Stingless Bees of Mexico

José Javier G. Quezada-Euán

Stingless Bees of Mexico

The Biology, Management and Conservation of an Ancient Heritage

Springer

José Javier G. Quezada-Euán
Departamento de Apicultura Tropical,
Campus de Ciencias Biológicas y Agropecuarias
Universidad Autónoma de Yucatán
Mérida, Yucatán, Mexico

ISBN 978-3-030-08539-1 ISBN 978-3-319-77785-6 (eBook)
https://doi.org/10.1007/978-3-319-77785-6

Printed on acid-free paper

This Springer imprint is published by the registered company Springer International Publishing AG part of Springer Nature.
The registered company address is: Gewerbestrasse 11, 6330 Cham, Switzerland

Preface

Stingless beekeeping (meliponiculture) in the Yucatan is an ancient activity. Nowhere else in the world meliponiculture reached the level of technical skills and productivity attained by the Maya in Mesoamerica with the stingless bee *Melipona beecheii*. However, changes in the economic system, together with severe deterioration of the habitat, caused a steady decline of this activity, almost to the verge of extinction. Fortunately, a series of events coincided in the mid-1980s that helped raising concern and renewed interest with important initiatives to rescue meliponiculture in the Maya region. In my opinion, 1986 was a critical year due to the arrival of the Africanized bee. The Africanization of European apiaries, drastically changed beekeeping in the whole of Mexico, but especially in the Yucatan Peninsula. In this area, with one of the largest densities of managed honey bee colonies in the Americas (17–21/km^2), hives were traditionally kept in the backyard of rural homes. To avoid stinging accidents, apiaries had to be relocated in the forests, further away from human settlements, leaving a vacant niche in rural homes. At around the same time Jorge González Acereto, a tireless promoter of stingless bees and meliponiculture in Mexico, started the characterization of nests of species from the Yucatán Península, with the objective of designing modern hives to keep them. Other students and I were involved in that project. Then, like most people, I was totally unaware of the existence of these insects and their ancient cultivation. We travelled the Yucatan, searching for colonies and species and got to know several stingless beekeepers from whom I learnt the importance and cultural value of this Mayan heritage. We visited remote rural villages in the forests of the Yucatan where stingless beekeeping had survived thanks to isolation from modern cultural influences. New species, like *M. yucatanica*, were discovered during our travel. With advice of various recognized researchers, as Professor Paulo Nogueira and Virgilio de Portugal Araujo, wooden hives of suitable size were produced for our different stingless bee species. This started the modernization of stingless beekeeping in the Yucatan, allowing better management and improvement on the production and quality of bee products. I obtained my bachelor's degree with a thesis investigating the development of colonies of *M. beecheii* in various types of hive and concluded that, colonies did not adapt to hives with excess volume because of thermoregulation problems. The development of modern husbandry, together with the promotion of stingless beekeeping as an alternative activity for rural villages, triggered a rebirth of the activity. With support from the government, several groups of farmers, especially women, got involved in meliponicuture, and started to take over the empty niche left by honey bees. Initiatives like the Seminario Mesoamericano sobre Abejas Nativas, devised by Margarita Medina and other stingless bee enthusiasts, helped to gather scientists and beekeepers around similar objectives.

Nowadays, stingless bees have become a phenomenon all over the Yucatan Península and other regions of Mexico. Indigenous groups and agencies are promoting *M. beecheii* and other species like never before. However, here lays a paradox: the recent popularity of stingless beekeeping is also becoming a major hazard. Uncontrolled selling of colonies because of high demand, and the negative impact of inexperienced instructors are threatening the activity, perhaps at a larger scale than ever. An avalanche of written and visual information on stingless bees can be found today in social networks, especially for Latin America. Unfortunately, much information found on these platforms lacks support and can be seriously mistaken, but accepted to the letter by inexperienced hobbyists, in detriment of colonies.

A great deal of scientific information has accumulated on stingless bees. However, it is generally dispersed and not easily available to the Spanish speaking public. One purpose of this book is to compile information on different aspects of the biology and management of stingless bees and (hopefully) make it accessible to students, academics, instructors, and the general public. The original work in Spanish, is aimed at readers in México, where information on these insects is limited. I dared to translate this work to English, intending that the information produced in Mexico (and Latin America), could be better known in other regions. I take full responsibility for the mistakes.

I would like to acknowledge many colleagues, with whom I have shared constructive scientific discussion, and pleasant times for several years. I particularly thank friends, who kindly reviewed and commented on earlier versions of the different chapters: Rodolfo Jaffé, Claus Rasmussen, Rob Paxton, and Adam Hart. My sincere thanks to all of you. My profound gratefulness to the agencies that have supported my research during all this time: the International Foundation for Science in Sweden, Fundación Produce Yucatán 2001–2007 (Manejo tecnificado de abejas sin aguijón y su uso en la polinización de cultivos), SISIERRA (Rescate de melponicultura en la Península de Yucatán), CONACyT Projects 2002–1556 (Rescate,conservación y mejoramiento genético de los recursos apícolas de México), 103341 (Conservación de las abejas sin aguijón de México), B237532 (Climate change and pollinators), and SAGARPA-CONACYT 291333 (Manejo sustentable de polinizadores).

I would also like to acknowledge my colleagues in the Department of Tropical Apiculture and my institution the Universidad Autónoma de Yucatán. Teresita Solis and Humberto Moo-Valle contributed with photographs and information in Chaps. 6 and 8, thank you for your help. Finally, I thank the many students and stingless beekeepers for their help and enthusiasm; you have made my work truly enjoyable. I dedicate this book to my beloved parents, Evangelina and José Guadalupe, my sisters Elia and Rosa, my dear friend and companion Alvaro Pat, my aunt Catalina, and my niece and nephew Mariana and Bernardo. Thank you all for your support and understanding.

Mérida, Yucatán, Mexico
February 2018

José Javier G. Quezada Euán

Contents

Chapter 1
Taxonomy and Diversity of the Stingless Bees

The two species of *Melipona* present in the Yucatan Peninsula. Left *M. (Melikerria) beecheii* and right *M. (Melipona) yucatanica.*

The bees are among the largest and most diverse group of insects. A worldwide estimation of 20,000 bee species is presently acknowledged (Roubik 1989). Mexico's bee fauna is considered mega-diverse encompassing an estimated 1800 species, representing approximately 10% of the world richness (Ayala 2006).

Together with their close relatives, the ants and wasps, the bees constitute the order Hymenoptera or membrane-winged insects (Michener 1974) (Fig.1.1). As a taxonomic group, the bees can be separated from their wasp relatives mainly by their feeding habits, being predominantly vegetarian (with a few exceptions). They feed upon nectar and pollen during their larval development and adulthood. The bees have evolved especialized structures for the collection of resources from plants, which are also useful in their taxonomic characterization (Michener 2000).

The bees or Apiformes are classified into seven families within the superfamily Apoidea, which also includes the sphecoid wasps (Michener 2000). One of these families of bees, the Apidae, is the most diverse and includes a small group in which the pollen-transporting apparatus of the females is the corbicula or pollen basket (Michener 1999; Hedtke et al. 2013; Fig. 1.2). The other Apidae as well as members

J. J. G. Quezada-Euán, *Stingless Bees of Mexico*,
https://doi.org/10.1007/978-3-319-77785-6_1

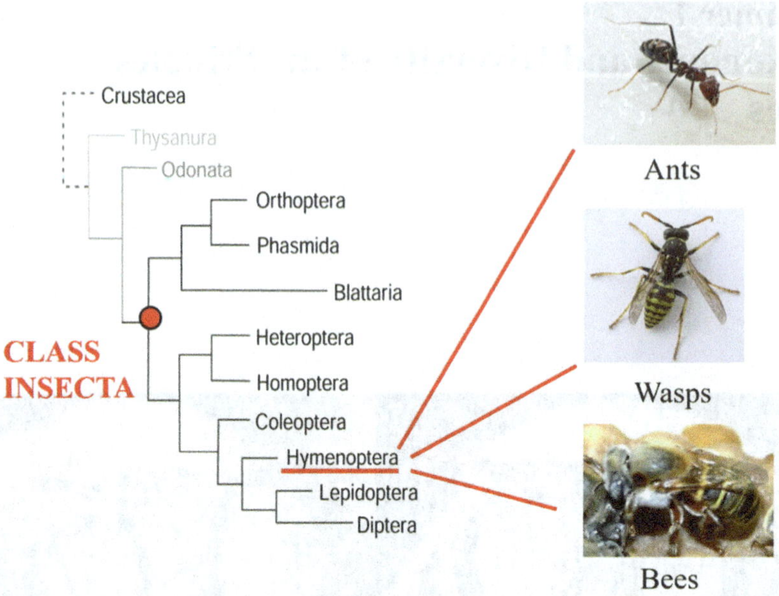

Fig. 1.1 Simplified phylogeny of the orders within the class Insecta. Underlined is the order Hymenoptera, which includes the wasps, ants, and bees

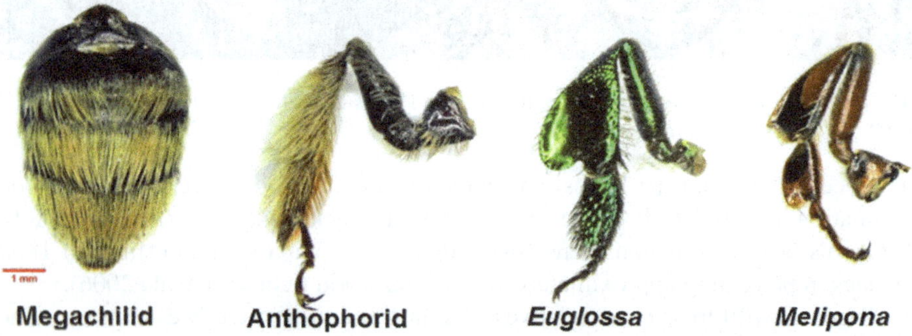

Fig. 1.2 Pollen-collecting apparatus of females of different bee groups. Most bees have hair brushes or scopas on the abdomen (Megachilid) or legs (Anthophorid). The corbiculate bees are characterized by the pollen basket or corbicula, a naked concavity of the tibia of their hind legs (examples, *Euglossa* and *Melipona*)

of the other six families (but not the parasitic ones) possess scopas (hair brushes) on the legs or abdomen, which are used for similar purpose (Fig. 1.2). The evolution of the pollen basket in the form of a naked concavity on the outer side of the hind tibia, allowed the corbiculate bees the transport of sticky materials and rapid dislodging of their load (Roubik 1989). The first corbiculate bees probably diverged 72–95 mya from New World Centridini ancestors preceding the evolution of eusociality (Martins et al. 2014).

Fig. 1.3 A simplified classification of the four tribes encompassing the corbiculate bees, a subset within the Apinae (after Hedtke et al. 2013)

The corbiculate bees are placed within the subfamily Apinae and comprise four tribes: Apini (honeybees), Bombini (bumble bees), Euglossini (orchid bees), and Meliponini (stingless bees) (Michener 2000) (Fig. 1.3).

Most bee species worldwide are solitary, but a small group of species have evolved different social lifestyles, from primitive to highly social behavior or eusociality (Michener 1974). At the highest level of eusociality two groups of bees are found: the honeybees and the stingless bees. In accordance to Michener (1974), the highly eusocial bees are distinguished by the following traits: morphological and physiological differentiation among females (true reproductive castes), division of labor with cooperative work, colonies in which individuals of different generations overlap, and reproduction by swarming.

1.1 General Features of the Meliponini

The Meliponini or stingless bees are by far the most diverse group within the corbiculate bees, not only for the current number of species (ca. 500), but also for the contrasting differences in morphology, nesting habits, and behavior between species (Roubik 2006). Nonetheless, it is possible to separate the stingless bees from the other corbiculate bees by some common features Wille (1979):

1. A vestigial sting.
2. The presence of a penicillium, a comb of rigid bristles on the internal margin of the corbicula. This structure is weaker or absent in the cleptobiotic genera *Lestrimelitta* and *Cleptotrigona*.

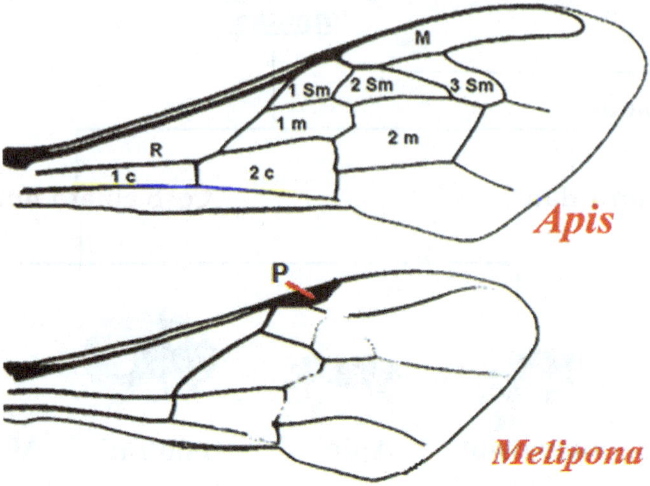

Fig. 1.4 Comparative forewing venation and cells of *Apis* and *Melipona*. A reduction of veins and cells, in particular the marginal cell (M) and the three submarginal cells (Sm), is observed in the latter. The P indicates the pterostigma

3. Weak and reduced forewing venation with a marginal cell relatively short, and submarginal cells weak or absent (Fig. 1.4).
4. A pterostigma of the forewing of moderate (*Melipona* species), to large size (non-*Melipona* species) (Fig. 1.3).
5. A reduced number of hamuli (hooks) on the hind wings (8 to 16 in *Melipona* and 4 to 10 in non-*Melipona*), compared with *A. mellifera* (ca. 20), and *Bombus* (14 to 28).
6. Absence of a pollen press (auricle) between the corbicula and the basitarsus. The basitarsus is small and slender compared with *Apis*.
7. Maxillary palps absent.
8. Wax-producing glands on the dorsal part of the metasoma.
9. Tibial spurs absent.
10. Female claws simple, bifurcate in the males (Fig.1.5).

1.2 Biogeography and Phylogeny

The stingless bees are exclusively found in the tropical and subtropical regions of the planet. In accordance to Moure et al. (2007), 60 genera are currently recognized. However, this figure may vary in accordance to the classification system used, of which two are presently accepted (Michener 2000; Camargo and Pedro 2007; Moure et al. 2007; Rasmussen and Cameron 2010).

The highest abundance and diversity of stingless bees are found in the neotropics with ca. 391 species and approximately 32 genera (Camargo and Pedro 2007; Fig. 1.6). Other regions of the world seem to have less diversity, although it is noted

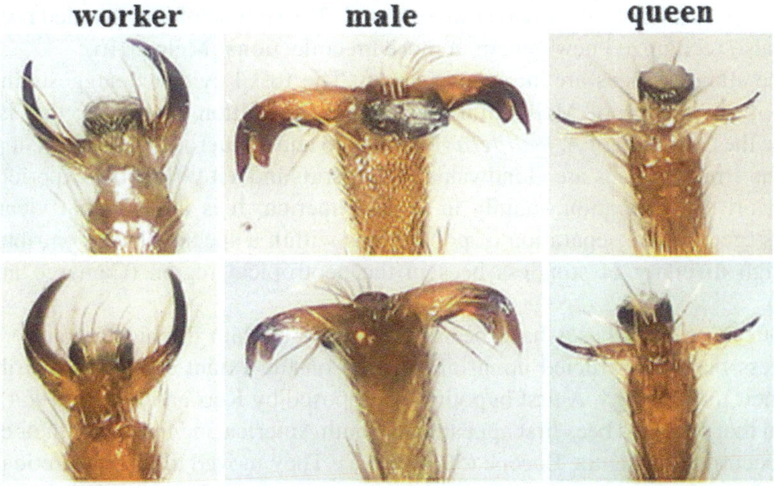

Fig. 1.5 Distal portion of the hind leg of *M. beecheii*: ventral (above) and dorsal view (below). The claws are simple in the worker and queen and bifurcate in the male

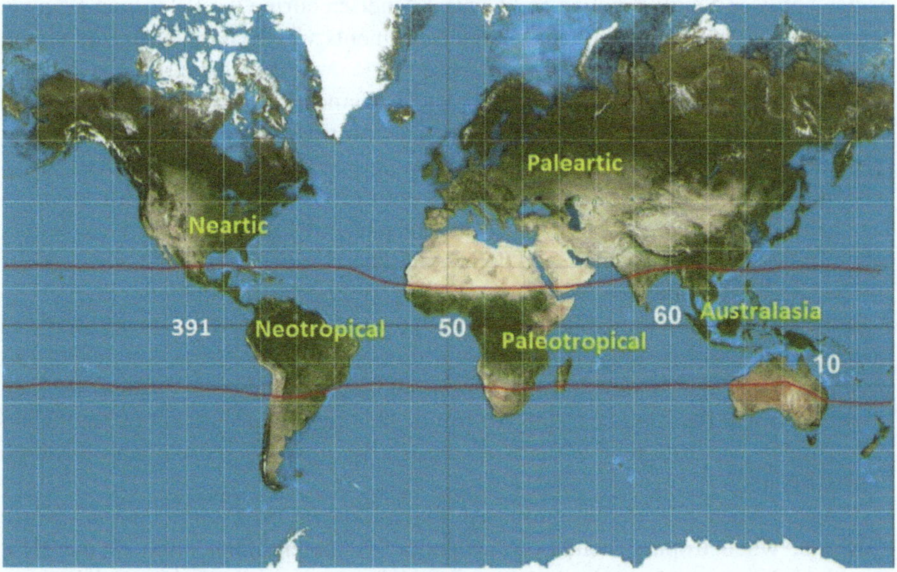

Fig. 1.6 Present geographic distribution of the stingless bees. The red lines represent the limits of their distribution confined to the tropics and subtropics. In white is the estimated number of estimated species per biogeographic region

that they have been less intensely studied. In Africa, for instance, recent revisions have reported some 50 species, and 70 species for Australasia and Indo-Malaysia (Eardley 2004; Rasmussen and Cameron 2007). The advent of accessible molecular methods has increased the number of studies on species complexes and the number of new species being identified (Halcroft et al. 2016). In addition, detailed revisions have also recognized new genera in museum collections (Melo 2016).

The stingless bees are an ancient group. The fossil evidence suggests that the time of emergence of Meliponini may be 70–90 million years ago; this is long before the evolution of *A. mellifera* (Rasmussen and Cameron 2010). Thus, it seems that the stingless bees are Gondwanian (Camargo and Vit 2013), but experienced a posteriori diversification, mainly in South America. It is thought that vicariance events (geographic separation of populations within a species) have contributed to the high diversity of stingless bees in the neotropical region (Camargo and Vit 2013).

Four main hypotheses have been proposed to explain the biogeography of the stingless bees, constructed upon information on the extant and fossil distribution and their morphology. A first hypothesis, proposed by Kerr and Maule (1964), suggested that stingless bees first appeared in South America and from here their expansion occurred during the Eocene (50–60 mya). They moved to the Palearctic region through the Bering Strait and from there they colonized the tropical regions of the Old World in the late Oligocene (20–30 mya). The main argument for Kerr and Maule's (1964) hypothesis is that the higher diversity in genera and species is presently found in South America. Their hypothesis assumes the existence of a land bridge between South America and Central America during the Paleocene-Eocene. However, it is now accepted that both subcontinents were separated for more than 100 mya, and that a connection with North America was established via the Panama Canal only 3–4 mya (perhaps only with patchy connections before).

Wille (1979), considering that the species with the more primitive or ancestral characteristics are presently found in Africa, suggested that stingless bees originated in this continent during the late Cretaceous (80–100 mya). From Africa stingless bees supposedly moved to the southern parts of Europe during the Eocene, when some land bridges connected both continents. From here, stingless bees eventually dispersed to the New World.

Later on, Camargo and Wittmann (1989) proposed a hypothesis suggesting the appearance of meliponines in Gondwana during the early Cretaceous (100–120 mya), and subsequent movement through land connections that existed between South America, the Antarctic, and Australia. This hypothesis attempted to explain the disjunct distribution of modern species in the genus *Plebeia* across all three subcontinents.

Finally, Michener (1990) proposed that stingless bees originated in North America, when tropical conditions extended to septentrional parts of the New World. However, this hypothesis does not explain a route of dispersion, and does not recognize that Africa and South America share taxa. Michener (1990) also proposed *Melipona* as a sister group to all the other stingless bees, and that *Trigona* sensu *lato (s.l.)* should include species of the neotropical and Indo-Malayan regions, but considered African species a separate clade.

In a further document, Camargo and Pedro (1992) elaborated on the original hypothesis by Camargo and Wittmann (1989), and suggested two main routes of dispersion of the meliponines from West-Gondwana via the Holartic and Panaustral regions. Camargo and Pedro (1992) proposed that *Trigona* from the neotropical, Indo-Malayan, and Australasian regions should form one group. However, these authors considered the Afrotropical species a sister group of all other stingless bees.

A first attempt at constructing a molecular phylogeny of the stingless bees, was undertaken by Costa et al. (2003). The authors built phylogenies based on the fragment 16S of the rRNA. However, due to the low number of species, and the use of only one fragment, not enough resolution for 34 species of stingless bees was obtained.

A larger phylogenetic analysis was conducted by Rasmussen and Cameron (2007, 2010) based on 9 genes for the basal framework, and including 79 taxa from the Old and New World (Fig. 1.7). The results revealed three groups clearly differentiated: Neotropical, Afrotropical, and Indo-Malayan-Australasian. The Neotropical group diverged early from the others. An important finding was that the species of *Trigona* from the New and Old World formed two separate clades implying different origin. The authors argue that since the first descriptions of *Trigona* were made for Neotropical specimens, the genus should be used for American species exclusively. In the analysis by Rasmussen and Cameron (2010) *Melipona* was more parsimoniously placed within the neotropical clade, and not as a sister group of all stingless bees as proposed by Michener (1990). In addition, the Neotropical clade appeared composed by three groups (Fig. 1.7): one including the species of *Melipona*, another for the minute species (e.g., *Trigonisca s.l.*), and a third one with the other New World taxa (Rasmussen and Cameron 2010).

The findings of Rasmussen and Cameron (2010) are useful to test hypotheses on the biogeography of stingless bees. The Gondwanian origin of the stingless bees seems accurate. Rasmussen and Cameron (2010) also estimated the time of divergence between Old and New World stingless bees at around 71–73 mya, the separation of the Afrotropical clade in 61 mya, and the clade Indo-Malayan-Australasia in 49 mya (Fig. 1.7).

Similar to *Trigona*, Old and New World *Plebeia* are not closely related, which proves that the hypothesis for the dispersion of this genus is incorrect (Camargo and Wittmann 1989). In the case of Kerr and Maule's (1964) hypothesis, it does not seem congruent with recent geological findings. Nonetheless, the other hypotheses, in the opinion of Rasmussen and Cameron (2010), cannot be ruled out due to a lack of evidence from fossils for a better phylogenetic picture.

Stingless bees are presently classified within a single tribe, the Meliponini (Michener 2000). However, contrasting differences within the Meliponini have long been recognized between the species of *Melipona* and the other genera (Wille 1983; Table 1.1; Fig. 1.8).

It is interesting to note that *Melipona* and *Trigona* derive from the Greek, meaning in the first case working incessantly for the honey (*meli* (honey) + *ponein*

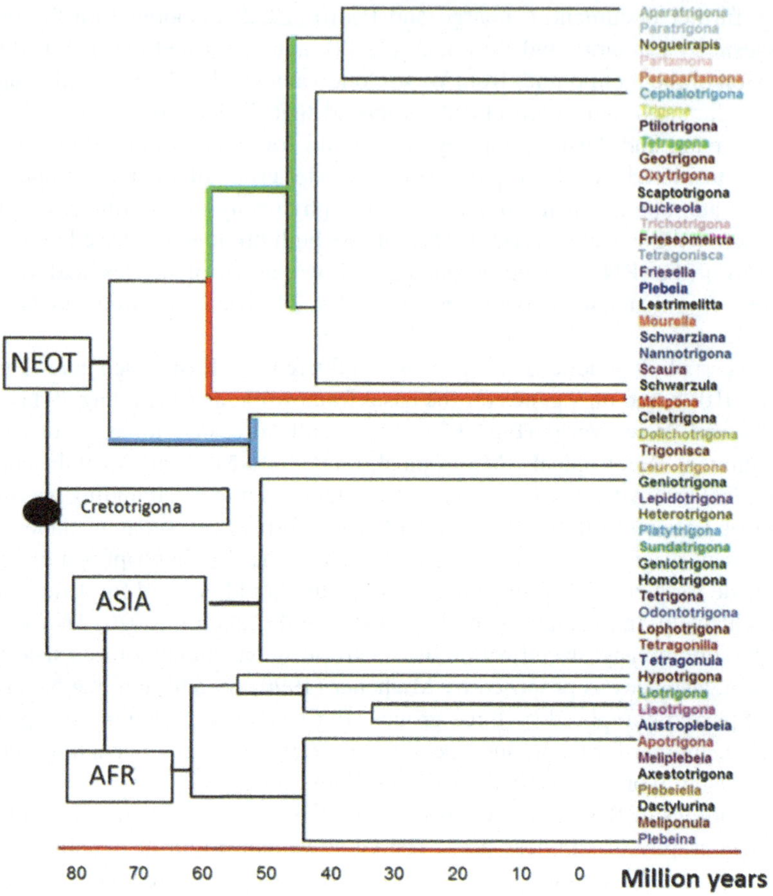

Fig. 1.7 Phylogenetic relationships of the stingless bees in accordance to the molecular study by Rasmussen and Cameron (2010). Three main clades are identified, NEOT = neotropical, ASIA = Indo-Malaya, AFR = Africa. Within the NEOT clade, three groups are found: the minute stingless bees (blue root), *Melipona* (red root), and the other taxa of New World (green root). *Cretotrigona*, the oldest fossil for the group, was used to provide an estimation of the time of emergence of each branch

(to toil)). *Trigona* makes reference to the triangular shape of the abdomen of these bees (*tri* (three) + *gono* (angle)) (Schwarz 1932).

In the neotropics, some 33 genera (Camargo and Vit 2013; Melo 2016), and approximately 418 species of stingless bees are recognized (Pedro 2014). The genus *Melipona* (Illiger, 1806) is exclusively Neotropical, and it encompasses an estimate of 74 species found between Mexico and Argentina (Camargo and Pedro 1992; Ramírez et al. 2010; Pedro 2014).

Melipona includes some of the most representative, economically and culturally, species in the Americas. A first comprehensive phylogeny of the genus *Melipona*

Table 1.1 Characteristics of the species in the genus *Melipona* and the other genera of the stingless bees

Characteristic	Meliponini	
	Melipona	Non-*Melipona*
Nest entrance	Made of mud, with a radiated shape	Made of cerumen or resin. Some species are distinctive for their tube or trumpet-like entrances
Worker body size	Robust, with a total length >1 cm	Wasplike, slender, total length <1 cm
Body pilosity	Abundant, the integument is furry, which gives an overall opaque appearance	Scarce—the integument is mostly glabrous, which gives a shiny appearance
Forewing	Total length not beyond the tip of the abdomen Pterostigma of moderate size	Total length well beyond the tip of the abdomen Pterostigma comparatively large in relation to wing size
Hind wing	8–16 Hamuli	Generally <8 hamuli
Dorsal vessel	Type II (see Chap. 3)	Type I
Digestive tract	Long	Short
Royal cells	There are no royal cells—cells where queens are reared are of the same size as workers and males	Royal cells are larger than those of males and workers
Distribution of queen cells	The cells where queens are reared are intermingled with those of males and workers	Queen cells are normally found in the periphery of the combs or the outer parts of clustered cells
Gyne size	Gynes are generally smaller than workers and males—thorax width notably smaller	Normally gynes are larger than males and workers
Ovaries of gynes at emergence	Partially developed	Well developed
Gyne rate production	Gynes are frequently produced	Gynes are produced at a lower rate
Caste determination	Tropho-genetic	Trophic

Fig. 1.8 Workers of *M. beecheii* and *Cph. zexmeniae* to show some contrasting features between *Melipona* and non-*Melipona* stingless bees. Body pilosity is more abundant in *M. beecheii*, and the forewing is comparatively larger in *Chp. zexmeniae* (Table 1.1)

was constructed using morphological characters by (Moure 1992), who recognized four subgenera.

The four subgenera recognized by Moure (1992) within *Melipona* are:

Melipona (*Melipona s. str.*) Illiger, 1806
Melipona (*Michmelia*) Moure, 1975
Melipona (*Melikerria*) Moure 1992
Melipona (*Eomelipona*) Moure 1992

Molecular studies have confirmed that *Melipona* is a modern clade within the stingless bees, of post-Gondwanian origin, which probably rose approximately 24 mya in the upper Eocene (Rasmussen and Cameron 2010). However, an extensive study of 35 *Melipona* species using mitochondrial, ribosomal, and nuclear DNA markers, suggests even a more recent origin (14–17 mya) of this genus (Ramírez et al. 2010). Interestingly, the phylogeny of *Melipona* by Ramírez et al. (2010), is largely in agreement with the phylogeny proposed by Moure (1992), and more recently by Camargo and Pedro (2007), both using morphological characters.

It is important to mention that in his most recent classification, Michener (2000) did not recognize the subgenera proposed by Moure (1992) for *Melipona,* due to the apparently subtle morphological differences between them. Michener (2000) neither recognized many of the genera used by Camargo and Pedro (2007) and Moure et al. (2007), for other stingless bees.

The *Melipona* clade seems largely monophyletic, with a common ancestor (Fernandes-Salomão et al. 2005; Ramírez et al. 2010; Rasmussen and Cameron 2010) (Fig. 1.9). Nevertheless, in the most recent study by Ramírez et al. (2010), it was found that three of the subgenera seem monophyletic (*Melikerria*, *Melipona s. str.*, and *Michmelia*), but in the case of *Eomelipona*, it seemed polyphyletic (more than one possible ancestor, Fig. 1.9). Moreover, there was a group of species, which were not assigned to any of the four proposed subgenera (*Incertae sedis*). These non-classified *Melipona* species, are considered a sister group to all others, and apparently form a monophyletic clade with *Michmelia* (Ramírez et al. 2010).

In Mexico, the most important species of *Melipona, M. beecheii*, belongs in the subgenus *Melikerria* (Roubik and Camargo 2012). Two derived (apomorphic) traits are characteristic of the subgenus *Melikerria* (Moure 1992):

1. The inner half of the apex of the mandibles is subdivided by strong emargination into two distinct contiguous teeth (the median tooth with sinuosity).
2. The anterolateral angles of the mesonotum show a rust-red patch of hair.

The males have a scythe-shaped gonostyle (Roubik and Camargo 2012).

In addition to the Mesoamerican *M. beecheii*, eight other species are included in the subgenus *Melikerria*, most of South American distribution. The following species with the present geographic distribution are recognized in *Melikerria* (Camargo and Pedro 2007; Roubik and Camargo 2012):

M. beecheii from Mexico to Costa Rica
M. triplaridis from Eastern Panama
M. salti, from the north of Colombia

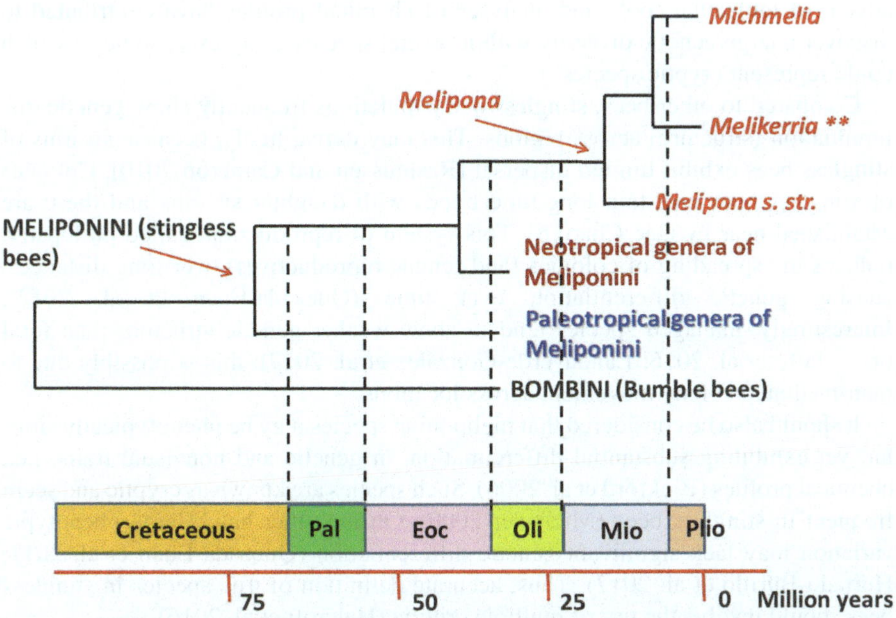

Fig. 1.9 Simplified phylogeny of the genus *Melipona* in accordance to Ramírez et al. (2010). The tribes are written in capital letters. The estimated time of emergence of each clade is presented, Pal (Paleocene), Eoc (Eocene), Oli (Oligocene), Mio (Miocene), Plio (Pliocene). The subgenus *Eomelipona* has a likely polyphyletic origin, and is not included in the figure. The subgenus *Melikerria* where *M. beecheii* is included is marked with a double asterisk

M. compressipes from Venezuela, Guyana, Suriname, French Guiana, and the north of Brazil and Colombia
M. interrupta from Guyana, Suriname, French Guiana, and the north of Brazil
M. grandis from South Western Amazonia
M. fasciculata from oriental Amazonia
M. quinquefasciata from North East Brazil and Argentina
M. insularis from the island of Coiba, province of Veraguas, Panama
M. ambigua from North East Colombia

The recent discovery of two new *Melikerria* species, namely *M. insularis* and *M. ambigua*, from Panama and Colombia, agrees with the idea that a land bridge united North and South America around 20 mya, long before the generally accepted estimation of 3 mya (Roubik and Camargo 2012). Thus, *M. beecheii* may have recently derived from the most northern species of *Melikerria*.

1.3 Diversity of Mexican Stingless Bees

The extant stingless bee fauna of Mexico seems to be the result of recent migrations of Central or South American taxa during the Pliocene and Pleistocene. Presently, 46 recognized species of stingless bees exist in Mexico (Ayala 1999). However, the advent of molecular tools and analyses of chemical profiles have contributed to discover a large genetic diversity within several species complexes, some of which could represent cryptic species.

Compared to other bees, stingless bee populations frequently show genetic differentiation (structure) across regions. This may derive firstly, because swarms of stingless bees exhibit limited dispersal (Rasmussen and Cameron 2010). Colonies of stingless bees maintain long-time bonds with daughter swarms and these are established near by (see Chap. 6). This system of reproduction, called philopatry, reduces the spreading of colonies (and female reproductives) over long distances, causing genetic differentiation over time (Quezada-Euán et al. 2012). Interestingly, managed species tend to show weaker genetic structure than feral ones (Jaffé et al. 2016; Landaverde-González et al. 2017); this is possibly due to man-mediated colony movement across locations.

It should also be considered that meliponine species may be phenotypically similar, yet exhibiting substantial differentiation, in genetic and nonvisual traits, i.e., chemical profiles (Bickford et al. 2006). Such species are known as cryptic and seem frequent in stingless bees (Michener 2000). On the other hand, large phenotypic variation may lack significant genetic differentiation (Quezada-Euán et al. 2012; Hurtado-Burillo et al. 2017). Thus, accurate definition of true species in stingless bees should involve the use of multiple criteria (Halcroft et al. 2016).

In Mexico, the genus *Melipona* is one of the most important. It is represented by seven species, which comprise 15% of the Mexican stingless bee fauna (Ayala et al. 2013; Table 1.2). Among the Mexican species of *Melipona*, two are found in highland forests: *M. colimana* and *M. fasciata*, both in the subgenus *Michmelia*. The workers of these species have the largest body size of them all. There are two other species of the subgenus *Michmelia* in Mexico; both are found in lowland tropics, *M. solani* and *M. belizae*. The workers of both species are clearly distinguished by the orange pilosity. The smallest Mexican *Melipona* are both endemic, namely *M. yucatanica* and *M. lupitae*. These species are phenotypically similar and are close phylogenetically (Ayala 1999; Table 1.2).

The species with the most extensive geographic distribution is *M. beecheii*, and also the most relevant due to its economic and cultural importance (Ayala 1999) (Fig. 1.9a). It is also the only species of the subgenus *Melikerria* out of South America (Roubik and Camargo 2012).

Molecular comparisons have shown that *M. beecheii* has probably experienced earlier divergence from the other *Melikerria* (Roubik and Camargo 2012).

Roubik and Camargo (2012) suggested that *M. beecheii* probably derives from the lineage of *M. insularis*, the most northern of all South American *Melikerria*. Its lineage may have gone extinct on mainland, but relict populations survived on islands of the Panama microplate, which eventually colonized the northern parts of Central America and the tropical regions of Mexico. This scenario may also explain the larger

Table 1.2 Description of the seven species of the genus *Melipona* found in Mexico (based on Ayala 1999)

Species		Distribution	Main features
Melipona beecheii Subgenus: *Melikerria*		The Yucatán Península, Gulf Coast (Tabasco, Veracruz, Tamaulipas) and Pacific Coast (Chiapas, Jalisco, Nayarit, Oaxaca). Also found in San Luis Potosí and Zacatecas	Integument black with variable yellow, red, or black marks on the legs. Body length 9.7–10.7 mm. Front of the antennal scape yellow. Scutum of the thorax with orange/yellow hair. Lateral tufts of dense orange-red hairs in strong contrast with the rest of the scutum. Black abdomen with well-defined yellow bands
Melipona solani Subgenus: *Michmelia*		Tropical forests of Chiapas and possibly Tabasco. More common in Central America	Orange and black integument with orange pilosity. Body length 8.0 mm. Hair on face orange-brown with some black hair. Pilosity red-orange with black hairs. Abdominal segments orange generally with no lines
Melipona belizeae Subgenus: *Michmelia*		Belize, Southern areas of Campeche and Quintana Roo	Integument black and orange. Body length 8.0 mm. Face without marks with grey-brown hair around the ocelli. Scutum with ocher and some black hairs. Sides of thorax with ocher or red hair only. Abdomen brown with yellow bands in last segments
Melipona fasciata Subgenus: *Michmelia*		One of the two highland *Melipona* species in México between 500 and 3000 m. From Guerrero and Michoacán, also Orizaba and Córdoba on the Gulf coast of Mexico.	Integument of head and thorax black, but orange in the abdomen with variable yellow lines. Pilosity clear yellow and orange. Body length 9.0 mm. Yellow marks below the eye (as in *M. colimana*). Sides of the thorax with orange tufts

(continued)

Table 1.2 (continued)

Species		Distribution	Main features
Melipona colimana Subgenus: *Michmelia*		Endemic to the highland forests on the west Coast in Jalisco above 1000 m (sierra de Manantlán, Volcán and Nevado de Colima and sierra del Tigre)	Black integument and yellow marks with orange pilosity no black hair. Body length 9.5 mm. Yellow marks below the eye. Side portions of the thorax with orange hairs and a tuft of dark orange hairs. Abdominal segments black with well-defined yellow apical bands
Melipona lupitae Subgenus: *Melipona*		Endemic to Mexico, only found in low tropical deciduous forests on the west side of the River Balsas Basin, in the state of Michoacán	Black integument with white and yellow marks. Light brown-orange fur with some whitish or black hairs. Body length 7.9–8.5 mm. Black scutellum. Tibia dark brown with black hairs. Abdominal segments with yellow stripes well defined with some black hair.
Melipona yucatanica Subgenus: *Melipona*		Endemic to the Southern forests of the Yucatan Península and the Tehuantepec Isthmus. Also present in Guatemala but the population is genetically different to the Mexican ones	Morphologically similar to *M. lupitae*. Black integument with yellow marks. White to orange hair on the face. Body length 8.2–8.5 mm. Scutum with orange hairs. Lateral portion of the thorax with a yellow thin line. Brown-red tibiae with yellow hairs. Corbicula with a black mark on its distal portion. Abdominal segments almost hairless and intensely colored yellow lines

Photographs kindly provided by Dr. Ricardo Ayala (UNAM), Dr. Terry Griswold, and Chelsey Ritner (USDA ARS—Utah State University)

genetic diversity found in *M. beecheii* from Central America compared to Mexico (May-Itzá et al. 2009), in agreement with a more recent colonization of the latter.

Presently, the distribution of *M. beecheii* in Mexico includes the low tropical and subtropical forests of the Pacific and Gulf coasts, the Isthmus of Tehuantepec, and the Yucatan Peninsula. Outside Mexico, it can be found in Guatemala, Belize, Honduras, Salvador, and Nicaragua. The southernmost distribution is the Guanacaste Peninsula of Costa Rica (Ayala 1999; Fig. 1.10A). This species is also one of the only two *Melipona* found in the Antilles, on the islands of Cuba and Jamaica (Genaro 2008). In the Yucatan Peninsula and some regions of Central America,

Fig. 1.10 (**A**) Geographic distribution (in orange) of *M. beecheii* in accordance with Ayala (1999) and Moure et al. (2007). (**B**) The two species of *Melipona* present in the Yucatan Peninsula: (**a**) *M. (Melikerria) beecheii* and (**b**) *M. (Melipona) yucatanica*

M. beecheii is sympatric with *M. yucatanica* (*ts'ets* in Maya), both recognized by the Maya as different species (González-Acereto 1984). Although externally both species may look similar, especially in the color of the integument (Fig. 1.10B), they are not related (Camargo et al. 1988). Indeed, in accordance with Moure et al. (2007), *M. yucatanica* belongs to the subgenus *Melipona s. str.* while *M. beecheii* is included in *Melikerria.*

Across its wide geographic distribution, *M. beecheii* is known with a diversity of names: abeja-real, *Yilkil-kab*, *Colel-kab*, *Xunan kab* (*Xunaan-kaab*), *Pool-kab*, colmena-kab, abeja-alazana, *pipioli*, *mimialcuatl*, *tsaspena*, gato, abeja-aluva, blanco-aluva, criolla, jicote-gato, jicote-estrella, abeja de la tierra, *Ajau-chab*, and *Suk-ajatié*, the latter two among the Maya, Chol, Chontal, and Tzeltal from Tabasco and Chiapas, in Mexico (González-Acereto 2008; Moure et al. 2007). The diversity fo names may be an indication of the extensive use of this species.

The first scientific description of *M. beecheii* was made by captain F.W. Beechey in his manuscript narrating a 3-year voyage to the Pacific coasts of Mexico: *Narrative of a voyage to the Pacific and Beering's strait, to cooperate with the polar expeditions: performed in his Majesty's Ship Blossom, under the command of Captain F. W. Beechey.* Vol. 2. (1831) London: H. Colburn & R. Bentley. In this volume's Appendix, from pages 357 to 365, the navigator includes a section called "*Mexican bees. Some account of the habits of a Mexican bee*" (Fig. 1.11).

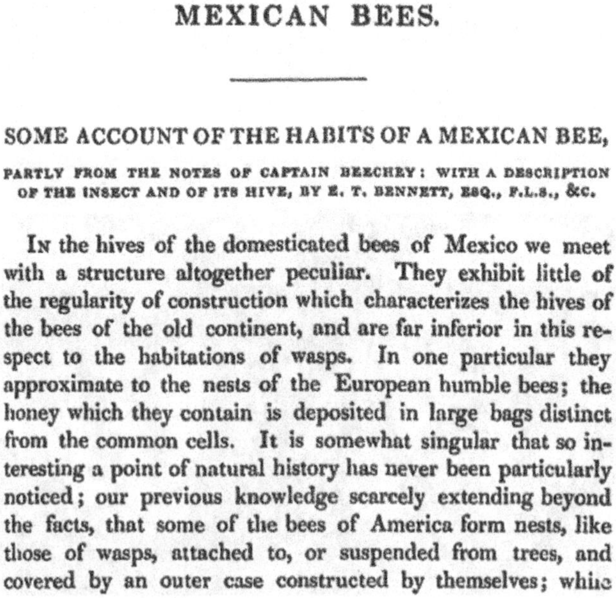

MEXICAN BEES.

SOME ACCOUNT OF THE HABITS OF A MEXICAN BEE,

PARTLY FROM THE NOTES OF CAPTAIN BEECHEY: WITH A DESCRIPTION OF THE INSECT AND OF ITS HIVE, BY E. T. BENNETT, ESQ., F.L.S., &c.

In the hives of the domesticated bees of Mexico we meet with a structure altogether peculiar. They exhibit little of the regularity of construction which characterizes the hives of the bees of the old continent, and are far inferior in this respect to the habitations of wasps. In one particular they approximate to the nests of the European humble bees; the honey which they contain is deposited in large bags distinct from the common cells. It is somewhat singular that so interesting a point of natural history has never been particularly noticed; our previous knowledge scarcely extending beyond the facts, that some of the bees of America form nests, like those of wasps, attached to, or suspended from trees, and covered by an outer case constructed by themselves; while

Fig. 1.11 A section of the first page of the Appendix of the narrative of the journey by captain Beechey (1831) along the Pacific coast of Mexico, where he described *M. beecheii* for the first time

Because captain Beechey was the person who first gave an account on this bee and its colonies, Bennett in 1831 used the captain's name for its species nomenclature, and assigned it to the genus *Melipona*, described by Illiger in 1806.

Beechey (1831: on the page 444 from the same Appendix) indicates the localities visited along the Mexican Pacific coast: San Blás and Tepic in Nayarit, as well as Mazatlán in Sinaloa and Acapulco in Guerrero. However, gives no precise locality where he collected his specimens. Paradoxically, the economic and cultural value of *M. beecheii* for the indigenous inhabitants of the west coast of Mexico is much less representative compared with the Yucatan Peninsula (Quezada-Euán et al. 2001; Fig. 1.10). It is believed that the description made by Bennett of *M. beecheii* was based on a single specimen whose present location is unknown (Moure et al. 2007).

In his revision of the stingless bees of Mexico, Ayala (1999) describes *M. beecheii* as a medium-sized bee, measuring between 9.7 and 10.7 mm, with a forewing span between 7.7 and 7.9 mm. The integument is black with yellow, brown, or black marks on the legs, depending on the geographic origin (Fig. 1.12). The metasomal tergi are black, with well-defined yellow bands. Noteworthy, there is a wide variation in size and coloration among specimens from different regions, particularly on the maculae on the clypeus and the dark marks on the hind legs (Ayala 1999).

Ayala (1999) assigned *M. beecheii* in what he called the *fasciata* group. The *fasciata* group of *Melipona*, have the lower part of the antennal scape yellow; the head vertex on frontal view slightly concave with no elevation at the level of the ocelli; the frontal-lateral sides of the scutum are covered by tufts of dense red hair, in contrast with the rest of the pubescence, which is clear and less dense.

There has long been controversy on the taxonomy of *M. beecheii* as a single species. This due to contrasting variation of color, body size, and occurrence of intergrades in specimens from different geographic regions (Schwarz 1932; Wille 1976; Quezada-Euán et al. 2001, 2007). Schwarz (1932) proposed the existence of two subspecies of *M. beecheii*. One such subspecies (*M. b. beecheii*), may involve populations from southern Mexico, Guatemala, Nicaragua, and Costa Rica, and the other (*M. b. fulvipes*), may include specimens from Cuba, Jamaica, Belize, and the Yucatan Peninsula in Mexico. Camargo et al. (1988) also acknowledged the possible existence of these two subspecies in *M. beecheii*. However, to date, *M. b. beecheii* and *M. b. fulvipes* are not accepted as true subspecies (Moure et al. 2007).

A first quantitative approach to analyze the morphological variation in *M. beecheii* was done by Carrillo et al. (2001). Their study included specimens from the Yucatan, Costa Rica, and Chiapas and compared body size and degree of maculation of the clypeus (the section on the head below the antennae and above the mandibles) and the malar region (the area beside the clypeus) (Fig. 1.12). The results showed that the bees from the Yucatan Peninsula were smaller and had significantly more yellow marks on the face than the bees from the other two regions.

In a following study Quezada-Euán et al. (2007) showed that morphological differentiation between bees from the Yucatan Peninsula and Costa Rica was genetically supported (*FST* value = 0.280). In spite of clear genetic differences, the sequencing of the mitochondrial *cox*1 revealed a differentiation between both populations of only 1.2%. Typically, sequence divergence of *cox*1 between insect species

Fig. 1.12 Intraspecific diversity in the pattern of coloration (maculation) of the face and body of *M. beecheii* workers from (**a**) Costa Rica, (**b**) Chiapas, Mexico, and (**c**) Yucatan, Mexico. There is a clinal distribution in the degree of face and leg maculation decreasing from North to South. The lower image shows workers from Chiapas (left) and Yucatan (right). The specimens from Chiapas are larger and darker than the specimens from the Yucatan

is in a range between 3 and 10% (Hebert et al. 2003), though sibling species may show far lower sequence divergence (Hebert et al. 2004). Although the observed divergence of 1.2% between populations of *M. beecheii* representing different species is not yet clear, distinct genotypes can be found in both regions (Quezada-Euán et al. 2007).

In subsequent studies, samples from Guatemala, El Salvador, and Nicaragua, together with samples from Mexico and Costa Rica were compared using the segments *ITS*2 and *ITS*1 of the ribosomal DNA (De La Rúa et al. 2007; May-Itzá et al. 2009). An interesting finding was that the *ITS*2 segment of *M. beecheii* is one of the longest found in insects, with 1788 bp (De la Rúa et al. 2007). After digestion, three different patterns were found for both *ITS*2 and *ITS*1, which are in general congruent with the geographic origin of the samples (De la Rúa et al. 2007; May-Itzá et al. 2009). A first pattern of *ITS*1 and *ITS*2 was obtained in specimens from Mexico (Yucatan and Chiapas) and northern de Guatemala, around the Petén region (Fig. 1.13). A second *ITS*1 pattern (B), was found in populations from southern Guatemala and El Salvador, and a third *ITS*1 pattern (C), was found in the bees from San Marcos, in the southwestern part of Guatemala (Fig. 1.13). In the case of *ITS*2, a characteristic pattern was found in specimens from southern Guatemala, Salvador, and Costa Rica, and third pattern was only found in the bees from San Marcos (De la Rúa et al. 2007). The largest genetic diversity of *M. beecheii* was found in Guatemala, where all the different patterns of *ITS*1 and *ITS*2 found, are present (Fig. 1.13).

The evidence from molecular and morphometric analyses, suggests that *M. beecheii* is a monophyletic species with two clearly differentiated groups (May-Itzá et al. 2012). Both groups are largely in agreement with the original subspecies proposed by Schwarz (1932). The distribution of the two main genetic lineages in *M. beecheii* seems defined by the Sierra Madre del Sur, which starts in Chiapas and extends down to Central America (Fig. 1.14). This large mountain chain may have

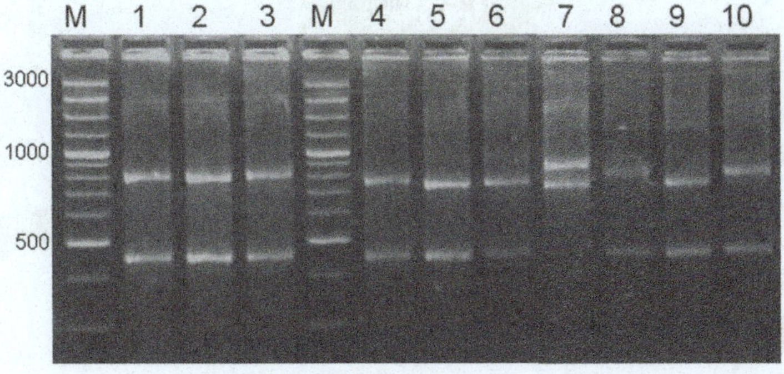

Fig. 1.13 The three different patterns found after digestion of the ribosomal segment *ITS*1 of *M. beecheii*. *ITS*1-A is obtained in specimens from Mexico (Yucatan Peninsula and Chiapas, lanes 1, 2, and 3, respectively) and the north of Guatemala (Petén, lanes 8 and 10). The *ITS*1-B pattern is found in bees from El Salvador (lane 4), and southern Guatemala (lanes 5, 6, and 9, respectively). Pattern *ITS*1-C (lane 7), has only been detected in specimens from San Marcos, in southwestern Guatemala

acted as a barrier for genetic flow between populations to the north and south. Speciation seems to frequently occur in stingless bees as a result of vicariance due to geographic isolation (Quezada-Euán et al. 2012; Jaffé et al. 2016).

There is also evidence for the existence of a cline in body size within the lineages of *M. beecheii* (Fig. 1.14). The bees from central Guatemala have the largest body size and this gradually decreases in the specimens from Salvador, Nicaragua and Costa Rica. Interestingly, a similar pattern occurs to the north, with the bees from the Yucatan Peninsula having the smallest body size of all populations studied (Fig. 1.14).

The mobilization of colonies is a potential threat for the genetic identity of the different lineages of *M. beecheii*, as this species is extensively used in Mexico and

Fig. 1.14 Phenotypic and genetic diversity in *M. beecheii*. There is a cline in body size represented by the mean length of the forewing (mm) across the different regions. Two clear *ITS* patterns are also evident (black and red dots), geographically separated by the Sierra Madre del Sur: (1) Yucatan Peninsula-northern Guatemala-Chiapas, and (2) Central America. There are also *ITS* differences within the Central American clade, suggesting one more genetic lineage in it (May-Itzá et al. 2009; May-Itzá et al. 2012)

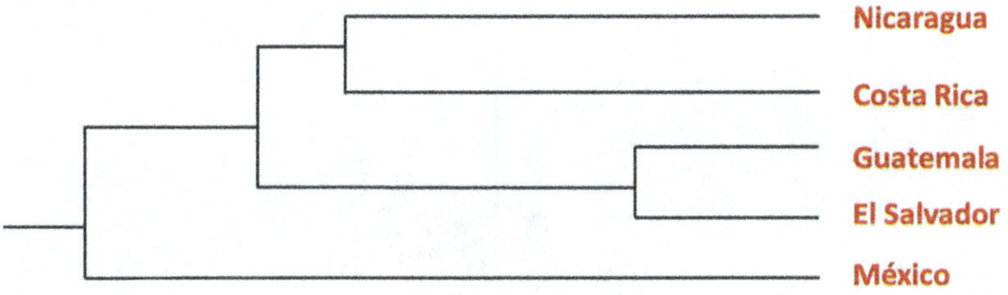

Fig. 1.15 UPGMA tree based on the geometric morphometrics of wing shape in *M. beecheii* specimens from different geographic regions (Francoy et al. 2011). The clades are in agreement with the molecular results of *ITS*1

Central America for honey production, and colonies can be exchanged among regions (May-Itzá et al. 2012). In order to find quick and reliable methods that allow geographic traceability of populations and lineages in *M. beecheii*, Francoy et al. (2011) analyzed the geometric morphometrics of wings of the samples used in the studies by May-Itzá et al. (2009) and May-Itzá et al. (2012). Their approach correctly assigned 87.1% of the colonies to their geographic region, and 92.4% to their haplotype (Fig. 1.15). Thus, geometric morphometrics can be a cheap and reliable method to accurately trace the geographic origin of specimens of *M. beecheii*, and serve as a tool to control the movement of specimens and colonies across geographic regions (Francoy et al. 2011).

Although a clearer picture is now evident for the biogeography of *M. beecheii* across the southern part of its distribution, nothing is yet known for populations along the Gulf Coast of Mexico (Veracruz, San Luis Potosí), and the Pacific coast, between Chiapas and Sinaloa. Stingless beekeeping is increasing in these areas, and it is important to develop studies on the genetic makeup of such populations in order to preserve them.

M. beecheii is one of the two species of *Melipona* found in Cuba (Fonte 2007; Genaro 2008), and the only one in the Greater Antilles. Given the limited ability for colonies to move across large bodies of water, a plausible explanation for its presence on these islands seems human introduction (Michener 1982; Genaro 2008).

Because of their similarity in size and color, Schwarz (1932) and Camargo et al. (1988), considered the bees from Cuba as *M. fulvipes* or a subspecies (*M. beecheii fulvipes*), and placed together with the bees from the Yucatan Peninsula. Recently, Lóriga-Peña and Quezada-Euán (2009) compared the body size and the maculation of the head and legs of *M. beecheii* samples from Cuba with bees from the Yucatan Peninsula and Costa Rica. Although the bees from Cuba and the Yucatan differed in individual characters, a multivariate comparison showed extensive overlapping in their overall body size. The bees from Costa Rica were significantly larger than those from Yucatan and Cuba (Fig. 1.16).

DNA microsatellite markers confirmed that the samples from Cuba were genetically more similar to those from the Yucatan Peninsula too (Lóriga-Peña et al. 2011;

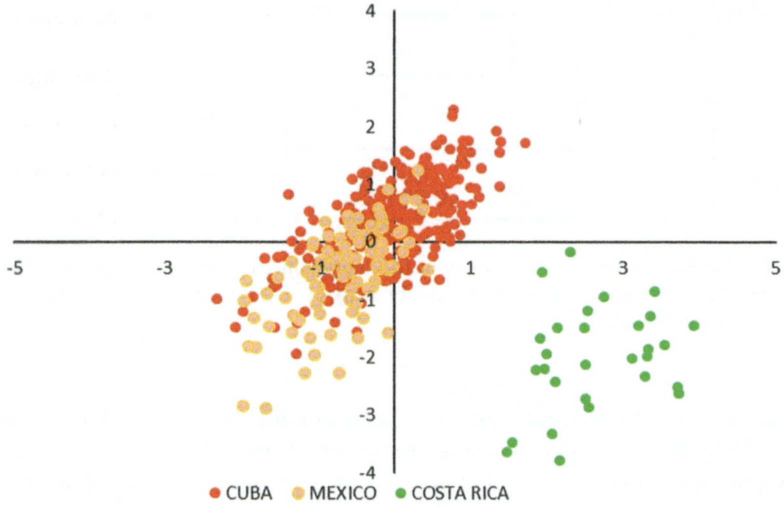

Fig. 1.16 Morphological comparison of *M. beecheii* from Cuba and populations from Yucatan (Mexico) and Costa Rica by means of a principal component analysis (abscissa = component 1 and ordinate = component 2). The specimens from Cuba extensively overlap with those from the Yucatan Peninsula

Hurtado-Burillo et al. 2014). Altogether, these results indicate that the "abejas de la tierra," as *M. beecheii* is known in Cuba (Fonte 2007), probably descend from continental ancestors taken to the island from the Yucatan. Although there does not seem to be records on this, it is likely that colonies were introduced to Cuba at some point during the Spanish colony (sixteenth to nineteenth centuries) as a result of the extensive trading between the Capitanía General de Cuba and the Nueva España's Yucatan Peninsula. It is important noticing that during the Spanish colony, *M. beecheii* was the only honey-producing bee found in the Yucatan, because *Apis mellifera* was not introduced until the first part of the twentieth century (Quezada-Euán et al. 2001). Thus, it sounds plausible that a population from the Yucatan may have been introduced to Cuba for honey production.

A sympatric species with *M. beecheii* in southern Mexico is *M. yucatanica.* This species could be considered endemic, although it is also present in Guatemala (Ayala 1999). Morphological and molecular analyses have shown that the populations of *M. yucatanica* from the Yucatan and Guatemala show significant differentiation (May-Itzá et al. 2010). Similarly to *M. beecheii*, workers of *M. yucatanica* from Guatemala are larger compared with the ones from the Yucatan. Analyses of the *cox*1 segment of the mitochondrial DNA showed two different clades corresponding to the geographic regions of Guatemala and the Yucatan Peninsula (May-Itzá et al. 2012).

Regarding other species of Mexican *Melipona*, recently a molecular comparison was conducted between populations of *M. fasciata* and *M. colimana* from the Pacific coast, using *cox*1 and *ITS*1. It was confirmed that both species are sister lineages (sequence divergence of only 0.6%), and belong within the subgenus *Michmelia* (Ruiz et al. 2013).

Another economically and culturally important group of stingless bee species in Mexico is the genus *Scaptotrigona*. In the last taxonomic revision of the stingless bees of Mexico, only three *Scaptotrigona* species had been recognized, namely, *Scp. mexicana, Scp. hellwegeri*, and *Scp. pectoralis* (Ayala 1999: Fig. 1.17). Recently, the genetic diversity of the Mexican *Scaptotrigona* has been investigated, and evidence of possible cryptic species has been found (Quezada-Euán et al. 2012; Hurtado-Burillo et al. 2013; Hurtado-Burillo et al. 2016).

The genetic diversity of *Scp. hellwegeri* was first studied using specimens from two populations, one on the Trans-Pacific volcanic belt in Jalisco and another some 500 km south in Guerrero. Significant differences were found in the genetic makeup and morphometrics between both populations indicating the existence of at least two genetic lineages (Quezada-Euán et al. 2012) (Fig. 1.17).

Later, Hurtado-Burillo et al. (2013) barcoded specimens of the three putative species of Mexican *Scaptotrigona* and found intraspecific divergence of 0.7%, whereas the interspecific value was 2.79%. Sequence analyses evidenced the close affinity of *Scp. mexicana* with *Scp. hellwegeri*. The results also suggested the existence of cryptic species within *Scp. mexicana*.

To evaluate the previous hypothesis, Hurtado-Burillo et al. (2016) and Hurtado-Burillo et al. (2017), performed a species delimitation analysis of the three putative Mexican *Scaptotrigona* species, based on the geometric morphometrics of the forewing and multilocus data. Their findings confirmed strong support for two clades in *Scp. mexicana*, one distributed along the Gulf of Mexico (Veracruz and Puebla) and another along the Pacific coast (Chiapas) (Fig.1.17). It had already been noted that specimens of *Scp. mexicana* from the East and West coast show contrasting color patterns (Ayala 1999); the findings of Hurtado-Burillo et al. (2017) suggest that they may be cryptic species. In the case of *Scp. pectoralis*, no evidence of genetic structure was found among populations from the Yucatan Peninsula, and the Pacific and the Gulf coasts. This species seems a single homogeneous population across Mexico (Hurtado-Burillo et al. 2017) (Fig. 1.17). However, it would be important to compare the genetic make up of Mexican *Scp. pectoralis* with populations from Central America south of the Sierra Madre del Sur. In contrast, the populations of *Scp. hellwegeri* from Jalisco and Guerrero do not seem to constitute separate species (Hurtado-Burillo et al. 2017). Nonetheless, these two groups show genetic differences which should grant their conservation as separate units (Quezada-Euán et al. 2012).

The pattern of geographic distribution of the 46 stingless bee species reported for Mexico (Annex 1) allows the identification of three ecological groups (Ayala 1999):

1. Wide distribution (montane, tropical, or subtropical)
2. Distribution associated with evergreen tropical forest
3. Endemic species

Fig. 1.17 Geographic distribution of the three species of *Scaptotrigona* in Mexico, in accordance with Ayala and Ortega-Huerta (2008). The analyses by Hurtado-Burillo et al. (2016) suggest the existence of two cryptic species in *Scp. mexicana (S.m.1* and *S.m.2)*

Most species present in the Yucatan Peninsula also exhibit wide distribution across Mexico (Ayala 1999). Representatives are *Pt. bilineata*, *M. beecheii*, *Pb. frontalis*, *Tr. fulviventris*, and *Tgca. pipioli*. Among them, the species with the most northern distribution on Eastern Mexico is *Pb. frontalis*, reaching the state of Nuevo Leon. Special cases of species with wide distribution are *N. perilampoides* and *Fr. nigra*. *N. perilampoides* is the species with the most northern distribution of all stingless bees; some specimens have been collected on the Pacific coast close to the 29° of latitude. This species is highly eclectic being found from lowland forests up to 1500 m of altitude in pine forests. Populations of *Fr. nigra* are mostly found along the Pacific coasts and the basin of the River Balsas reaching south in the Yucatan Peninsula, but in contrast with *N. perilampoides*, it is not common along the Gulf.

Among the Mexican stingless bee fauna, the genus *Plebeia* stands out for the number of species. The 11 species of this genus of minute stingless bees represent nearly a quarter of the meliponines in Mexico (Ayala 2016). To date, no detailed studies have been conducted to analyze morphological differences and the genetic structure of any Mexican *Plebeia*; however, in accordance to the results in other species and the more limited capacity of dispersion of *Plebeia* colonies, the existence of cryptic species in this genus is very likely. Ayala (2016) presented an updated revision of the Mexican *Plebeia*, and separated them in at least three ecological groups. A first group, commonly found in tropical forests, is comprised by *Pb. frontalis*, *Pb. llorentei*, and *Pb. jatiformis*. A second group, restricted to the temperate forests of central and southern Mexico, includes *Pb. fulvopilosa* and *Pb. manantlensis*. Another group is formed with *Pb. cora*, *Pb. parkeri*, and *Pb. mexica*, due to morphological similarities. On the other hand, morphologically similar species, like the dark form of *Pb. fulvopilosa* and *Pb. melanica*, are considered different species because of their allopatric distribution, restricted to the mountains and the lowland evergreen tropics of Michoacán and Sierra Madre del Sur, respectively. In contrast with the dark integument of the former two species, *Pb. pulchra* and *P. jatiformis* are the species with lighter integument of all Mexican *Plebeia*. The smallest species is *Pb. mourena*, whose workers are only 3 mm long.

The distribution of many stingless bee species in Mexico is being defined with the help of collecting works (Ayala 1999). Using such data, recently, Yurrita et al. (2017) presented a model of distribution for the *Melipona* in Mexico and Central America. Interestingly, only 20% of the distribution of several species includes protected areas. Clearly, studies of such kind should be encouraged for the development of future conservation strategies and to better plan collecting efforts.

To finish this chapter, general information on the geographic distribution, nest characteristics, and external anatomy of the worker is presented in Figs. 1.18, 1.19, 1.20, 1.21, 1.22, 1.23, 1.24, 1.25, 1.26, 1.27, and 1.28 for some stingless bees from the Yucatan Peninsula.

Fig. 1.18 General information on *Melipona beecheii* with images of the body, leg, and wing of a worker. The map includes those localities where this species has been collected in accordance with Ayala (1999) and Moure et al. (2007). The typical locality and the ubication of the holotype are unknown for this species

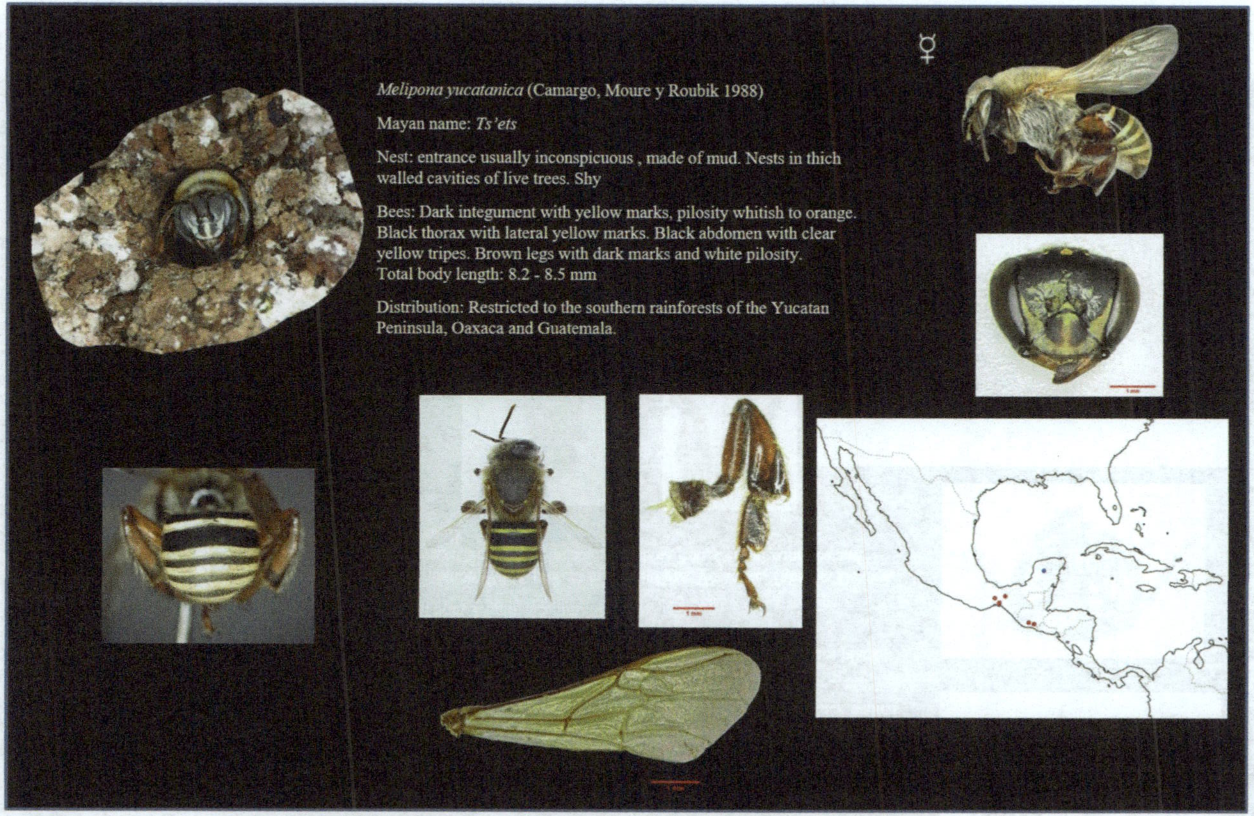

Fig. 1.19 General information on *Melipona yucatanica* with images of the body, leg, and wing of a worker. The map includes those localities where this species has been collected in accordance with Ayala (1999) and Moure et al. (2007). In blue the typical locality for this species

Fig. 1.20 General information on *Cephalotrigona zexmeniae* with images of body, wing, and leg of a worker. The map includes those localities where this species has been collected Ayala (1999) and Moure et al. (2007). In blue the typical locality for this species

Fig. 1.21 General information *on Partamona bilineata* with images of body, wing, and leg of a worker. The map includes those localities where this species has been collected Ayala (1999) and Moure et al. (2007). In blue the typical locality for this species

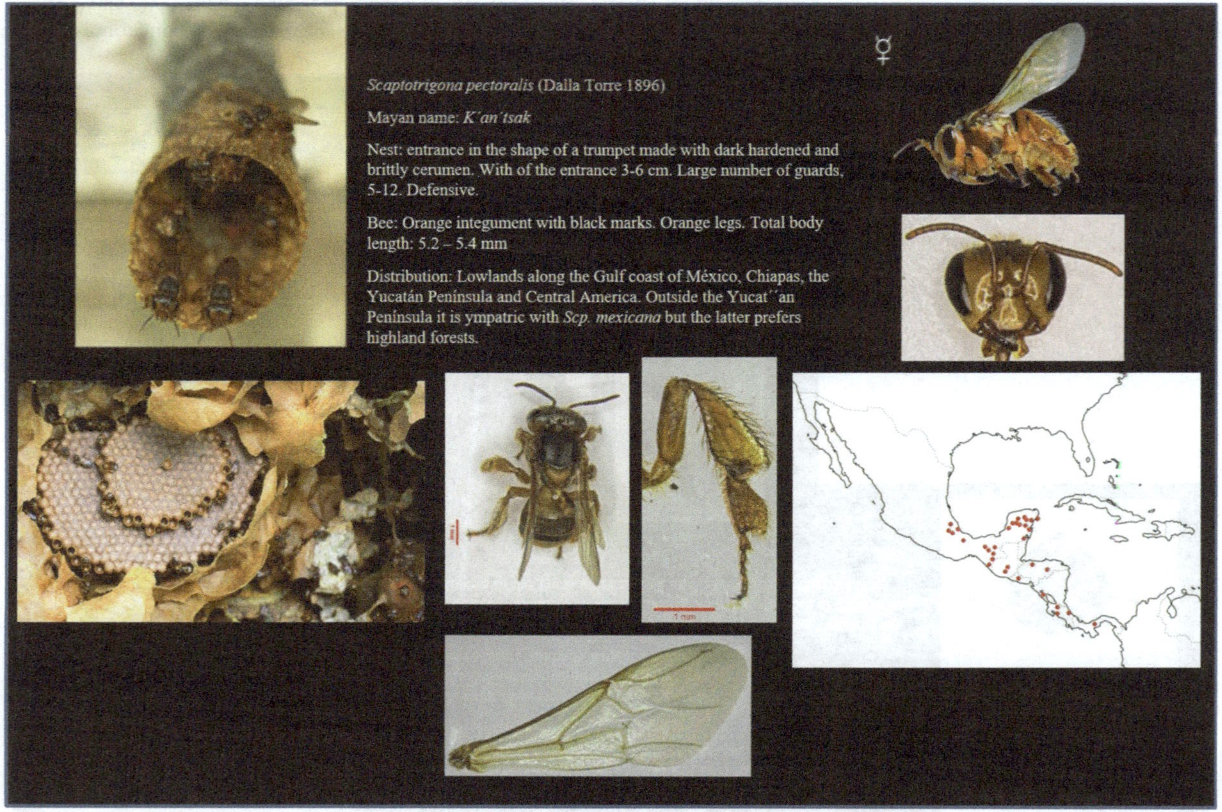

Fig. 1.22 General Information on *Scaptotrigona pectoralis* with images of body, wing, and leg of a worker. The map includes those localities where this species has been collected Ayala (1999) and Moure et al. (2007)

Fig. 1.23 General information on *Nannotrigona perilampoides* with images of body, wing, and leg of a worker. The map includes those localities where this species has been collected Ayala (1999) and Moure et al. (2007). In blue the typical locality for this species

Fig. 1.24 General information on *Frieseomelitta nigra* with images of body, wing, and leg of a worker. The map includes those localities where this species has been collected Ayala (1999) and Moure et al. (2007)

Fig. 1.25 General information on *Plebeia frontalis* with images of body, wing, and leg of a worker. The map includes those localities where this species has been collected Ayala (1999) and Moure et al. (2007). The typical locality is marked in blue

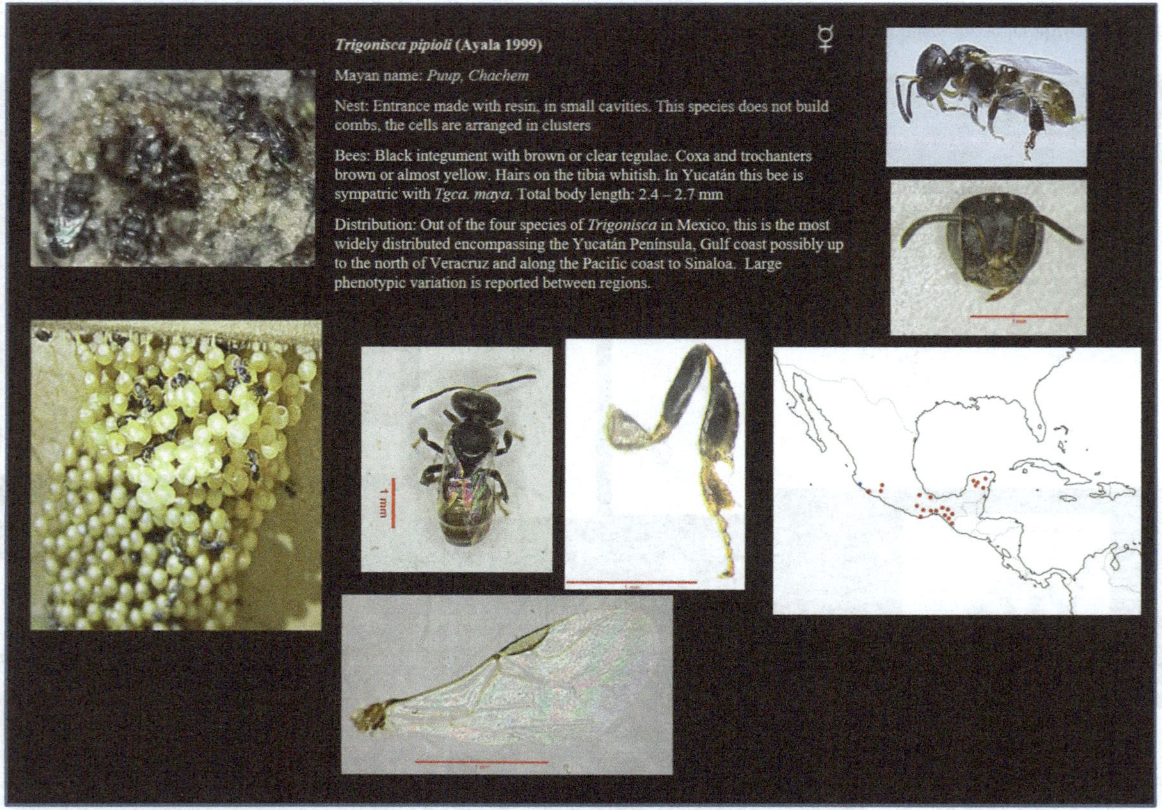

Fig. 1.26 General information on *Trigonisca pipioli* with images of body, wing, and leg of a worker. The map includes those localities where this species has been collected (Ayala 1999; Moure et al. 2007). In blue the typical locality for this species

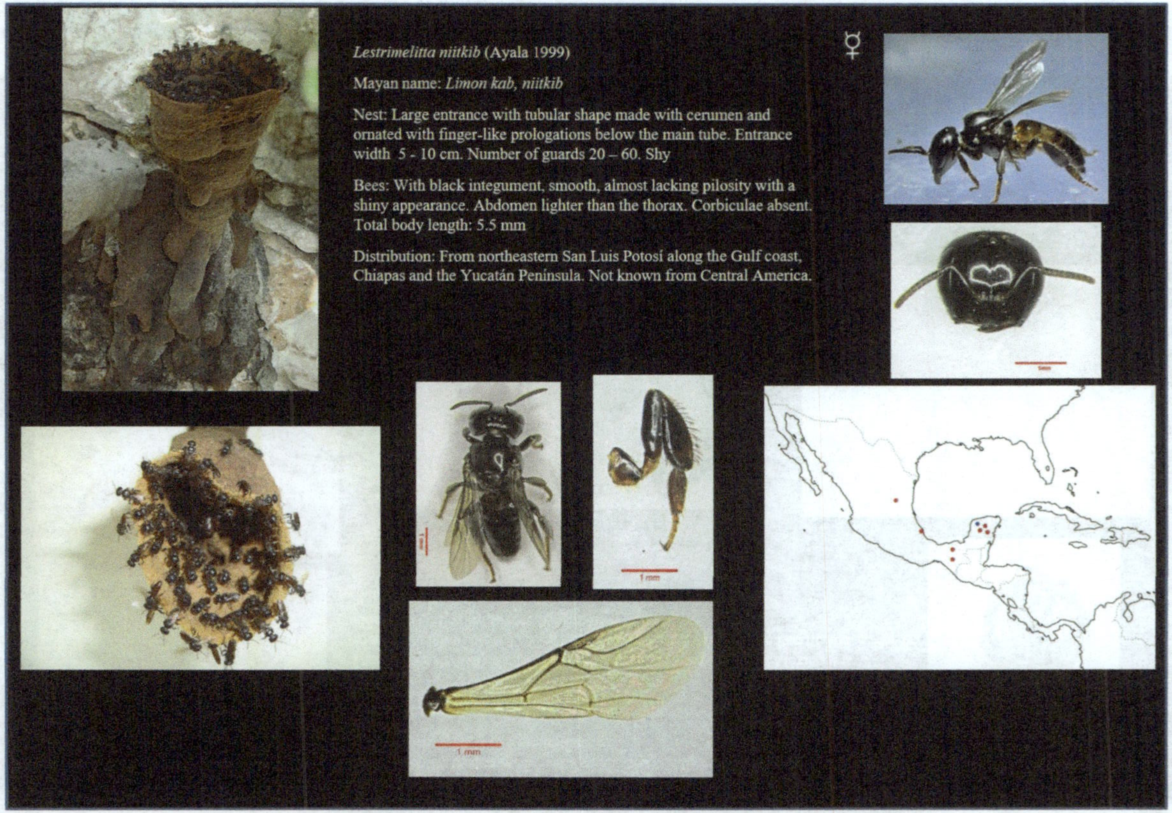

Fig. 1.27 General information on *Lestrimelitta niitkib* with images of body, wing, and leg of a worker. The map includes those localities where this species has been collected (Ayala 1999; Moure et al. 2007). In blue the typical locality for this species

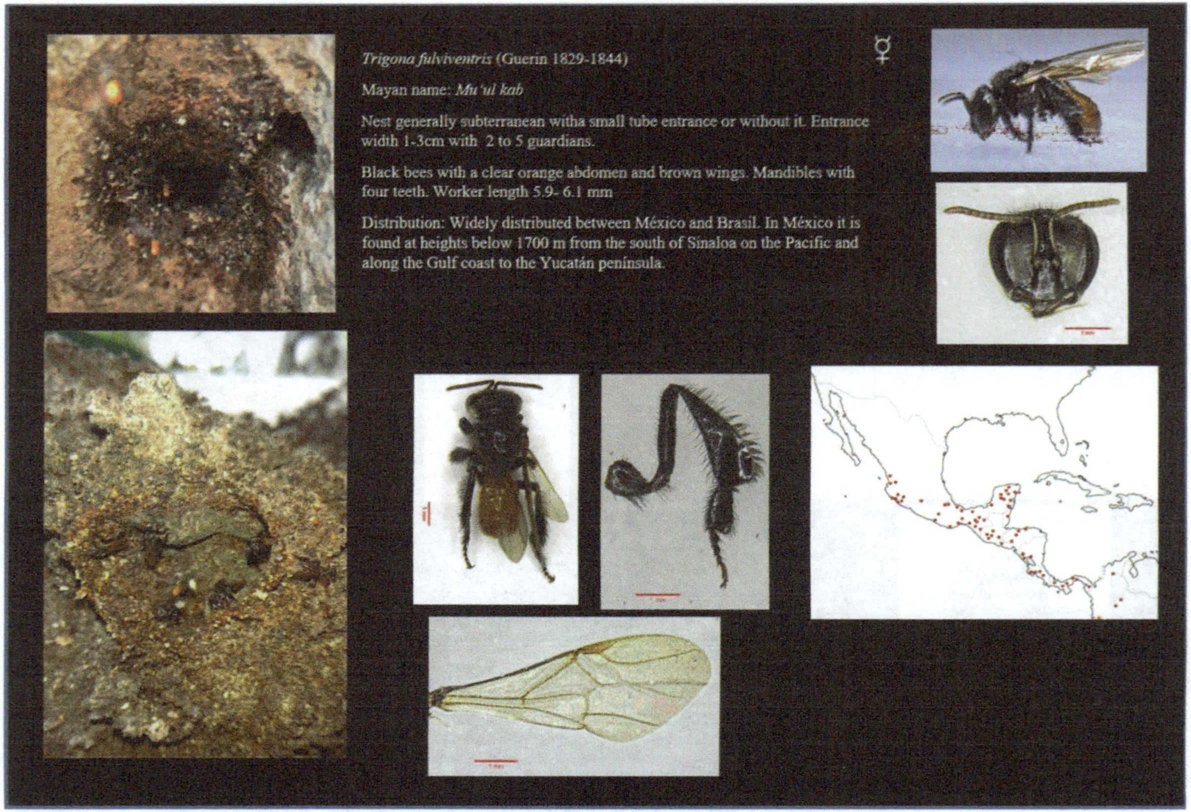

Fig. 1.28 General information on *Trigona fulviventris* with images of body, leg, and wing of a worker. The map includes the localities where the species has been collected in accordance with Ayala (1999) and Moure et al. (2007)

References

Ayala R (1999) Revisión de las abejas sin aguijón de México (Hymenoptera: Apidae: Meliponini). Folia Entomológica Mexicana 106:1–123

Ayala R (2006) Bee diversity in Mexico. In: Kevan P, Imperatriz-Fonseca VL (eds) Pollinating bees the conservation link between agriculture and nature. Ministry of Environment, Brasilia, p 318

Ayala R (2016) Las abejas del género *Plebeia* Schwarz (Apidae: Meliponini) de México. Entomología Mexicana 3:937–942

Ayala R, Ortega-Huerta M (2008) Distribución potencial de las abejas sin aguijón (Hymenoptera: Meliponini) de México. In: V Congreso Mesoamericano sobre abejas sin aguijón. Universidad Autónoma de Yucatán, Mérida, pp 74–75

Ayala R, González VH, Engel MS (2013) Mexican stingless bees (Hymenoptera: Apidae): diversity, distribution, and indigenous knowledge. In: Vit P, Pedro SRM, Roubik DW (eds) Pot-honey: a legacy of stingless bees. Springer, New York, pp 135–152

Beechey FW (1831) Narrative of a voyage to the Pacific and Beering's Strait, Vol. 2 of 2: to co-operate with the polar expeditions; performed in His Majesty's Ship Blossom, under the command of Captain F. W. Beechey, R. N. F. R. S. &C., in the years 1825, 26, 27, 28. H. Colburn and R. Bentley, London

Bickford D, Lohman DJ, Sodhi NS, Ng PKL, Meier R, Winker K, Ingram KK, Das I (2006) Cryptic species as a window on diversity and conservation. Trends Ecol Evol 22:148–155

Camargo JMF, Pedro SRM (1992) Sistemática de Meliponinae (Hymenoptera, Apidae): sobre a polaridade e significado de alguns caracteres morfológicos. Naturalia (São Paulo) (número especial): Anais do encontro brasileiro de biología de abelhas e outros insetos sociais Special Issue, p 45–49

Camargo JMF, Pedro SRM (2007) Meliponini. In: Moure JS, Urban D, Melo GAR (eds) Catalogue of bees (Hymenoptera, Apoidea) in the Neotropical region. Sociedade Brasileira de Entomologia, Curitiba, pp 272–578

Camargo JMF, Wittmann D (1989) Nest architecture and distribution of the primitive stingless bee, *Mourella caerulea* (Hymenoptera, Apidae, Meliponinae): evidence for the origin of *Plebeia* (s. lat.) on the Gondwana continent. Stud Neotropical Fauna Environ 24:213–229

Camargo JMF, Vit P (2013) Historical biogeography of the Meliponini (Hymenoptera, Apidae, Apinae) of the neotropical region. In: Vit P, Pedro SRM, Roubik DW (eds) Pot-honey: a legacy of stingless bees. Springer, New York, pp 19–34

Camargo JMF, Moure JS, Roubik DW (1988) *Melipona yucatanica* new species (Hymenoptera: Apidae: Meliponinae); stingless bee dispersal across the Caribbean Arc and post-Eocene vicariance. Pan Pac Entomol 64:147–157

Carrillo A, Quezada-Euán JJG, Moo-Valle JH (2001) Estudio preliminar sobre la variabilidad morfológica de *Melipona beecheii* (Apidae: Meliponini) en su rango de distribución de México, América Central y el Caribe. In: Memoarias II (ed) Seminario Mexicano sobre abejas sin aguijón. Universidad Autónoma de Yucatán, México, pp 73–78

Costa MA, Del Lama MA, Melo GAR, Sheppard WS (2003) Molecular phylogeny of the stingless bees (Apidae, Apinae, Meliponini) inferred from mitochondrial 16S rDNA sequences. Apidologie 34:73–84

De la Rúa P, May-Itzá W d J, Serrano J, Quezada-Euán JJG (2007) Sequence and RFLP analysis of the ITS2 ribosomal DNA in two neotropical social bees, *Melipona beecheii* and *Melipona yucatanica* (Apidae: Meliponini). Insect Soc 54:418–423

Eardley CD (2004) Taxonomic revision of the African stingless bees (Apoidea: Apidae: Apinae: Meliponini). Afr Plant Protect 10:63–96

Fernandes-Salomão TM, Rocha RB, Campos LAO, Arafflјo EF (2005) The first internal transcribed spacer (ITS-1) of *Melipona* species (Hymenoptera, Apidae, Meliponini): characterization and phylogenetic analysis. Insect Soc 52:11–18

Fonte L (2007) Las "abejas de la tierra" en zonas de las provincias occidentales de Cuba: las colmenas, la miel que producen y los "meliponicultores". Trabajo de diploma. Universidad Agraria de La Habana, Cuba, p 77

Francoy TM, Grassi ML, Imperatriz-Fonseca VL, May-Itzá WdJ, Quezada-Euán JJG (2011) Geometric morphometrics of the wing as a tool for assigning genetic lineages and geographic origin to *Melipona beecheii* (Hymenoptera: Meliponini). Apidologie 44:499–507

Genaro JA (2008) Origins, composition and distribution of the bees of Cuba (Hymenoptera: Apoidea: Anthophila). Insecta Mundi, Paper 583

González-Acereto JA (1984) A cerca de la regionalización de la nomenclatura Maya de las abejas sin aguijón (*Melipona* sp.) en Yucatán. Revista de Geografía Agrícola Chapingo 6

González-Acereto JA (2008) Cría y manejo de abejas nativas sin aguijón en México. Ed. UADY-Secretaría Fomento Agropecuario y Pesquero Estado de Yucatán, México

Halcroft MT, Dollin A, Francoy TM, King JE, Riegler M, Haigh AM, Spooner-Hart RN (2016) Delimiting the species within the genus *Austroplebeia*, an Australian stingless bee, using multiple methodologies. Apidologie 47:76–89

Hebert PDN, Ratnasingham S, de Waard JR (2003) Barcoding animal life: cytochrome *c*oxidase subunit I divergences among closely related species. Proc R Soc Lond B 270(Suppl):S96–S99

Hebert PDN, Penton EN, Burns JM, Janzen DH, Hallwachs W (2004) Ten species in one: DNA barcoding reveals cryptic species in the neotropical skipper butterfly *Astraptes fulgerator*. Proc Natl Acad Sci USA 101:14182–14187

Hedtke SM, Patiny S, Danforth BN (2013) The bee tree of life: a supermatrix approach to apoid phylogeny and biogeography. BMC Evol Biol 13:138

Hurtado-Burillo M, Ruiz C, May-Itzá WdJ, Quezada-Eúan JJG, De la Rúa P (2013) Barcoding stingless bees: genetic diversity of the economically important genus *Scaptotrigona* in Mesoamerica. Apidologie 44:1–10

Hurtado-Burillo M, Martinez J, May-Itzá WJ, Quezada-Euán JJG, de la Rúa P (2014) Estudio de microsatélites en las abejas sin aguijón *Melipona colimana* y *M. beecheii* de Mesoamérica. Archivos de Zootecnia 63:145–151

Hurtado-Burillo M, Jara L, May-Itzá WJ, Quezada-Euán JJG, Ruiz C, De la Rúa P (2016) A geometric morphometric and microsatellite analyses of *Scaptotrigona mexicana* and *S. pectoralis* (Apidae: Meliponini) sheds light on the biodiversity of Mesoamerican stingless bees. J Insect Conserv 5:753–763

Hurtado-Burillo M, May-Itzá W d J, Quezada-Euán JJG, De la Rúa P, Ruiz C (2017) Multilocus species delimitation in Mesoamerican *Scaptotrigona* stingless bees (Apidae: Meliponini) supports the existence of cryptic species. Syst Entomol 42:171–181

Jaffé R, Pope N, Acosta AL, Alves DA, Arias MC, de la Rúa P, Francisco FO, Giannini TC et al (2016) Beekeeping practices and geographic distance, not land use, drive gene flow across tropical bees. Mol Ecol 25:5345–5358

Kerr WE, Maule V (1964) Geographic distribution of stingless bees and its implications (Hymenoptera: Apidae). J NY Entomol Soc 72:2–18

Landaverde-González P, Enriquez E, Ariza MA, Murray T, Paxton RJ, Husemann M (2017) Fragmentation in the clouds? The population genetics of the native bee *Partamona bilineata* (Hymenoptera: Apidae: Meliponini) in the cloud forests of Guatemala. Conserv Genet 18(3):631–643

Lóriga-Peña W, Quezada-Euán JJG (2009) Afinidades morfométricas de *Melipona beecheii* B. de la Isla de Cuba y poblaciones del continente. In: Memorias VI (ed) Congreso Mesoamericano sobre abejas nativas. Universidad de San Carlos, Antigua, pp 197–205

Lóriga-Peña W, May-Itzá WdJ, De la Rúa P, Quezada-Euán JJG (2011) *Melipona beecheii* de Cuba: afinidades genéticas y posible origen biogeográfico. In: Memorias VII seminario Mesoamericano sobre abejas nativas. El Colegio de la Frontera Sur, Tosepan

Martins AC, Melo GAR, Renner SS (2014) The corbiculate bees arose from new world oil-collecting bees: implications for the origin of pollen baskets. Mol Phylogenet Evol 80:88–94

May-Itzá WdJ, Quezada-Euán JJG, De la Rúa P (2009) Intraspecific variation in the stingless bee *Melipona beecheii* assessed with PCR-RFLP of the ITS1 ribosomal DNA. Apidologie 40:549–555

May-Itzá WDJ, Quezada-Euán JJG, Medina-Medina LA, Enríquez E, De la Rúa P (2010) Morphometric and genetic differentiation in isolated populations of the endangered Mesoamerican stingless bee *Melipona yucatanica* (Hymenoptera: Apoidea) suggest the existence of a two species complex. Conserv Genet 11:2079–2084

May-Itzá WDJ, Quezada-Euán JJG, Ayala R, De La Rúa P (2012) Morphometric and genetic analyses differentiate Mesoamerican populations of the endangered stingless bee *Melipona beecheii* (Hymenoptera: Meliponidae) and support their conservation as two separate units. J Insect Conserv 16:723–731

Melo GAR (2016) *Plectoplebeia*, a new Neotropical genus of stingless bees (Hymenoptera: Apidae). Fortschr Zool 33:e20150153

Michener CD (1974) The social behavior of the bees: a comparative study. Belknap Press, Harvard University, Cambridge

Michener CD (1982) A new interpretation of fossil social bees from the Dominican Republic. Sociobiology 7:37–45

Michener CD (1990) Classification of the Apidae (Hymenoptera). Appendix: *Trigona genalis* Friese, a hitherto unplaced New Guinea species. Univ Kansas Sci Bull 54:75–163

Michener CD (1999) The corbiculae of bees. Apidologie 30:67–74

Michener CD (2000) The bees of the world. Johns Hopkins University Press, Baltimore

Moure JS (1992) *Melikerria* e *Eomelipona*, dois subgeneros novos em *Melipona* Illiger, 1806. (Hymenoptera—Apidae). In: Cruz Landim CC, Chaud NJ (eds) Anais do Encontro Brasileiro de biologia de abelhas e outros insetos sociais. Homenagem aos 70 Anos de Warwick Estevam Kerr (Naturalia, special tome). UNESP, Rio Claro, pp 45–49

Moure JS, Urban D, Melo GAR (2007) Catalogue of bees (Hymenoptera, Apoidea) in the Neotropical region. Sociedade Brasileira de Entomologia, Curitiba

Pedro SRM (2014) The stingless see fauna in Brazil (Hymenoptera: Apidae). Sociobiology 61:348–354

Quezada-Euán JJG, May-Itzá W d J, González-Acereto JA (2001) Meliponiculture in México: problems and perspective for development. Bee World 82:160–167

Quezada-Euán JJG, Paxton RJ, Palmer KA, May-Itzá WdJ, Tay WT, Oldroyd BP (2007) Morphological and molecular characters reveal differentiation in a Neotropical social bee, *Melipona beecheii* (Apidae: Meliponini). Apidologie 38:247–258

Quezada-Euán JJG, May-Itzá WdJ, Rincón M, De la Rúa P, Paxton RJ (2012) Genetic and phenotypic differentiation in endemic *Scaptotrigona hellwegeri* (Apidae: Meliponini): implications for the conservation of stingless bee populations in contrasting environments. Insect Conserv Divers 5:433–443

Ramírez SR, Nieh JC, Quental TB, Roubik DW, Imperatriz-Fonseca VL, Pierce NE (2010) A molecular phylogeny of the stingless bee genus *Melipona* (Hymenoptera: Apidae). Mol Phylogenet Evol 56:519–525

Rasmussen C, Cameron SA (2007) A molecular phylogeny of the Old World stingless bees (Hymenoptera: Apidae: Meliponini) and the non-monophyly of the large genus *Trigona*. Syst Entomol 32:26–39

Rasmussen C, Cameron SA (2010) Global stingless bee phylogeny supports ancient divergence, vicariance, and long distance dispersal. Biol J Linn Soc 99:206–232

Roubik DW (1989) Ecology and natural history of tropical bees. Cambridge University Press, Cambridge

Roubik DW (2006) Stingless bee nesting biology. Apidologie 37:124–143

Roubik DW, Camargo JMF (2012) The Panama microplate, island studies and relictual species of *Melipona* (*Melikerria*) (Hymenoptera: Apidae: Meliponini). Syst Entomol 37:189–199

Ruiz C, May-Itzá WdJ, Quezada-Euán JJG, De la Rúa P (2013) Presence of nuclear copies of mitochondrial origin (NUMTs) in two related species of stingless bee genus *Melipona* (Hymenoptera: Meliponini). J Zool Syst Evol Res 51:107–113
Schwarz FH (1932) The genus *Melipona*. The type genus of the Meliponidae or stingless bees. Bull Am Mus Nat Hist 63:231–460
Wille A (1976) Las abejas Jicotes del género *Melipona* (Apidae: Meliponini) de Costa Rica. Rev Biol Trop 24:123–147
Wille A (1979) Phylogeny and relationships among the genera and subgenera of the stingless bees (Meliponinae) of the world. Rev Biol Trop 27:241–277
Wille A (1983) Biology of the stingless bees. Annu Rev Entomol 28:41–64
Yurrita CL, Ortega-Huerta MA, Ayala R (2017) Distributional analysis of *Melipona* stingless bees (Apidae: Meliponini) in Central America and Mexico: setting baseline information for their conservation. Apidologie 2:247–258

Chapter 2
Nesting Biology

Queen of *Scp. pectoralis* wandering among workers and combs.

The nest is central to the lives of stingless bees; here they develop and pass through different life stages, feed, and reproduce. Stingless bee nesting biology is also tightly linked with defense strategies and interaction with other organisms (Roubik 2006). Notably, large variation exists in the size, ornaments, and substrates used in the nests of stingless bees (Wille and Michener 1973; Roubik 1983). In the forests, different species may nest in tree trunks and stems, crevices within rocks, burrows made by other animals, and termite nests (Roubik 2006). In general, the space required for nesting is a function of the size of the colony and the body size of the

J. J. G. Quezada-Euán, *Stingless Bees of Mexico*,
https://doi.org/10.1007/978-3-319-77785-6_2

individuals; species with large body size and large number of individuals usually require larger nesting spaces (Wille and Michener 1973). Accordingly, large variation in nest volume can be found across species, from 0.1 in some *Plebeia* up to 35 L in the largest *Melipona* and *Trigona* (Roubik 1983, 2006).

Adaptability or flexibility of different species to varied substrates and environments seem to play an important role in the conformation of stingless bee communities (Eltz et al. 2002, 2003). For instance, in the Yucatan Peninsula, some species (e.g., *M. yucatanica*) only nest in large living trees, a type of substrate mostly found in pristine forests. Such specific requirements for nesting, can restrain distribution outside these environments. Less selective species may be more frequently found in areas with some degree of alteration, such as *M. beecheii* and *Cph. zexmeniae*. Even though these two species prefer pristine rainforests, they may use alternative spaces in rocks or buildings, allowing them to nest in comparatively less conserved areas. At the other extreme, there are eclectic species like *N. perilampoides* and *Fr. nigra* showing ample flexibility in the use of different substrates and environments for nesting. Such species can be common even in highly disturbed areas (Fig. 2.1).

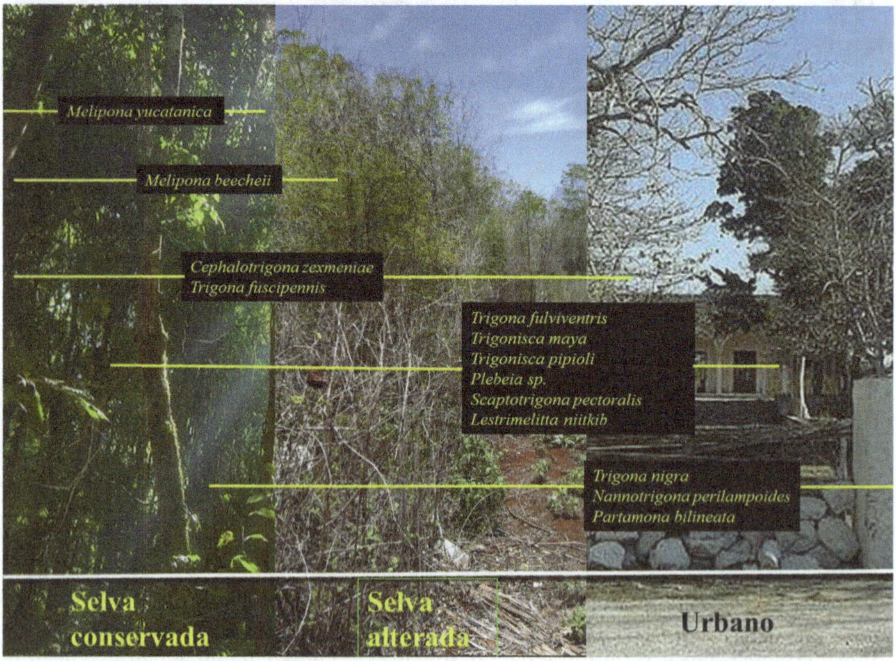

Fig. 2.1 Relative frequency of stingless bee species across three different landscapes in Yucatan. The highest frequency for each species is represented by the black rectangles; the yellow lateral lines show the relative presence across habitats. Selva conservada = pristine forest; Selva alterada = fragmented forest; Urbano = Urban (data from González-Acereto et al. 2006)

Although adequate and sufficient nest substrates are strong determinants of stingless bee communities, the availability of pollen and nectar can also affect nest density and diversity (Eltz et al. 2002). Interestingly, considerable variation in nesting habits can be found across different regions for the same species (Roubik 1983), but no formal intraspecific comparison of nest characteristics has been conducted to identify factors affecting such variation (Roubik 2006).

2.1 Nest Structure

In spite of the great variation in the size and accessories found in nests of different stingless bees, the basic design is remarkably homogenous across species (Fig. 2.2; Roubik 2006).

The basic material used for nest buildup for all species of stingless bees is the cerumen. The bees make cerumen mixing the wax they produce on their dorsal abdominal glands, with resins collected from various plants. In stingless bees the use of pure wax, as used by honey bees, is practically nonexistent.

The nest connects with the outside world through the entrance. Entrances can be very simple, in the shape of a simple hole scarcely surrounded by resin, to some very large and ornate structures (for nest entrance diversity check Chap. 1 Figs. 1.18 to 1.28). The relative size (length and width) of the entrance seems to be under conflicting selective pressures. Smaller entrances are easier to defend, but a larger entrance can accommodate heavier forager traffic, so there may be a trade-off between foraging capacity and security (Couvillon et al. 2008). Deposits of resin near the entrance can be common in several species (e.g., *Fr. nigra*) which are used in rapid defense against some predators like ants (Roubik 1983).

A layer of solid cerumen mixed with propolis (sometimes also mixed with mud) known as bitumen is used to cover the interior of the nest. The bitumen is used to line up the cavity and to isolate the nest from temperature changes. It also helps in

Fig. 2.2 Basic structure of a stingless bee nest (*Scp. pectoralis*): (**a**) vertical cells in horizontal combs in the brood area. (**b**) Involucrum. (**c**) Food pots

the control of the volume of the cavity. The bitumen can be removed to accommodate growth during the flowering season and contracted during dearth. Most structures inside the nest are attached to the bitumen by means of cerumen pillars and connectives (Nogueira-Neto 1997).

Stingless bees, like honey bees, store food for long periods. For this, meliponines use specially built containers, not the brood combs, as in the case of the honey bee (Roubik 1989). Stingless bees build cerumen structures called pots where the honey and pollen are kept (Roubik 2006; Camargo and Pedro 2007). Pollen and honey are stored separately, meaning there are pollen and honey pots. An exception is found in cleptobiotic species which store a mixture of pollen and honey in the same pot. Pollen and honey pots are usually of similar size and they are intermingled (Fig. 2.2). The shape of pots is usually ovoid, but may also vary across species, sometimes appearing conical or cylindrical (Wille and Michener 1973).

In many species, there are one or few layers of thin cerumen called the involucrum. The involucrum separates the food pots from the brood area and serve as protection to the latter. The involucrum also helps to control the temperature in the brood area (Fig. 2.2). The brood area is the nest section containing the cells with individuals in development. Callow workers and the queen are more frequently found in the brood area. Similar to honey bees, stingless bees develop in individual cells. However, the cells are in all, but two, species of stingless bees vertical (in honey bees they are oblique, almost horizontal) (Wille 1983). In contrast with the honey bees, the brood cells are built with cerumen and are used only once. The cells can be attached to each other by small cerumen threads (called connectives) forming clusters or built wall to wall forming horizontal combs (Fig. 2.3). Interestingly, the arrangement in clusters or combs seems to relate to the body size of the species; minute species more frequently build brood cells in clusters (Michener 2001). Noteworthy, some comb-building species can also develop them in a spiral shape as *N. perilampoides* (Fig. 2.3).

Wille (1983) divided the nest of stingless bees in accordance with the type of cavity and the arrangement of the cells (combs or clusters) in five categories:

1. Nests in cavities with horizontal combs (examples are *Melipona, Scaptotrigona, Nannotrigona*) (Fig. 2.3).
2. Exposed nests (*Partamona, Tr. fuscipennis*).
3. Subterranean nests (*Tr. fulviventris*).
4. Nests with cells in clusters (*Plebeia, Trigonisca, Fr. nigra*) (Fig. 2.3).
5. Nests with vertical combs (*Scaura*—one species in South America, and *Dactylurina*—one species in Africa).

Invariably, when individuals enter the pupal phase, in all stingless bees the workers remove the external layer of cerumen from the cell, exposing the cocoon (Fig. 2.4). Since the cocoon has no color, the cells and combs containing pupae are paler than the combs with eggs and larvae, made with dark cerumen.

When the adult emerges, the workers remove the cocoon remains, which are then taken out of the colony. The removal of empty cells normally starts from the center

Fig. 2.3 Two basic types on brood cell arrangement in the Meliponini: above cells in clusters of *Fr. nigra,* below horizontal combs of *M. beecheii* and *N. perilampoides*. A spiral shape is evident in the horizontal comb of the latter

of the combs as these were the first cells produced. Therefore, an empty space starts to show in the center of the comb, which increases gradually as individuals emerge and eventually disappearing (Fig. 2.4). In contrast with honey bees that rear many generations of workers in the same combs, stingless bees use brood cells once. Perhaps this is an adaptation for an economic use of cerumen (Fig. 2.4, Wille and Michener 1973).

Some species, like *Cph. zexmeniae*, do not eliminate the cocoons and waste from their colonies; they store them in a separate chamber, usually in the lower part of the cavity, forming a mass called scutellum. The scutellu may be used for temperature control (Sakagami 1982). The scutellum is usually separated from the main nest by a thin layer of bitumen (Zwaal 1992; Fig. 2.5). In many species like *M. beecheii*, temporary waste dumps can be distributed on different parts of the nest, and eventually some specialized workers remove them from the colony (Medina-Medina et al. 2014).

Fig. 2.4 Dynamics of comb construction in stingless bees. The combs start with a few cells (brown rectangles) that will eventually occupy the center, as more cells are added (red arrows), and the comb expands. The new cells are made of cerumen and can be found at different stages of construction. When the queen lays an egg, the cell is sealed, and remains with its cerumen cover during the egg and larval stages of individuals. When the bee enters the pupal stage, and builds up a cocoon, the outer layer of cerumen is removed by the workers exposing the light-colored cocoons (cream rectangles). When the imago emerges, the workers remove the remains of the cocoon and the cells start to disappear (blue arrows). In the empty space new cerumen cells start to be constructed (red arrows at the bottom and image on the right). The image below shows *M. beecheii* workers removing the cerumen from cells with individuals already in the pupal stage and a cell where a bee has already emerged in the center of the comb

Some stingless bees are associated with termite or ant nests (Wille and Michener 1973), although this association seems flexible (Roubik 1983). It seems that species that nest in live termite mounds may use the holes made by birds or other predators to start building their nest (Roubik 2006). Some species like *Paratrigona peltata* and *Scr. latitarsis* may dig holes in the nests of *Nasutitermes* and *Camponotus* from which the workers can start expanding a cavity and eventually build a nest (Wille 1983). Other species, e.g., *Tr. fuscipennis*, use termite nests usually when these are almost empty (Wille 1983).

Fig. 2.5 Nest of *Cph. zexmeniae* showing the position of the scutellum (E), the brood chamber (C), and food pots (P)

2.2 Homeostasis

One of the greatest achievements of eusociality is homeostasis, the capacity to maintain a stable environment in the nest with relative independence from environmental fluctuations (Michener 1974). Nest thermoregulation or temperature control is one basic aspect of homeostasis in eusocial species. Solitary insects are ectotherms; that is, they lack, or have, negligible physiological processes to control body temperature. This means that their body temperature fluctuates with the environment. Highly eusocial insects show a progress towards nest temperature control using a series of passive and active mechanisms resulting in various degrees of endothermy. Notably, some eusocial species can keep nest temperature within narrow range variations (Seeley 1985; Heinrich 1993).

Despite the advantages of an optimal environment for the development of the brood, homeostasis represents a big challenge for eusocial colonies. In the case of the honey bee, producing heat in the winter or cooling down the nest in the summer, implies an energy expenditure that can be similar to that of mammals (Moritz and Southwick 1992). In social insects Colony heating is a product of individual heat production involving the contraction of thoracic muscles. To produce heat over long periods, workers need a source of energy in the form of carbohydrates. This involves the accumulation of a large reserve of honey which requires a great deal of energy to collect (Seeley 1985; Winston 1987).

The economy of heat production is contrasting between honey bees and stingless bees. Generally, it seems that stingless bees, in spite of living in highly populated eusocial colonies are not quite capable of controlling nest temperature at narrow ranges (Roubik 1983). They seem to rely more on the use of passive mechanisms to maintain nest temperatures, than the active production of heat (Heinrich 1993). Such an incipient control of nest temperature may explain the distribution of stingless bees confined to the tropics, in comparison with the honey bee which has colonized diverse environments (Engels et al. 1995). Stingless bees have probably taken advantage of the warm temperatures, usually between 18 and 35 °C found in the tropics, and have not evolved more efficient and active mechanisms of temperature control (Roubik 1989). Indeed, the immature of various species develop well at such wide range of warm temperatures (Macías-Macías et al. 2011).

Because of the lack of active mechanisms to produce heat, stingless bee colonies seem particularly sensitive to cold. Individual and group heat production seem generally poor. When exposed to low temperature, stingless bees do not show the instinctive clustering observed in workers of *A. mellifera.* Stingless bees move closer to each other, but the cluster is not tight, leaving wide spaces among individuals, which result in comparatively high heat loss in the group (Zucchi and Sakagami 1972). The workers of several species are also less efficient in producing heat by means of the contraction of their thoracic muscles (Roubik and Peralta 1983; Engels et al. 1995). In addition, meliponine colonies do not store large amounts of honey and this is more humid (with less sugar per weight unit), compared to that of the honey bee (Roubik 1983). This means that their energy reserves are comparatively poor in quantity and quality, thus reducing the capacity to produce heat over long periods. As a result, stingless bee colonies strongly depend on local environmental conditions and well-protected nest spaces to preserve the temperatures inside their colonies (Roubik 1989). Interestingly, with adequate nest conditions (good insulation) stingless bees can maintain nest temperatures within narrow ranges. For instance, species like *Tr. spinipes* can keep the temperature of the brood between 34 and 36 °C (Zucchi and Sakagami 1972). Some species in the Yucatan such as *S. pectoralis* can also keep the nest within 3–4°C variation during the day (Table 2.1).

Noteworthy, most species of stingless bee nest in cavities with thick walls further increasing the insulation with structures like the bitumen. Certainly, most structures in stingless bee nests may serve as a means of insultation. The surrounding layout of food pots and layers of involucrum could maintain stable temperatures in the brood. Structures like the scutellum may produce additional heat through fermentation of debri (Zucchi and Sakagami 1972).

Actually, Temperature economy inside stingless bee colonies centers on the brood. The brood chamber is the area experiencing the lowest temperature variation

Table 2.1 Mean and range temperatures in the brood area of various species of stingless bees

Species	Mean (°C)	Range (°C)	Locality	Reference
Frieseomelitta nigra	–	21.0–31.0	Colombia	Torres et al. (2009)
Frieseomelitta varia	23.74	18.0–29.6	Riberão Preto, Brazil	Zucchi and Sakagami (1972)
Lestrimelitta niitkib	27.6	23.5–31.9	Yucatan, Mexico	Moo-Valle et al. (unpublished)
Leurotrigona mulleri	23.31	16.5–29.2	Riberão Preto, Brazil	Zucchi and Sakagami (1972)
Melipona rufiventris	31.3	31.04–31.85	Manaus, Brazil	Roubik and Peralta (1983)
Melipona beecheii	30.2	25.4–34.0	Yucatan, Mexico	Moo-Valle et al. (2000)
Melipona colimana	26	24–28	Jalisco, Mexico	Macías-Macías et al. (2011)
Melipona fasciculata	29.6	26–33.5		Kerr et al. (1996)
Melipona interrupta	29.84	27.3–31.8	Manaus, Brazil	Becker (2014)
Melipona quadrifasciata	29.3	25.0–31.7	Riberão Preto, Brazil	Zucchi and Sakagami (1972)
Melipona rufiventris	31.91	30.3–34.0	Riberão Preto, Brazil	Zucchi and Sakagami (1972)
Melipona scutellaris (winter)	26.05	24.1–28.2	Riberão Preto, Brazil	Roldão (2011)
Melipona scutellaris (summer)	29.8	29.15–30.75	Riberão Preto, Brazil	Roldão (2011)
Melipona seminigra	31.73	31.03–32.36	Manaus, Brazil	Roubik and Peralta (1983)
Melipona subnitida	–	27.0–33.0	Mossoró, Brazil	Ferreira (2014)
Nannotrigona perilampoides	29.08	23.3–32.3	Yucatan, Mexico	Moo-Valle et al. (unpublished)
Plebeia droryana	26.14	20.0–29.5	Riberão Preto, Brazil	Zucchi and Sakagami (1972)
Scaptotrigona postica	32.0	29.0–35.0	Riberão Preto, Brazil	Engels et al. (1995)
Scaptotrigona depilis	30.03	27.6–31.8 26.5–35.0	Riberão Preto, Brazil	Zucchi and Sakagami (1972); Vollet-Neto et al. (2015)
Tetragonisca angustula	30.3	27.5–31.5	Colombia	Torres et al. (2007)
Trigona spinipes	35.06	34.1–36.0	Riberão Preto, Brazil	Zucchi and Sakagami (1972)

in the nest (Moo-Valle et al. 2000). Detailed recordings showed that the temperature of the brood in *M. fuliginosa* and *Partamona cupira* was relatively constant, in spite of a high variation in the environment. Temperature fluctuation in the environment was 4–26 °C, while the temperature in the brood only varied between 26.5 and 30 °C (average 28.3 °C) (Wille 1976).

The brood area is also one major source of heat. In nests of *M. rufiventris* and *M. seminigra* the brood produced most of the heat compared with the food reserves (Roubik and Peralta 1983). The pupae and larvae produce heat and some features of their body structure contribute to retain it (Sung et al. 2008; Dantas 2016).

For instance, the non keratinized body of immatures seems more capable of heat exchange. However, immatures also contain a larger proportion of water compared to adults, and can absorb proportionally more heat through convection (Dantas 2016).

The involucrum is relevant in the heat balance, and species that do not build involucrum experience comparatively wider temperature variations in the brood area (Zucchi and Sakagami 1972).

Interestingly, nest temperatures of stingless bees are not constant throughout the day; large variation can be found between day and night. In Yucatan, Moo-Valle et al. (2000) studied the thermoregulation of *M. beecheii* in natural nesting cavities (hollow logs), and found that colony temperature was more stable during the day (9:00–20:00 h) than at night, when the environmental temperature decreased. It seemed that workers did not engage in heat production at night to compensate low temperatures in the environment. Indeed, night temperatures inside the colonies varied more closely to the fluctuation outside the nest. The authors suggested that stingless bees may economize on the metabolic heat produced during the day by means of the insulating structures, when it decreases at night. This pattern reveals the close relationship between environmental and internal nest temperatures in stingless bees, and their need of sufficiently isolated cavities to avoid drastic reduction of temperature at night, or during the "colder" season. This is one basic aspect to consider when trying to adapt the different species to artificial hives (Moo-Valle et al. 2000; Macías-Macías et al. 2011).

All stingless bees defecate inside the nests and this material could produce additional heat, especially if accumulated in strategic places. Some species build temporal accumulations of feces and debris, including dead bees (waste dumps) across the nest (Rasmussen and Camargo 2008). In species that build a scutellum, like *Cephalotrigona,* the organic waste ferments, presumably producing heat that can act as a large radiator (Zwaal 1992). In the Yucatan, *Cph. capitata* is a species that builds a large scutellum, and this structure is frequently close to the brood area (Fig. 2.5; Zwaal 1992). The nests of this species are mostly found subterranean or in cavities of large trees with thick walls. Although *Cph. capitata* can produce large reserves of honey, its adaptation to artificial boxes is problematic. If the environmental temperature decreases below 21 °C, a significant mortality of the brood occurs. Thus, thermal constraints are possibly the main factor to consider in the management of this bee and perhaps others that also build a scutellum (Zwaal 1992). Interestingly, in nests of *Tr. spinipes*, the temperature of the scutellum was 10 °C below the temperature of the brood suggesting little participation in heat generation (Zucchi and Sakagami 1972). These contrasting findings show the need to develop more studies on the role of the scutellum in the thermoregulation of stingless bee colonies.

Pollen reserves are proportionally large in stingless bees and can be maintained for long periods (Roubik 1982, 1983). It is known that stingless bee pollen can contain a large diversity of fermenting microorganisms producing heat (Roubik 1989). The location of pollen reserves around the brood may be an additional strategy for thermoregulation.

Although stingless bees live in the tropics, not much is known about the mechanisms they use to deal with high environmental temperatures. This type of information is relevant in the light of a potential rise in global temperatures and an increased frequency of heat waves.

In the nest of most species, the excess of heat and humidity is eliminated by means of ventilation. In hot days it is possible to see workers around the entrance fanning their wings to create air currents in and out the nest (Roubik and Peralta 1983). Similar to the honey bee, part of the adult population may evacuate the nest to disperse heat excess (Winston 1987; Engels et al. 1995). The dynamics of water balance has not been studied in stingless bees (Roubik 2006). Normally, humidity is high in tropical areas, and water condensation occurs inside colonies. Some species can be seen eliminating water droplets at the entrance, possibly representing water excess in the colony (Pers. Obs.) Although water deficit is perhaps not common in stingless bees, some species can avidly collect water at times of hot environmental temperatures. Nonetheless, the link between water collection and thermoregulation has not been clearly established. In *N. perilampoides*, workers actively collect water when exposed to high environmental temperatures (Cauich et al. 2004); the same occurs in *M. beecheii* (pers. Obs.). Recently, it was reported that in *Scp. aff. depilis* from Brazil, water collection and fanning increased as the temperature of the brood was artificially raised. The authors suggested that these bees can actively reduce nest overheating by fanning and water evaporation but no definite proof was found (Vollet-Neto et al. 2015). On the other hand, it was noted that water could only have a nutritional value and used to dilute food or obtain minerals (Vollet-Neto et al. 2015). More research is needed to clear up if stingless bees use water evaporation for colony thermoregulation, and the contribution of this mechanism to control high nest temperatures.

In subterranean species, nest temperatures can be comparatively more stable even at extreme environmental variations (Moritz and Crewe 1988). In such species, the main challenge seems the elimination of excess carbon dioxide. This gas needs to ascend to the surface to be expelled, usually through very narrow and long subterranean tubes. In the African species *Tr. denoiti,* as a response to gas accumulation, the nests engage in active ventilation producing alternating phases of inspiration and expiration each 2–3 min. This causes a periodic air exchange and elimination of carbon dioxide (Moritz and Crewe 1988).

Activity outside the nests stops at night and some species close their entrance. This behavior could help in thermoregulation and defense (Fig. 2.6). It seems that highly defensive species do not close the entrance at night.

Most data on stingless bee thermoregulation have been obtained on tropical species, dwelling at relatively stable environmental temperature and humidity. However, some stingless bees inhabit tropical highlands, cloud forests, and boundaries between tropical and temperate regions with broad temperature oscillation (Ortiz-Mora et al. 1995). A study was conducted to compare temperature tolerance of highland *M. colimana* (a species of pine forests above 500 m in Jalisco, Mexico) with lowland *M. beecheii* and *Scp. hellwegeri.* The results showed that *M. colimana* better tolerated extreme changes in environmental temperatures (Macías-Macías et al. 2011). Pupae mortality in *M. colimana* kept at 40 °C for 24 h was 30% lower than

Fig. 2.6 Entrances of the nests of *N. perilampoides* (left) and *Fr. nigra* (right) at night. In the first species, cerumen is used to build a tight mesh on the entrance, while in the latter resin is used to build threads to block it

in the other two species. Interestingly, the highest mortality was registered in workers (40%) and in pupae (90%) of *M. beecheii* at 40 °C. The workers of *M. colimana* also tolerated low temperatures of 5 °C while 40% of *M. beecheii* and *Scp. hellwegeri* workers died at that temperature. In the three species, the brood was more sensitive to low temperatures compared with adults (Macías-Macías et al. 2011). At 5 °C, all *M. beecheii* and *Scp. hellwegeri* pupae died, but only 80% those of *M. colimana.*

An interesting finding is that the workers of *M. colimana* seem to engage in active mechanisms to compensate environment temperature extremes. At low temperatures in the lab, *M. colimana* workers increased the consumption of sugar syrup, which correlated with an increase in their thoracic temperature, a behavior that was not observed in lowland species. Moreover, when submitted to high temperatures, *M. colimana* workers engaged in water regurgitation and its evaporation through wing fanning (Macías-Macías et al. 2011). As a result of living in areas with daily and seasonal extreme temperatures *M. colimana* may have evolved some level of active control of nest temperatures. Nonetheless, the relative contribution of passive and active mechanisms for nest thermoregulation in stingless bees from highland forests needs to be assessed in whole colonies in their natural habitats (Macías-Macías et al. 2011).

In insects, as in other organisms, inter- and intraspecific phenotypic variation can occur in response to habitat temperatures (Laiolo 2013). Individuals at higher altitudes or latitudes can be larger than those at lower ones. However, it has been found that in insects, such a pattern defined as Bergmann's rule is not even across species or taxa (Shelomi 2012). Although no formal comparison has been made, in stingless bees, there are examples of species from the highland tropics (*M. colimana, M. fasciata*) that have comparatively larger body size than their tropical counterparts (*M. beecheii, M. yucatanica*) (Ayala 1999). *M. colimana* is also darker and with more pilosity on the thorax (Fig. 2.7), suggesting a better adaptation for maintaining

Fig. 2.7 Comparative body size, color, and pilosity of three species of *Melipona* from Mexico. *M. colimana* from the highlands of the west Mexican coast has an average larger body size, darker coloration, and more pilosity than *M. beecheii* and *M. yucatanica* from the lowlands of the Yucatan Peninsula (photos courtesy of Ricardo Ayala)

body heat at high altitudes. Evidence of an inverse relationship between body color and heat loss has also been found in stingless bees; that is, dark-colored species warm up more rapidly and can keep higher body temperature than light-colored ones (Pereboom and Biesmeijer 2003).

Although some aspects of nest temperature regulation in stingless bees are presently understood, much information is yet needed to comprehend the importance of active and passive mechanisms and the variation across species and environments. In the advent of global warming, the effect of high temperatures and the ability of different species to deal with them are relevant, albeit only little explored, areas of study.

References

Ayala R (1999) Revisión de las abejas sin aguijón de México (Hymenoptera: Apidae: Meliponini). Folia Entomológica Mexicana 106:1–123

Becker T (2014) Desenvolvimento de colmeias de abelhas Melipona interrupta Latreille, 1811 (Hymenoptera: Meliponini) sob diferentes temperaturas em condições de laboratorio. Dissertação PósGraduação em Entomologia. Instituto Nacional de Pesquisas da Amazônia, Manaus, Amazonas

Camargo JMF, Pedro SRM (2007) Notas sobre a bionomia de *Trichotrigona extranea* Camargo & Moure (Hymenoptera, Apidae, Meliponini). Revista Brasileira de Entomologia 51:72–81

Cauich O, Quezada-Euán JJG, Macias-Macias JO, Reyes-Oregel V, Medina-Peralta S, Parra-Tabla V (2004) Behavior and Pollination Efficiency of Nannotrigona perilampoides (Hymenoptera: Meliponini) on Greenhouse Tomatoes (Lycopersicon esculentum) in Subtropical Mexico. J Econ Entomol 97(2):475–481

Couvillon MJ, Wenseleers T, Imperatriz-Fonseca VL, Nogueira-Neto P, Ratnieks FLW (2008) Comparative study in stingless bees (Meliponini) demonstrates that nest entrance size predicts traffic and defensivity. J Evol Biol 21:194–201

Dantas MRT (2016) Thermogenesis in stingless bees: an approach with emphasis on brood's thermal contribution. J Anim Behav Biometeorol 4:101–108

Eltz T, Brühl CA, van der Kaars S, Linsenmair KE (2002) Determinants of stingless bee nest density in lowland dipterocarp forests of Sabah, Malaysia. Oecologia 131:27–34

Eltz T, Brühl CA, Imiyabir Z, Linsenmair KE (2003) Nesting and nest trees of stingless bees (Apidae: Meliponini) in lowland dipterocarp forests in Sabah, Malaysia, with implications for forest management. For Ecol Manag 172:301–313

Engels W, Rosenkranz P, Engels E (1995) Thermoregulation in the nests of the neotroprical stingless bee *Scaptotrigona postica* and a hypothesis on the evolution of temperature homeostasis in highly eusocial bees. Stud Neotropical Fauna Environ 30:193–205

Ferreira NdS (2014) Temperatura colonial e toleranca térmica de Melipona subnitida, uma especie de abelha sem ferrão (Hymenoptera: Apidae: Meliponini) da Caatinga. Dissertação Mestre em Ciencia Animal. Universidade Federal do Semiárido, UFERSA, Mossoró

González-Acereto JA, Quezada-Euán JJG, Medina-Medina LA (2006) New perspectives for stingless beekeeping in the Yucatan: results of an integral program to rescue and promote the activity. J Apic Res 45:234–239

Heinrich B (1993) The hot blooded insects: strategies and mechanisms of thermoregulation. Springer, Berlin

Kerr WE, Carvalho GA, Nascimento VA (1996) Abelha uruçu: biologia, manejo e conservação. Ed. Fundação Acangaú, Paracatu, MG, 144 pp

Laiolo P (2013) From inter-specific behavioural interactions to species distribution patterns along gradients of habitat heterogeneity. Oecologia 171:207–215

Macías-Macías JO, Quezada-Euán JJG, Contreras-Escareño F, Tapia González JM, Moo-Valle H, Ayala R (2011) Comparative temperature tolerance in stingless bee species from tropical highlands and lowlands of México and implications for their conservation (Hymenoptera: Apidae: Meliponini). Apidologie 42:679–689

Medina-Medina LA, Hart AG, Ratnieks FLW (2014) Waste management in the stingless bee *Melipona beecheii* Bennett (Hymenoptera: Apidae). Sociobiology 61:435–440

Michener CD (1974) The social behavior of the bees: a comparative study. Belknap Press, Harvard University, Cambridge

Michener CD (2001) Comments on minute Meliponini and the male of the genus *Pariotrigona* (Hymenoptera: Apidae). J Kansas Entomol Soc 74:231–236

Moo-Valle H, Quezada-Euán JJG, Navarro J, Rodriguez-Carvajal LA (2000) Patterns of intranidal temperature fluctuation for *Melipona beecheii* colonies in natural nesting cavities. J Apic Res 39:3–7

Moritz RFA, Crewe RM (1988) Air ventilation in nests of two African stingless bees *Trigona denoiti* and *Trigona gribodoi*. Experientia 44:1024–1027

Moritz RFA, Southwick EE (1992) Bees as superorganisms, an evolutionary reality. Springer, Berlin

Nogueira-Neto P (1997) Vida e criação de abelhas indígenas sem ferrão. Editora Nogueirapis, São Paulo

Ortiz-Mora A, van Veen JW, Corrales-Moreira G, Sommeijer MJ (1995) Influence of altitude on the distribution of stingless bees (Hymenoptera: Apidae: Meliponinae). Apiacta 30:101–105

Pereboom JJM, Biesmeijer JC (2003) Thermal constraints for stingless bee foragers: the importance of body size and coloration. Oecologia 137:42–50

Rasmussen C, Camargo JMF (2008) A molecular phylogeny and the evolution of nest architecture and behavior in *Trigona s.s.* (Hymenoptera: Apidae: Meliponini). Apidologie 39:102–118

Roldão YS (2011) Termorregulação colonial e a influência da temperatura no desenvolvimento da cria em abelhas sem ferrão, *Melipona scutellaris* (Hymenoptera, Apidade, Meliponini). Dissertação, Faculdade de Filosofia, Ciências e Letras, Ribeirão Preto

Roubik DW (1982) Seasonality in colony food storage, brood production and adult survivorship: studies of *Melipona* in tropical forest (Hymenoptera: Apidae). J Kansas Entomol Soc 55:789–800

Roubik DW (1983) Nest and colony characteristics of stingless bees from Panamá (Hymenoptera: Apidae). J Kansas Entomol Soc 56:327–355

Roubik DW (1989) Ecology and natural history of tropical bees. Cambridge University Press, Cambridge

Roubik DW (2006) Stingless bee nesting biology. Apidologie 37:124–143

Roubik DW, Peralta FJA (1983) Thermodynamics in nests of two *Melipona* species in Brasil. Acta Amazon 13:453–466

Sakagami SF (1982) Stingless bees. In: Hermann HR (ed) Social insects, vol III. Academic Press, London, pp 361–423

Seeley TD (1985) Honey bee ecology, a study of adaptation in social life. Princeton University Press, Princeton

Shelomi M (2012) Where are we now? Bergmann's rule *sensu lato* in insects. Am Nat 180:511–519

Sung IH, Yamane S, Hozumi S (2008) Thermal characteristics of nests of the Taiwanese stingless bee *Trigona ventralis hoozana* (Hymenoptera: Apidae). Zool Stud 47:417–428

Torres A, Hoffman W, Lamprecht I (2007) Thermal investigations of a nest of the stingless bee *Tetragonisca angustula* Illiger in Colombia. Thermochim Acta 458:118–123

Torres A, Hoffman W, Lamprecht I (2009) Thermal investigations of a nest of the stingless bee *Trigona (Frieseomelitta) nigra paupera* Provancher in Colombia. J Therm Anal Calorim 95:737–741

Vollet-Neto A, Menezes C, Imperatriz-Fonseca VL (2015) Behavioural and developmental responses of a stingless bee (*Scaptotrigona depilis*) to nest overheating. Apidologie 46:455–464

Wille A (1976) Las abejas Jicotes del género *Melipona* (Apidae: Meliponini) de Costa Rica. Rev Biol Trop 24:123–147

Wille A (1983) Biology of the stingless bees. Annu Rev Entomol 28:41–64

Wille A, Michener CD (1973) The nest architecture of stingless bees with special reference to those of Costa Rica (Hymenoptera: Apidae). Rev Biol Trop 21:1–278

Winston ML (1987) The biology of the honey bee. Harvard University Press, Cambridge

Zucchi R, Sakagami SF (1972) Capacidade termo-reguladora em *Trigona spinipes* e em algumas outras especies de abelhas sem ferrão (Hymenoptera: Apidae: Meliponinae). In: Cruz-Landim C (ed) Homenahem a Kerr, W.E. Ribeirão Preto, Brazil, pp 301–309

Zwaal N (1992) Investigation on the stingless bee *Cephalotrigona capitata* in the Yucatan Peninsula. Tesis FMVZ-UADY & Faculty of Biology, University of Utrecht, Utrecht

Chapter 3
Anatomy and Physiology

Workers of *M. beecheii* collecting pollen from a water lily.

A great deal of variation exists on the external features of stingless bees; this is in strong contrast with the rather uniform body pattern found in the Apini, the other eusocial group of bees (Rasmussen and Cameron 2010). The stingless bees show notorious interspecific differences not only in size, but also in shape and color Ayala 1999). The size and number of mandibular teeth, the degree of body pilosity, the pattern of wing venation, and the size and shape of the corbiculae are

J. J. G. Quezada-Euán, *Stingless Bees of Mexico*,
https://doi.org/10.1007/978-3-319-77785-6_3

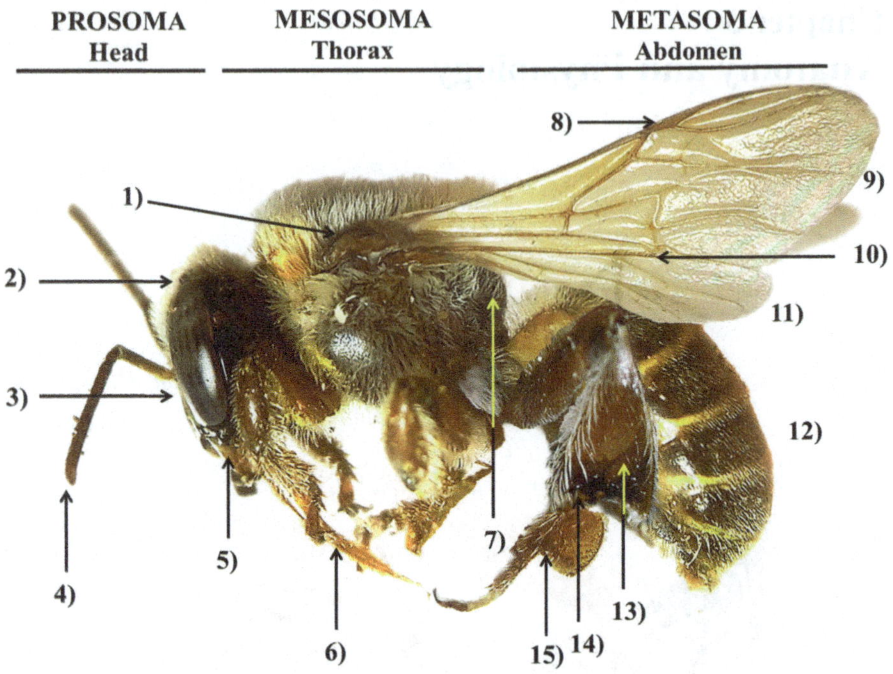

Fig. 3.1 External anatomy of a *M. beecheii* worker: (1) tegula; (2) ocellum; (3) compound eyes; (4) antenna; (5) mandibula; (6) proboscis; (7) propodeum; (8) pterostigma; (9) forewing; (10) hamuli on the hind wing; (11) hind wing; (12) abdominal segments A2 a A7; (13) corbicula; (14) penicillum; (15) basitarsum

notorious examples of such variability. Nonetheless, stingless bees exhibit a basic pattern of the body, which is similar to that of other Hymenoptera.

In this chapter a general description of the anatomy and physiology of stingless bees is presented. Descriptions proceed in accordance to the main segments conforming the Hymenopteran's body: the prosoma or head, the mesosoma or thorax, and the metasoma or abdomen (Fig. 3.1).

For some particular structures, differences are pointed between *Melipona* and other genera. Aspects related to the anatomy and physiology of the reproductive system are presented in Chap. 6, in the section referring to individual reproduction of gynes and males.

3.1 Prosoma

Structures found on the head or prosoma are important for the interaction of the insect with its environment, most acting as centers for feeding and the senses. The main organs here are the eyes, antennae, and proboscis.

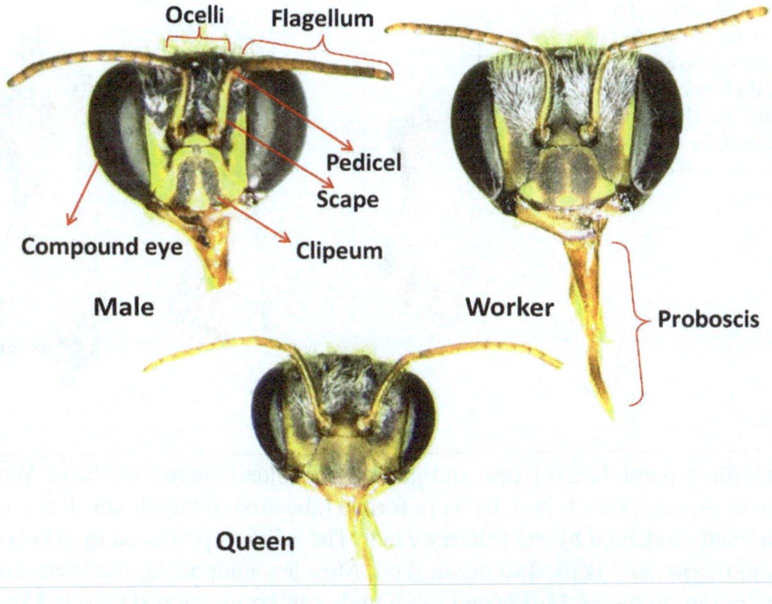

Fig. 3.2 Frontal image of the head of *M. beecheii* male, worker and queen. The ocelli and parts of the antennae are labelled on the male and the proboscis on the worker

Bees have two types of eye. One, is a set of three simple eyes called ocelli that are located at the frontal apex of the head (Fig. 3.2). The ocelli are not involved in the process of image production, but in the perception of light intensity. There is also a pair of large compound eyes located at both sides of the head. The large compound eyes are formed by small facets called ommatidia. Each ommatidium is a simple but complete optic unit with a corneal lens, a crystalline cone, and a light-sensitive pigment, the rhodopsin. It seems that the ommatidia act by capturing simple images in their visual area and that these are processed in the brain to obtain a more complex and wider image (Dade 1985).

At the front of the head there is a pair of antennae. Each antenna is formed by three segments; the larger one attaching the antenna to the head is the scape. After the scape, the following small spherical segment is the pedicel, which allows free movement of the distal section of the antennae called the flagellum (Fig. 3.2). The flagellum is divided into small segments called flagellomeres, of which females (workers and queens) have 10 and males 11. The surface of the flagellomeres is covered with chemical and mechanical receptors called sensorial plates or *sensilla*. There are different kinds of *sensilla*, each involved in the detection of the levels of CO_2, pressure, and temperature, as well as taste and smell, among others (Dade 1985). Inside the antennae, the Johnston's organ is involved in the reception of airborne sound. In *M. scutellaris* this organ seems capable of detecting the vibrations produced by forager bees through the movement of air particles at distances of up to 2 cm (Hrncir et al. 2008).

Underneath the head is the proboscis, an articulated structure used by the bee as a straw in the suction of nectar and liquids. When not in use, the parts of the proboscis

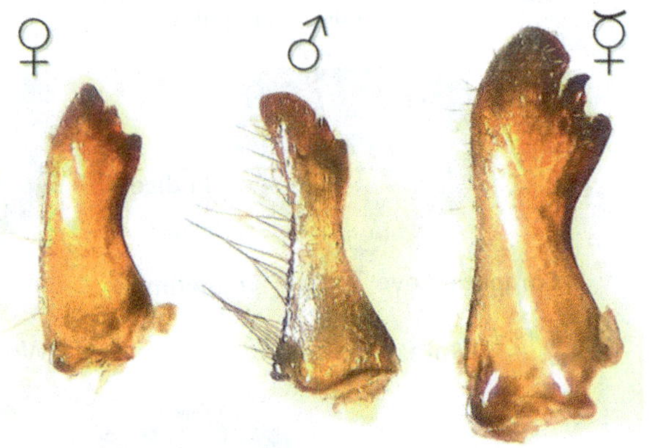

Fig. 3.3 Mandibles of a queen, male, and worker of *M. beecheii*. There is a striking difference in the size of the mandible and sharpness of the teeth between the worker and both sexuals

(the maxillary palpi, labial palpi, and glossa) are folded behind the head. When in use, the proboscis extends and the palpi form a tube around the glossa. The glossa is longitudinally indented by the salivary canal. The saliva is produced in glands in the head and thorax, and is used to dilute food. Muscles, underneath the head, create a sucking pump on the proboscis and the liquids can be ingested (Dade 1985). Taste seems to be mainly sensed at the tip of the glossa (the flabellum) and the palpi.

The mandibles are found in the lower part the head. These appendages are the main tools for processing food (pollen) and building materials (wax, cerumen, and resin), but also the most important weapons for defense (Fig. 3.3). Given that meliponines lack a sting, the mandibles can be especially developed in some species. Accordingly, the mandibles can vary in size, the number of teeth, and their sharpness. Interestingly, highly defensive species have larger mandibles with more numerous and sharper teeth (Shackleton et al. 2015). Sexuals generally make little use of the mandibles, and these are significantly smaller (Fig. 3.3).

3.2 Mesosoma

The mesosoma or thorax is the second largest body segment. The organs involved in the locomotion of the insect (walking and flight) are found here. The mesosoma is formed by segments that have been reorganized in the transition from the larval to the adult stage during metamorphosis. The first abdominal segment of the larva (A1) moves into the mesosoma to form the last segment of the adult's thorax, and is called the propodeum (Fig. 3.1). Between the propodeum and the first abdominal segment of the adult (originally the larval A2), a narrowing of the body (wasp-waist) called the petiole unites the thorax with the metasoma (abdomen). The petiole allows great flexibility of the metasoma (Dade 1985).

The thorax carries two pairs of wings and three pairs of legs. Both types of appendage are propelled by groups of muscles. The bee's legs are three pairs

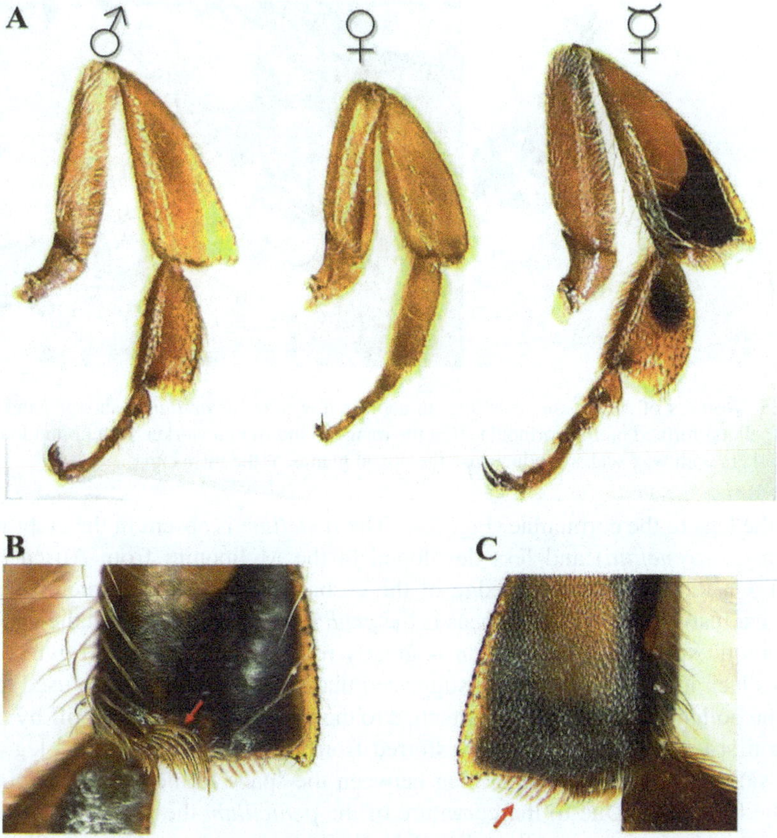

Fig. 3.4 (**a**) Hind leg of *M. beecheii* male, queen, and worker. The corbícula and details of the *penicillum* and *rastellum* (only found in the worker), are shown in images **b** and **c**, respectively

of segmented appendages. The segments of the legs, from the most proximal to the more distal are, the coxa (the first segment attaching the leg to the mesosoma), followed by the trochanter, femur, tibia, tarsus, and pretarsus. The tarsus is formed by five segments called the tarsomeres of which the first, the basitarsus, is wider and larger than the others. The basitarsus of the first pair of legs has a notch with hair that is used to clean the antennae (antenna cleaner). The third pair of legs is of particular importance in the workers of stingless bees (and other corbiculate Apinae), because the tibia forms the *corbicula* or pollen basket. The corbicula is a broad expansion of the tibia; it is slightly concave with curved hairs along the edge to help keeping the load in place (Fig. 3.4). The inner face of the basitarsus on the third leg is covered with thick short bristles (the pollen brush) that groom the pollen from the bee's body and transfer it to the corbiculae.

On the distal edge of the corbicula a group of short stiff bristles are known as the pollen rake or *rastellum*. The pollen rake is used during the transference of pollen

Fig. 3.5 Workers of *M. beecheii* visiting an achiote flower (*Bixa orellana*) showing the pollen loads (pellets) formed on the corbiculae. On the image to the right a worker with pollen loads and two workers with wax scales produced on the dorsal glands of the metasoma

from the legs to the corbiculae (Fig. 3.4). The *rastellum* is absent in the cleptobiotic genera (*Lestrimelitta*) and less developed in the Meliponini from Africa (Wille 1979). On the apical internal edge of the corbicula, a brushlike structure of stiff setae, exclusive of the stingless bees is the *penicillum* (Wille 1979). In the genera of cleptobiotic species the *penicillum* is greatly reduced; the males and gynes of all species lack it too (Fig. 3.4). It is suggested that the *penicillum* is also used to process the pollen loads. The pollen attached to the bee's body is brushed off by means of the first two legs. Then it is transferred from the second to the third leg of the same side by passing the second leg between the space formed by the *penicillum* and the basitarsus. Due to the curvature of the *penicillum* the pollen is pushed up into the corbicula to form pellets (Fig. 3.5). Stingless bees do not have a pollen press apparatus on the basitarsus like the honey bees (Wille 1979).

The wings of the bees are membranous structures, found in pairs on each side of the body. Each pair is formed by a large forewing and a smaller hind wing (Fig. 3.6). The wings have veins that act as mechanical support for their fine structure. The spaces among the different veins are called wing cells (Fig. 3.6). The pattern of forewing veins and cells is important in the taxonomic identification of species and lineages of stingless bees (Francoy et al. 2011).

Flight is produced by the action of two sets of muscles found in pairs inside the mesosoma, i.e., the longitudinal and dorsoventral (or vertical) muscles. These muscles act to generate the energy and movement required for flight. Thanks to the elastic cuticle between its segments, the mesosoma can expand and contract alternatively. The longitudinal muscles pull the mesosoma front to back and the vertical muscles pull it dorso-ventrally. The wing bases are inserted into a slit of the thorax (longitudinal slit) and by the alternative movements of the muscles, the wings are pushed up and down. The dorsoventral contraction elevates the wings, and the longitudinal contraction flips them down. Muscles contract and relax at a very high speed (200–250 beats per second), elevating and pushing the wings down in alternative movements, creating the necessary stroke for flight (Dade 1985).

Fig. 3.6 Fore and hind wings of three Meliponini from Yucatan to compare the relative size of the pterostigma (P) and hamuli number (H), the latter in the amplified section. The species *Scp. pectoralis* and *N. perilampoides* have seven and five hamuli, respectively; *M. beecheii,* with a larger body mass, has ten

The mesosomal muscles produce a great amount of energy during contraction (Winston 1987). During flight the forewing and hind wing form a functional unit. This is achieved by means of a canal in the posterior margin of the forewing and a set of hooks (hamuli) on the frontal edge of the hind wing. When the bee expands its wings in preparation for flight, the hamuli on the hind wing clasp into the forewing canal, effectively becoming one structure.

Through this mechanism, the bee can increase the surface and strength of the wings, for a more powerful flight (Dade 1985). In the Meliponini, it has been found that the number of hamuli in the different species is correlated with body mass rather than wing length (Schwarz 1948). Non-*Melipona* stingless bees usually have between 4 and 10 hamuli while *Melipona* species have between 8 and 16 (Schwarz 1948). Queens have less hamuli compared with males and workers (Schwarz 1948). This is probably due to their low requirements for continuous and long flights, compared to the other individuals.

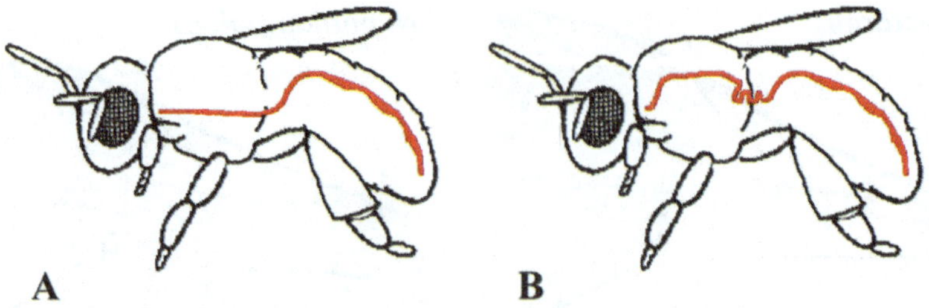

Fig. 3.7 Shape of the dorsal vessel (in red) in stingless bees: (**a**) type I aorta present in non-*Melipona* species and (**b**) type II aorta present in *Melipona* and *Meliponula*

Some fossil stingless are small, like the present *Plebeia.* Wille (1979) suggests that small size of their ancestors may explain some of the present anatomical features of the stingless bees, for instance, the reduced wing venation. Weak veins like those presently found in Meliponini, can be a hurdle for the strong flight needs of highly social bees. It seems that non-*Melipona* and *Melipona* stingless bees evolved different methods to compensate for the relative weakness of the veins. In the case of non-*Melipona* species, the forewing evolved to a comparatively larger size and the pterostigma is proportionally larger than in other corbiculate bees (Fig. 3.6). Large pterostigmas are characteristic of small Hymenoptera which provide additional support to the wings during flight (Michener 2001). In comparison, *Melipona* have comparatively shorter wings with a smaller pterostigma, but the number of wing hamuli has increased, probably resulting in a better union between fore and hind wings increasing flight capacity (Wille 1979; Fig. 3.6).

One portion of the circulatory system, the dorsal vessel, or aorta, is found in the thorax. This vessel shows differences between *Melipona* and non-*Melipona* species. In non-*Melipona* bees the thoracic portion of the aorta is normally straight, and it passes below the longitudinal thoracic muscles not between them; this type of dorsal vessel is known as type I (Fig. 3.7). In *Melipona* the thoracic portion of the aorta forms an arch between the longitudinal thoracic muscles and is known as type II (Fig. 3.7). It is thought that Type II aorta probably evolved as response for a better irrigation of the muscles or temperature regulation during flight in larger bees (Wille 1979; Winston 1987).

3.3 Metasoma

Most organs and systems of the bee are found in the metasoma or abdomen. The metasoma is formed by nine segments, each with a dorsal (tergite) and a ventral plate (sternite). The segments of the metasoma lack the pleurite or lateral sclerite between the terga and sterna found in the mesosomal segments. In contrast with the thorax, the bee's abdomen can expand notably due to the elastic membranes between the plates.

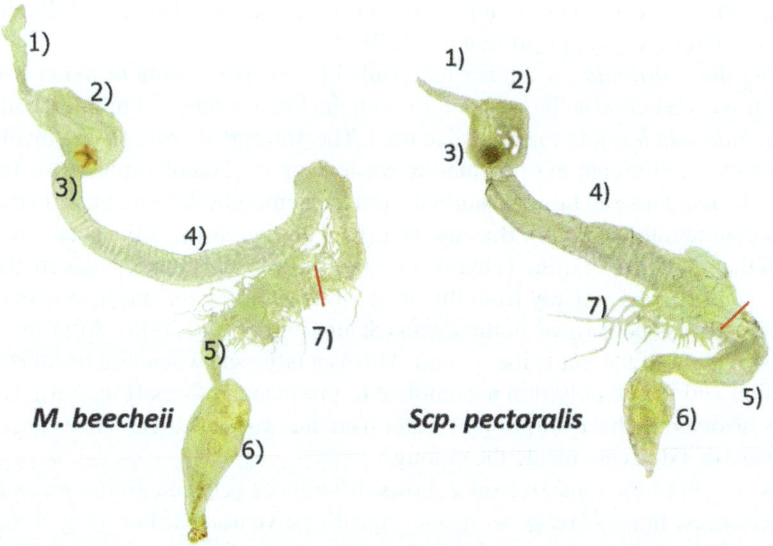

Fig. 3.8 Digestive tract of *M. beecheii* and *Scp. pectoralis*workers: (1) esophagus, (2) honey sac, (3) proventriculus, (4) ventriculus, (5) short intestine or ileum, (6) rectum, (7) Malpighian tubules. Elongation of the tract has occurred in *Melipona,* compared to other stingless bees, in particular of the ileum (5). The red lines indicate the start of the ileum

In the adult stingless bee, only segments A2–A7 are externally visible. Segment A1 has moved into the mesosoma to form the propodeum and segments A8 to A10 are located internally in the abdomen, some of them part of the vestigial sting.

The digestive tract extends from the proboscis in the head to the metasoma. In the abdomen, the digestive tract is continuation of the esophagus, a long and slender tube that extends from the head through the thorax. In the abdominal cavity, the esophagus expands to form a sac called the honey stomach or crop. When empty, the honey stomach is folded but its thin walls can expand notoriously to accommodate a large nectar volume (Fig. 3.8). The honey stomach does not have a digestive function; it acts like a reservoir for the nectar. The honey stomach is separated from the true stomach by a sphincter valve called the proventriculus which controls the passage of pollen and other solids from the honey sac. The true stomach or *ventriculus* is a thick muscular segmented tube where the digestion takes place (Fig. 3.8). A series of membranes inside the *ventriculus* (the peritrophic membranes), wrap around the pollen grains as they enter this part of the tract (Dade 1985). The peritrophic membranes may possibly act to protect the delicate lining of the *ventriculus* against the coarse or spinous exines of some types of pollen. In the *ventriculus* the pores of the pollen grains are broken by osmotic shock (Crailsheim 1990). In addition, the cells lining the *ventriculus* produce digestive juices and proteolytic enzymes penetrate the pores of the pollen. Inside the pollen, the enzymes digest the large molecules and a broth can exit the pores. Once in the *ventriculus* lumen, the nutrients can absorbed into the hemolymph. Nutrients can be stored in the abdominal fat bodies as fat, glycogen, or protein (Dade 1985).

After the *ventriculus*, a thinner tube called the short intestine or *ileum* is found. In the insertion between the *ventriculus* with the *ileum* a series of tubules, called the *Malpighian tubules*, join the digestive tract. The *Malpighian tubules* are involved in the process of elimination of metabolic waste from the hemolymph. In the tubules, metabolic residues and liquid absorbed from the hemolymph form the primary urine that is poured into the ileum. Because bees feed on liquids (mostly nectar), it is suggested that the primary urine is hypo-osmotic. To keep an osmotic balance, the bees need to absorb ions actively from the urine while avoiding the absorption of excess water; a process performed in the *ileum* (Santos and Serrão 2006). After the *ileum,* the last section of the gut is the rectum. This is a large sac where the residues of the digestion and the hemolymph accumulate to produce the feces (Fig. 3.8). The rectum is involved in the absorption of water from the waste. Stingless bees in contrast to honey bees defecate inside the colony.

The length of the digestive tract shows differences between *Melipona* and other stingless bees. In non-*Melipona* species, the digestive tract is short (Fig. 3.8), compared with that of *Melipona*. A long digestive tract is considered a derived characteristic (Wille 1979). Notably, in the cleptobiotic *Lestrimelitta,* the ventriculus in addition to the ileum is also shortened (Wille 1979). This may be a result of a lifestyle involving the consumption of larval food and pollen, predigested by the hosts.

The bee's circulatory system is open; there are no vessels in the form of arteries or capillaries. Basically, there is one large vessel with two parts, the larger posterior part or heart, is found in the metasoma. The first part, the aorta, already described in Sect. 3.2, continues into the abdomen to form theheart. The heart is a muscular organ that pumps the hemolymph from the abdomen to the mesosoma and the head. The bee's hemolymph is a yellowish liquid composed of the plasma and cells called hemocytes. Hemocytes are involved in the immunity of the insect. The insect's hemolymph reaches all the organs and muscles carrying the nutrients and removing excretions. However, the hemolymph does not contain red cells and is not involved in the transport of oxygen. In the abdomen, the hemolymph enters the heart through a series of holes called the ostia. The walls of the heart pump the liquid to the aorta and to the head where the hemolymph is poured onto the brain from which it flows backwards to the abdomen (Dade 1985).

The bees, as other insects, breathe through the body cuticle by means of lateral orifices called *spiracles*. From these, the incoming air flows through ducts called *tracheae* to the different organs. The exchange of oxygen takes place in a series of microscopic tubes called *tracheoles* which are in close contact with the cells. The contractions of the abdomen pump the air in and out of the body (Dade 1985).

The bee's nervous system comprises the brain and a set of ventral nerve cumulus, called *ganglia*. The ganglia act as small satellite brains. It is important to note that in stingless bees the abdominal *ganglia* have migrated towards the front (cephalization) of the body (Wille 1979). This is more evident in *Melipona* (Fig. 3.9).

In some species, the abdominal *ganglion* three has migrated into the mesosoma. Thus, the thorax in these species carries two *ganglia* of mesosomal origin and one of metasomal origin (Wille 1961). It is believed that the cephalization of the abdom-

Fig. 3.9 Ventral nervous chain of a *M. beecheii* worker in which the four abdominal *ganglia* are observed

inal *ganglia* could have occurred as a result of the loss of a functional sting which would reduce the activity of the *ganglia* associated with this structure. In males, for instance, there is a trend in all bee taxa for a reduction in the number of abdominal ganglia (Wille 1979). Nonetheless, abdominal ganglion three also has a thoracic position in females of some bumble bees and orchid bees (Cruz-Landim et al. 1972), both having a functional sting.

The workers and males of some *Melipona,* such as *M. quadrifasciata* have five abdominal *ganglia*, while the queens have only four (Kerr and Nielsen 1966). In such species, the number of abdominal *ganglia* has been used to differentiate individuals with a queen genotype (Kerr 1950), but that due to nutritional factors are phenotypically workers (see Chap. 4). In *M. beecheii,* however, both queens and workers have four abdominal *ganglia* (Fig. 3.9; Darchen and Delage-Darchen 1975). In the adult bee, the last abdominal *ganglion* results from the fusion of the larval abdominal *ganglia* 7, 8, and 9 and is found in the fourth abdominal segment (Wille 1961).

The vestigial sting is found in the last portion of the abdomen. In the most primitive genera, it is still possible to find a small sac, probably homologue to the acid gland that exists in stinging bees. Interestingly, this sac is also found in the queens of Meliponini (Kerr and Lello 1962). Parts of the vestigial sting include the *valvulae* 1 and 2, which are the remains of the stylet of stinging bees (Packer 2003; Fig. 3.10). The *valvulae* 3 or *gonostylus* are the remains of the sheath. In stingless bees, valvulae 3 are robust, and straight, profusely covered with bristles, compared with the concave and smooth structures of stinging bees (Radović 1981; Fig. 3.10). In the stingless bees the bristles of the *gonostylus* may act as mechanical receptors. Interestingly, the contact of the *gonostylus* with non-sclerotized surfaces like the skin, triggers some defensive responses, like biting and abdominal movements of the workers that resemble the act of stinging (Radović 1981).

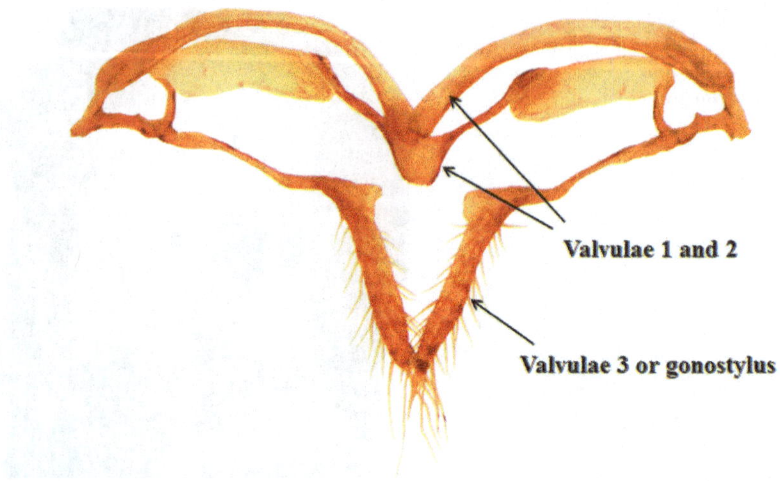

Fig. 3.10 Structure of the vestigial sting of a worker of *M. beecheii* in anterodorsal view. *Valvulae* 1 and 2 are the remains of the sting stylets and *valvulae* 3 or *gonostylus* are the remains of sting sheath. The *gonostylus* is covered with abundant bristles that may act as mechanoreceptors triggering defense responses

References

Ayala R (1999) Revisión de las abejas sin aguijón de México (Hymenoptera: Apidae: Meliponini). Folia Entomológica Mexicana 106:1–123

Crailsheim K (1990) The protein balance of the honey bee worker. Apidologie 21:417–429

Cruz-Landim C, Höfling JF, Zaniboni MC (1972) Estudo comparativo do cordão nervoso ventral em abelhas. In: Homenagem à Warwick E. Kerr. Rio Claro, Brasil, pp 113–134

Dade HA (1985) Anatomy and dissection of the honeybee. International Bee Research Association, London

Darchen R, Delage-Darchen B (1975) Contribution a l'étude d'une abeille du Mexique *Melipona beecheii* B. (Hymenoptére: Apidae). Le dèterminisme des caste chez les Mélipones. Apidologie 6:295–339

Francoy TM, Grassi ML, Imperatriz-Fonseca VL, May-Itzá Wd J, Quezada-Euán JJG (2011) Geometric morphometrics of the wing as a tool for assigning genetic lineages and geographic origin in *Melipona beecheii* (Hymenoptera: Meliponini). Apidologie 42:499–507

Hrncir M, Schorkopf DLP, Schmidt VM, Zucchi R, Barth FG (2008) The sound field generated by tethered stingless bees (*Melipona scutellaris*): inferences on its potential as a recruitment mechanism inside the hive. J Exp Biol 211:686–698

Kerr WE (1950) Genetic determination of castes in the genus *Melipona*. Genetics 35:143–152

Kerr WE, Lello E (1962) Sting glands in stingless bees—a vestigial character. J NY Entomol Soc 70:190–214

Kerr WE, Nielsen RA (1966) Evidences that genetically determined *Melipona* queens can become workers. Genetics 54:859–866

Michener CD (2001) Comments on minute Meliponini and the male of the genus *Pariotrigona* (Hymenoptera: Apidae). J Kansas Entomol Soc 74:231–236

Packer L (2003) Comparative morphology of the skeletal parts of the sting apparatus of bees (Hymenoptera: Apoidea). Zool J Linnean Soc 138:1–38

Radović IT (1981) Anatomy and function of the sting apparatus of stingless bees (Hymenoptera: Apidae: Apinae). Proc Entomol Soc Wash 83:269–273

Rasmussen C, Cameron SA (2010) Global stingless bee phylogeny supports ancient divergence, vicariance, and long distance dispersal. Biol J Linn Soc 99:206–232

Santos CG, Serrão JE (2006) Histology of the ileum in bees (Hymenoptera, Apoidea). Braz J Morphol Sci 23:405–413

Schwarz HF (1948) The stingless bees (Meliponinae) of the Western Hemisphere. Bull Am Mus Nat Hist 90:1–546

Shackleton K, Toufailia A, Balfour NJ, Nasciento FS, Alves DA, Ratnieks FLW (2015) Appetite for self-destruction: suicidal biting as a nest defense strategy in *Trigona* stingless bees. Behav Ecol Sociobiol 69:273–281

Wille A (1961) Evolutionary trends in the ventral nerve cord of the stingless bees (Meliponini). Rev Biol Trop 9:117–129

Wille A (1979) Phylogeny and relationships among the genera and subgenera of the stingless bees (Meliponiane) of the world. Rev Biol Trop 27:241–277

Winston ML (1987) The biology of the honey bee. Harvard University Press, Cambridge

Chapter 4
Sex Determination and Caste Production

J. J. G. Quezada-Euán, *Stingless Bees of Mexico*,
https://doi.org/10.1007/978-3-319-77785-6_4

4.1 Ontogenetic Development

Like all Hymenopterans, bees are holometabolous; they pass through a series of anatomical and physiological changes, from larva to adult, called metamorphosis (Savard et al. 2006; Fig. 4.1). In holometabolous insects, there are five main stages in the ontogenetic development: egg, larva, prepupa, pupa, and adult or imago. After the larva hatches, it experiences a rapid growth. Because the rigid outer cuticle prevents the expansion and growth of the body, the larva needs to shed it in a series of molts or *ecdysis*. The regulation of growth and the *ecdysis* is under the effect of various hormones, the most important being the juvenile hormone (JH), insulin, and ecdysone (Hartfelder et al. 2006; Nijhout and Callier 2015). During the larval phase of the bee, there are four molts and five stages (Dade 1985). There is also a fifth molt between the prepupal and pupal phases, plus a final one between the pupal and the imago phases, making a total of six molts throughout the ontogenetic development.

In general, there is little information on the duration of ontogenetic development in stingless bees. However, although the different phases are the same, there seem to be substantial differences in duration across species and also between sexes and castes (Table 4.1). For instance, in two Mexican species of *Melipona*, *M. beecheii* and *M. colimana,* there are significant differences in the time of development of males, gynes, and workers (Moo-Valle et al. 2004; Macías-Macias and Quezada-Euán 2015). The total time for the development of the workers in *M. beecheii* is 53 ± 1 days. The queens develop faster (51 ± 1 days) and the males slower (53.5 ± 1 days) than the workers (Table 4.1). Marked differences among individuals are found in the duration of the third larval stage and the pupal stages 2 and 3 (Table 4.1).

In comparison with *Melipona,* the duration of development of the workers in non-*Melipona* bees seems comparatively shorter: *Ttr. angustula* (36.5 days, Nates-Parra et al. 1989), *Friesella* (35–36 days, Sakagami 1982), *Ttgn. minangkabau* (42 days), *Sundatrigona moorei*, and *Heterotrigona itama* (46.5 days for both species) (Salmah et al. 1996). However, the workers of *Austroplebeia* have a similar development time to that of *Melipona:* 54–55 days (Halcroft et al. 2013). It has been suggested that interspecific differences in the length of development may be due to variation in the ability to maintain stable nest temperatures. Species with less developed thermoregulation may develop less rapidly (Halcroft et al. 2013; Macías-Macias and Quezada-Euán 2015). Other factors like body size may also affect the length of development, with larger individuals normally taking longer to develop (Jarošík and Honekel 2007). In the case of stingless bees, the larval food is formed mainly by pollen and honey which may also contribute to delay development. In contrast, honey bee larval food is a product of the worker glands and more readily digestible, possibly accelerating maturation of individuals in this species (Velthuis et al. 2003; Quezada-Euán et al. 2011).

An interesting contrast regarding ontogenetic development seems to exist in stingless bees. While *Melipona* gynes develop faster than workers and males, in non-*Melipona* there seems to be an opposite trend; the workers and males develop faster than gynes (Wille 1983). In a preliminary investigation on *N. perilampoides*, we found that gynes take approximately 41 days to develop, and workers 36.5 days. Similarly, gynes of *S. pectoralis* emerge around 44 days, while workers do so in 37 days (Quezada-Euán, pers. Obs.). Information on other non-*Melipona* species is required to confirm this observation.

Fig. 4.1 Ontogenetic development of a *M. beecheii* worker. Capital letters indicate the ontogenetic phase with relevant features of the phase in brackets. The number of days below the title indicates the total duration of each phase, and the number of days above each image indicates the age of the individual

Table 4.1 Duration in days of the different phases of ontogenetic development of two species of *Melipona* from Mexico

	M. beecheii[a]			*M. colimana*[b]		
Phase	Workers	Males	Gynes	Workers	Males	Gynes
Egg	8	8	8	11	11	11
Larva 1	5	5	5	4	4	4
Larva 2	4	4	4	4	4	4
Larva 3	9	10	8	8	11	8
Prepupa	2	2	2	2	2	2
Pupa (white)	3	3	3	5	5	4
Pupa (pink eye)	8	8	6	5	5	4
Pupa (pigmented ocelli)	6	6	4	10	10	6
Pupa (base of the wings and antennae pigmented wings folded on the body)	3	3	3	3	3	3
Pupa (dark body, wings extended)	5	6	7	5	4	6
Average duration	**52.72**	**53.48**	**50.8**	**55.44**	**57.14**	**52.62**

[a]Moo-Valle et al. 2004
[b]Macías-Macias and Quezada-Euán 2015

Melipona bees, similar to other Apidae, emerge with a pigmented cuticle. However, the intensity of pigmentation and the thickness of the cuticle are still incipient compared with solitary species. There is a process of gradual melanization and maturation of the cuticle known as cuticular heterochrony (Elias-Neto et al. 2014). Interestingly, non-*Melipona* bees exhibit the highest degrees of heterochrony; they lack pigmentation and the cuticle is soft at emergence. Melanization occurs gradually as they age and is completed usually after 2 weeks (Salmah et al. 1996; Elias-Neto et al. 2014).

On the combs, the distribution of cells of different individuals varies across species. In the case of *M. beecheii* and *M. colimana*, male and gyne cells are randomly distributed (Moo-Valle et al. 2004; Macías-Macias and Quezada-Euán 2015). However, in other species, like Brazilian *M. quadrifasciata* and *M. favosa*, cells containing males are usually placed in the center of the combs and gyne cells in the periphery (Koedam 1999). It is suggested that in species in which workers participate in male production, such as *M. quadrifasciata* and *M. favosa*, the workers lay eggs in batches in periods of "worker reproductive dominance" resulting in clustered male cells (Chinh et al. 2003; Sommeijer et al. 2003).

When worker and male *Melipona* complete their development, they emerge from the cells with assistance from old workers. In contrast, gynes cut the cap by themselves, rapidly exit the cell and run to hide from workers.

4.2 The Sex and Caste of Individuals

In the highly eusocial colonies there are two sexes (males and females) and two castes, workers and queens. In social insects, the term caste is more frequently used to refer to the division of labor among females based on reproductive specialization

(Michener 1974). Reproductive specialization is reflected in workers and queens being morphologically and physiologically different (Michener 1974). Queens are devoted to reproduction and have lost all structures for the collection of food and for building a new nest. Workers, on the other hand, are dedicated to the collection of food and activities inside the nest. In some species, workers are capable of reproduction, but as they are uncapable of mating, they can only produce unfertilized eggs.

In the Hymenoptera, sex determination (how males and females are produced), is the result of a haplodiploid system. In this type of system, the males normally develop from unfertilized eggs, meaning that they only have one set of the chromosomes of the species (*n*), they are haploid. In contrast, females derive from fertilized eggs and own two sets of chromosomes (2*n*), thus, being diploid. Males result from the recombination of their mother's genes but have no genetic contribution from the males which sired the queen, they have no father (Fig. 4.2). This mode of reproduction in which males are produced from unfertilized eggs is known as arrhenotokous parthenogenesis (Zayed 2009). In exceptional circumstances, males can develop from fertilized eggs, as will be explained later.

In the Hymenoptera, there is no sex chromosome. A gene called *single-locus complementary sex determiner* or sl-*csd*, of which there are several alleles, is responsible of sex determination. The sex of the individual will depend on whether it inherits one or two sl-*csd* copies, and if these are in homozygous or heterozygous condition (Zayed 2009).

The product encoded by the sex locus sl-*csd* is necessary for feminization. At molecular level, the sl-*csd* locus acts upon a specific target, identified as the feminizing gene (*fem*) (Hasselmann et al. 2008). When sl-*csd* is artificially inactivated, individuals develop a whole set of masculine characters (Beye et al. 2003). In heterozygotes (diploid individuals with two different sl-*csd* alleles), the effect of both

Fig. 4.2 Production of sexes and castes in stingless bees. Males normally originate from unfertilized eggs and females from fertilized ones. Fertilized eggs can give rise to workers or queens, depending on the amount of food supplied by workers, or the genotype of individuals, as in *Melipona*

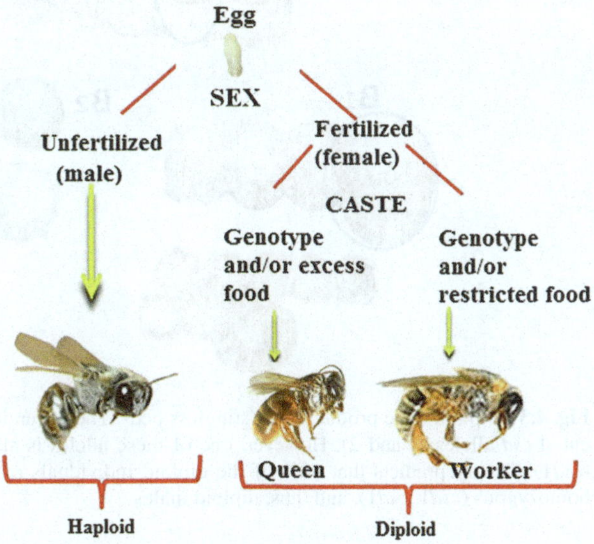

loci is additive and activation of the gene *fem* occurs. The activation of *fem* triggers a system for determining feminine characteristics. In individuals with only one copy (haploid), or two identical (homozygotes) copies of sl-*csd*, there is insufficient activation of the *fem* gene, and a male is produced (Zayed 2009; Fig. 4.3). Although complementary sex determination mediated by a sex locus has been frequently demonstrated in Hymenoptera, some exceptions have also been found (Ma et al. 2013).

The number of sl-*csd* alleles varies across population and species. In stingless bees, for instance, up to 25 alleles have been found for sl-*csd* in *M. scutellaris* from Brazil (Araujo Alves et al. 2011).

Diversity in the number of sl-*csd* alleles in Hymenoptera populations is important because as explained, when individuals receive two identical copies, a male with two sets of chromosomes is produced, a diploid male (Zayed 2009) (Fig. 4.3). Diploid males are commonly sterile; they cannot father diploid daughters and reduce the reproductive success of the queens (Zayed and Packer 2005; Vollet-Neto et al. 2017), representing dead ends for reproduction. A decrease in sl-*csd* allelic richness can result in the production of more diploid males and, in turn, reduced population growth, triggering what is known as the diploid male vortex. Thus, drivers reducing the size of Hymenopteran populations (isolation or genetic drift) are expected to increase the frequency of diploid males by means of increased homozygosity at the sl-*csd* locus (Zayed and Packer 2005; Zayed 2009).

Interestingly, some stingless bees have mechanisms that seem to reduce the burden of diploid males. In *M. seminigra* and *M. interrupta*, workers can detect diploid males (probably by some kind of chemical as in honey bees) when they emerge and

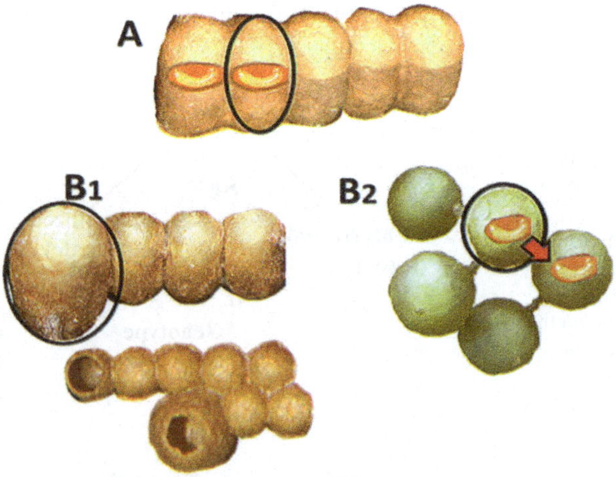

Fig. 4.3 Diploid male production in stingless bees. The queen in this example carries two different sl-*csd* alleles (1 and 2). However, one of these alleles is also present in the male siring her (*csd*1). Theory predicts that 50% of the diploid individuals resulting from this cross would be homozygous (*csd*1, *csd*1), and thus, diploid males

eliminate them (Winston 1987; Francini et al. 2012). Queens producing diploid males are also rapidly replaced (Francini et al. 2012; Vollet-Neto et al. 2017). Surprisingly, in some Meliponini, diploid males can reach adulthood (for instance in *M. quadrifasciata*; Borges et al. 2012), and can even join congregation areas, as in the case of *N. perilampoides*, *Ttr. angustula*, and *Scp. depilis* (Quezada-Euán et al. 2013; Santos et al. 2013; Vollet-Neto et al. 2015). Nonetheless, it is not known if stingless bee diploid males joining congregation areas can mate.

It is intriguing why in Brazil, some managed populations of stingless bees started with only a few colonies, and thus, highly inbred, show no negative effects on health and survival (Nogueira-Neto 2003; Alves et al. 2011). How these species and populations deal with the problem of diploid males is yet unknown, but surely understanding it should have important repercussions for the conservation of stingless bees and other Hymenopterans.

Stingless bees are highly diverse and cytogenetics can contribute greatly to the understanding the evolution of this Tribe (Rocha and Pompolo 1998; Tavares et al. 2017). Several studies have analyzed chromosome number variation and its distribution in Meliponini (Lopes et al. 2008; Fracini et al. 2011; Lopes et al. 2011).

Using data on 104 species of stingless bees, Tavares et al. (2017) recognized three main groups, in accordance to the most frequent chromosome number. The first group ($2n = 18$), comprises species of the genus *Melipona*, whereas karyotypes with $2n = 30$ and $2n = 34$, have been detected in species other than *Melipona*. However, it is noted that in the latter, the number of chromosomes can range between $2n = 16$ and $2n = 36$ (Tavares et al. 2012). In the case of *M. beecheii* and other Mexican species, the karyotype has not been analyzed yet.

In highly eusocial insects, the female larvae can develop either as worker or as queen. Caste determination (how a female larva becomes queen or worker) could be of two forms in stingless bees, trophic or tropho-genetic (Michener 1974; Sakagami 1982) (Fig. 4.4). Before describing each form, it is important to remember that in contrast with the honey bee, in which larvae are gradually fed by nurse workers in open cells, stingless bee larvae develop in isolation. The brood cells of stingless bees are mass provisioned and after the queen lays an egg, they are sealed. This system prevents contact between workers and larvae (Sakagami 1982). Rearing larvae in closed cells has a collateral effect in Meliponini; it reduces the possibility of workers controlling the fate of their sisters through feeding, as it occurs in the honey bee (Bourke and Ratnieks 1999).

In stingless bees, one form of caste determination is trophic. The trophic determination of caste is found in genera other than *Melipona* (Sakagami 1982). In this type of determination, the fate of the female larvae depends on differential nutrition; larvae to become queens are reared in cells with more food than workers.

There are two variants in the system of trophic determination of caste. The first variant includes the construction of royal cells in advance, that is, cells are expressly built to rear a queen (Sakagami 1982). In species building combs, royal cells are normally found in the periphery. In a second variant of the trophic determination, queens are reared from worker cells through emergency queen rearing (Faustino

	Queen (*n*=2) Xa_1 Xa_2 Xb_1 Xb_2 Egg genotypes			
Male (*n*=1) Xa_1 Xb_1	Xa_1 Xb_1	Xa_1 Xb_2	Xa_2 Xb_1	Xa_2 Xb_2
sperm genotype Xa_1, Xb_1	Xa_1 Xa_1 Xb_1 Xb_1 	Xa_1 Xa_1 Xb_1 Xb_2 	Xa_1 Xa_2 Xb_1 Xb_1 	Xa_1 Xa_2 Xb_1 Xb_2 
	Worker= Homozygote for both loci Xa and Xb	Worker= Homozygote for locus Xa	Worker= Homozygote for locus Xb	Queen= Heterozygote for both loci Xa and Xb

Fig. 4.4 The different systems of caste production in Meliponini (in the black circle, a cell with a future gyne): (**a**) Tropho-genetic in *Melipona,* the larva receives equal amount and quality of food to her sisters. (**b**) Trophic, with two variants: B1 in non-*Melipona* that build combs, queen cells are built expressively before the mother queen lays an egg; B2 occurs in non-*Melipona* species that build cells in clusters. Queen cells in this case are built from individual worker cells that are united

et al. 2002; Teixeira 2012). In this variant, enlarged queen cells are produced by merging two contiguous worker cells (Faustino et al. 2002). The workers unite the cells and help building a passage between, so that both larvae can get in contact. One larva prevails and consumes the food of the other cell. Alternatively, a support cell containing only food could be built for the same purpose. In both cases, the larva effectively increases its nutrient consumption, which will induce transformation into a queen (Faustino et al. 2002; Teixeira 2012). Trophic caste production through emergency queen rearing has been found in stingless bees that build clustered cells, like those in the genera *Frieseomelitta, Leurotrigona* (Terada 1974; Faustino et al. 2002), and *Austroplebeia* (Teixeira 2012).

A contrasting form of caste determination is found in the genus *Melipona.* Caste in this genus depends on the genotype of the individual, but the effect of food is important for its expression (tropho-genetic caste determination) (Fig. 4.3). A proximate model for caste determination in *Melipona* was originally proposed by Kerr (1948, 1950). The model considers that two non-linked loci, each with a pair of alleles, interact to produce the specific genotype of a queen. Only when both loci are found in heterozygous condition the individual can become a queen, but the expression of queen phenotype will require an adequate amount and/or quality of food.

On the other hand, homozygosity for one or both loci would result in a worker genotype (Fig. 4.4). Originally, Kerr (1948) also contemplated the possibility of a third locus involved (which would result in a proportion of queens to workers of one in eight), but this idea was later dismissed (Kerr 1969).

When female larvae with a queen genotype receive proper amount or quality of food, the *corpora allata* (a pair of endocrine glands at the base of the brain) increase the production of JH necessary for the transformation into a queen (Kerr et al. 1975). However, undernourished double-heterozygous female larvae would only express a worker phenotype (Darchen and Delage-Darchen 1975; Velthuis and Sommeijer 1991). JH seems to control hypomethylation of the genome in individuals to become queens. In contrast, in worker larvae, hypermethylation of the genome occurs (Cardoso-Júnior et al. 2017).

In accordance to the two-loci model of caste determination in *Melipona,* 25% would be the maximum possible queens. This means that potentially one queen could result from each of four female larvae produced in the colony (Fig. 4.4). However, the natural production of queens recorded in several species of *Melipona* is below that prediction, ranging between 14 and 21% (Moo-Valle et al. 2001; Sommeijer et al. 2003; Wenseleers et al. 2004; Morais et al. 2006). However, food conditions in colonies seems to influence the rate of queen production across time (Moo-Valle et al. 2001; Fig. 4.5). Interestingly, the experimental addition of 10 μg of geraniol (a compound produced in the labial glands of workers) to larval food, increased the production of queens in *M. beecheii* closer to the predicted 25% (Jarau et al. 2010). These findings suggest that geraniol can probably act as a precursor of compounds that regulate JH titers in female larvae (Jarau et al. 2010). Workers in malnourished conditions may produce less geraniol, explaining differences in queen production across colonies too.

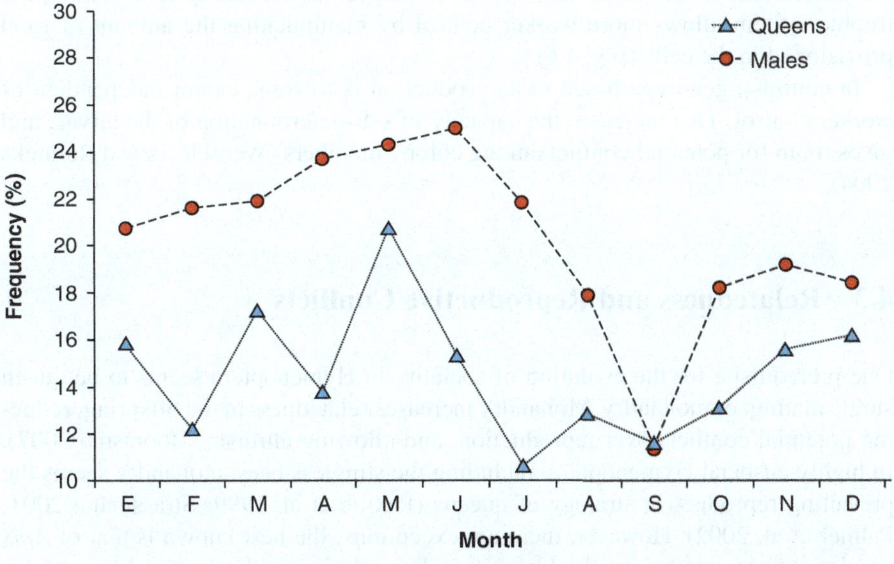

Fig. 4.5 Graphic representation of the genetic theory of caste production in *Melipona* (Kerr 1950). The queen produces four types of egg, each with a combination of the two caste determining loci. The male produces sperm of the same genotype given his haploid condition. The expected probability of an individual being double heterozygous (the queen genotype) is 25%. Workers are produced when the female is homozygous for one or both loci

Fig. 4.6 Monthly production of gynes and males of *M. beecheii* in Yucatan (Moo-Valle et al. 2001)

The two systems of queen production in stingless bees affect the possibility for self-determination of female larvae, i.e. its chances to become queen. Clearly, the trophic system allows more worker control by manipulating the amount of food provisioned in the cells (Fig. 4.6).

In contrast, genotype-based caste production is to some extent independent of worker control. This increases the capacity of self-determination of the larvae, and gives room for potential conflict among colony members (Wenseleers and Ratnieks 2004).

4.3 Relatedness and Reproductive Conflicts

One prerequisite for the evolution of sociality in Hymenoptera seems to be queen single mating, or monandry. Monandry increases relatedness in the offspring, reducing potential conflicts over reproduction, and allowing altruism (Boomsma 2007). In highly eusocial Hymenoptera, including the stingless bees, monandry seems the prevailing reproductive strategy of queens (Paxton et al. 1999; Strassmann 2001; Palmer et al. 2002). However, there are exceptions; the best known is that of *Apis* species whose queens are highly polyandric, mating with over a dozen males (Winston 1987).

The number of queen matings affects relatedness among colony members. In colonies headed by monandrous queens, as in stingless bees, the workers are supersisters (all daughters of the same father) and subfamilies are practically nonexistent (Peters et al. 1999). The daughters of single-mated Hymenopteran queens are highly related, because the male siring the queen is haploid (*n*) and the totality of his genes are passed on to his female offspring. Indeed, the coefficient of relatedness (*r*) between father and daughters is *r* = 1.00, and the he estimated coefficient of relatedness among supersisters is *r* = 0.75 (Peters et al. 1999; Palmer et al. 2002).

Relatedness among members of stingless bee colonies has important repercussions, because although workers are unable to mate, they can participate in male production. As a result of queen single mating, a stingless bee supersisters, is related more with her nephews (*r* = 0.375), than with her brothers (the queen sons, *r* = 0.25) (Fig. 4.7). It is thought that because relatedness is higher with their sisters sons than with their brothers, workers do not destroy eggs laid by other workers, and do not show aggression towards them. The acceptance of reproductive workers and their eggs in Meliponini is in strong contrast with the well-known behavior of "worker policing" in the honey bee. Presumably, worker policing is consequence of higher worker relatedness with brothers than nephews, due to queen polyandry (Ratnieks 1988).

In stingless bees, it is common that workers share in various degrees with the queen in the production of males (Tóth et al. 2002; Chinh et al. 2003; Koedam et al. 2005; Fig. 4.7). However, the sharing of worker reproduction varies among species

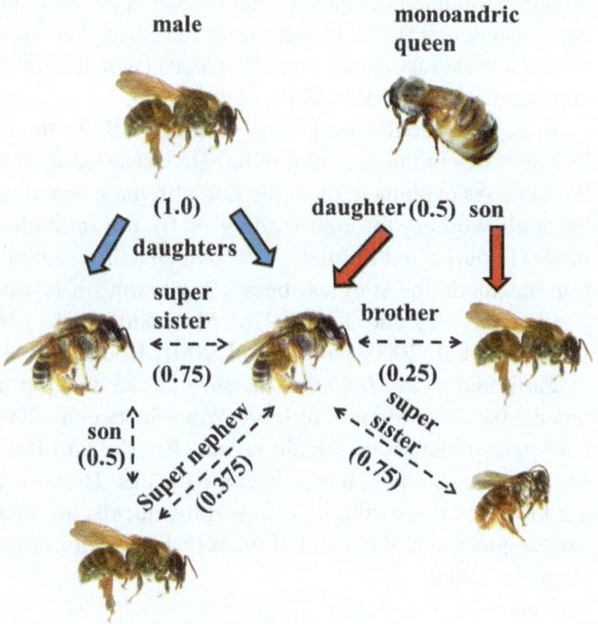

Fig. 4.7 Coefficients of relatedness (*r*) in stingless bees where queens are predominantly monandric. The male is haploid and the relationship with his daughters is 100% (*r* = 1.0). The queen only passes 50% of her genotype to sons and daughters (*r* = 0.5). The relatedness of a worker with her nephews is higher (*r* = 0.375) than with her brothers, sons of the queen (*r* = 0.25)

and among colonies of the same species (Tóth et al. 2002). There are also species in which workers do not seem to reproduce whatsoever. In *Fr. varia*, and probably other species of this genus, ovaries experience cell death and workers are permanently sterile (Boleli et al. 1999). In other species like *M. beecheii* workers have developed ovaries, but there is no molecular or behavioral evidence of worker reproduction, even during prolonged queenless periods (Paxton et al. 2001). For a yet not understood reason, workers in this species do not seem capable of producing fertile eggs. The eggs produced by *M. beecheii* workers lack nuclei, which is consistent with being trophic eggs (Quezada-Euán, unpublished data). No worker reproduction has been documented in *M. colimana* either (Macías-Macias and Quezada-Euán 2015). However, in this species no molecular evidence has been collected to assess the contribution of workers in male production. Some hypotheses suggest that the queen induces the production of trophic eggs by her workers (Cruz-Landim 2000), so it would be interesting to find out why, in the absence of the queen, the workers of some species produce only trophic eggs. Workers of species like *Sch. quadripunctata* are capable of producing males, but normally refrain from reproduction, probably because of the costs involved or because of strong reproductive control by the queen (Tóth et al. 2003). Ovarian supression of the workers via chemical control of the queen has been found in *Frs. schrotkyii* (Nunes et al. 2014). On the other extreme, there are species in which workers produce most of the males, as in *M. favosa* (Sommeijer et al. 1999).

Although workers can participate in male production, they would benefit more as queens, because of higher reproductive benefits. Adult workers are indifferent to which individual becomes worker or queen, because they are equally related to both supersisters ($r = 0.75$). However, for the colony's benefit there must be a balance, as not all female larvae can become queens (which would risk colony function). Thus, conflict arises (Bourke and Ratnieks 1999).

In accordance to Bourke and Ratnieks (1999), the root of caste conflict in stingless bees lies in the potential of female larvae (totipotentiality) to become queens or workers. As explained, the caste fate of female brood can be more or less controlled by adult workers through feeding or by the individual itself (the genetic-trophic mode) (Bourke and Ratnieks 1999). Such a differential outcome of caste determination has made the stingless bees a key taxon for testing arguments related to caste conflict (Contel and Kerr 1976; Machado et al. 1984; Sommeijer et al. 1999; Drumond et al. 2000; Paxton et al. 2001; Koedam et al. 2005).

Particularly, in *Melipona*, because larvae develop in cells of the same size and receive the same amount of food (Wenseleers et al. 2004), there is larger margin for auto-determination of female larvae (Bourke and Ratnieks 1999; Ratnieks 2001). Indeed, many female larvae become queens. However, excess queens in *Melipona* are killed by the workers as no surplus queens are generally needed. It seems that excess queens are the result of the selfish behavior of female larvae (Wenseleers and Ratnieks 2004).

In contrast, in stingless bee genera other than *Melipona,* the larvae do not have much margin for auto-determination, because through feeding, the workers can have more control of their future (Wenseleers et al. 2005). Nonetheless, it is interesting that in some genera like *Cephalotrigona, Nannotrigona, Plebeia,* and *Schwarziana,* there is evidence for some degree of female auto-determination. Females of those species can become queens with the amount of food provisioned for a worker, but their size is smaller than normal queens; they become "dwarf" queens. Dwarf queens can be functional, they can mate and head a colony (Ribeiro et al. 2006), representing a viable option for female larvae. It has been argued that the production of dwarf queens may be under a genetic mechanism like *Melipona* (Wenseleers et al. 2005; Ribeiro et al. 2006).

In a situation where the larvae can control their own caste fate, the probability of a larva becoming a queen has been calculated in a model proposed by Wenseleers et al. (2003):

$$\left(1-R_f\right)/\left(1+R_m\right)$$

Here, R_f is the coefficient of relatedness among sisters, and R_m the coefficient of relatedness of females with their brothers. In *Melipona,* because females are produced by monandric queens, the first coefficient is $R_f = 0.75$. However, the relatedness to males (R_m) in single-queen colonies can range from 0.25 to 0.75, as a result of variable degrees of worker reproduction (Tóth et al. 2002). Therefore, the production of gynes in species where all the males are the sons of the queen ($R_m = 0.25$) should be

$$\left(1-0.75\right)/\left(1+0.25\right)=20\%$$

Alternatively, in species in which the males are all sons of the workers ($R_m = 0.75$) the frequency of queen production should be

$$\left(1-0.75\right)/\left(1+0.75\right)=14\%$$

In accordance to those calculations, queen production in *Melipona* should vary between 14 and 20% depending on the extent of sharing between the queen and the workers in the production of males (Wenseleers et al. 2003). The model predicts that in species in which workers participate actively in the production of males, female larvae would gain less by becoming a queen, resulting in fewer queens produced.

Theoretical predictions of queen production are supported with empirical evidence in some *Melipona* species (Ratnieks 2001; Wenseleers and Ratnieks 2004) (Fig. 4.8 and Table 4.2). In species where queens and workers share on male production, like *M. subnitida* (Contel and Kerr 1976; Koedam 1999; Koedam et al. 2005), *M. quadrifasciata* (da Silva 1977; Tóth et al. 2002), and *M. favosa* (Sommeijer et al. 1999; Chinh et al. 2003), fewer female larvae become queens (5.1–8.6%, Kerr

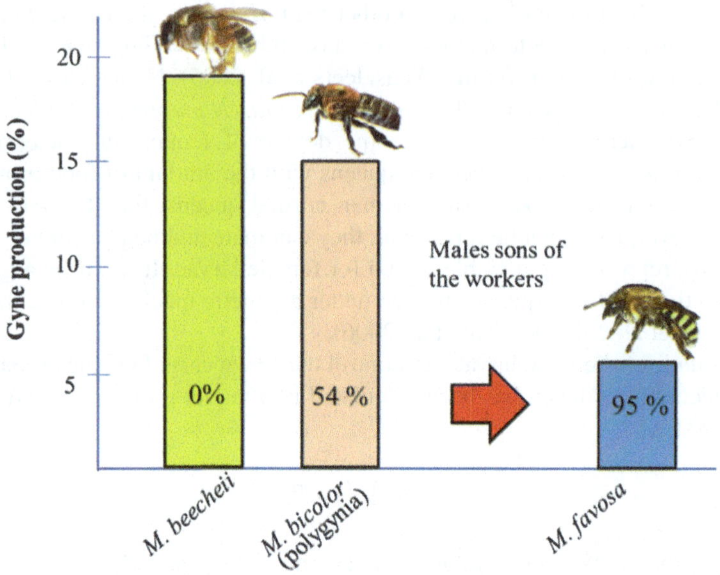

Fig. 4.8 In monandric *Melipona*, female relatedness is *Rf* = 0.75, theory predicts an inverse relationship between male production by workers and colony gyne production (from Wenseleers and Ratnieks 2004)

Table 4.2 Coefficients of female relatedness (*Rf*) and their relationship with the percentage of male sons of the workers, and the predicted proportion of female larvae that become gynes in various species of *Melipona*

Species	*Rf*	% Male sons of the workers	% gynes produced in the colony
M. beecheii[a]	0.75	0	21
M. subnitida[b]	0.75	34	8.5
M. quadrifasciata[c]	0.75	41	8.6
M. bicolor[d,e] (Polygynia)	<0.75	27–82 Average = 54	15.6
M. favosa[f]	0.75	95	5
M. colimana[g]	(?)	0	11.5

In *M. favosa,* males are mostly sons of workers, and queen production is lowest, while in *M. beecheii* all males are produced by the queen and gyne production is highest. In the case of *M. bicolor* where *Rf* < 0.75, the percentage of gynes produced also increases as predicted (Wenseleers et al. 2003; Wenseleers and Ratnieks 2004)

(?) = unknown

[a]Moo-Valle et al. (2001)

[b]Koedam et al. (2005)

[c]Tóth et al. (2002)

[d]Alves et al. (2012)

[e]Ferreira et al. (2013)

[f]Chinh et al. (2003)

[g]Macías-Macias and Quezada-Euán (2015)

1950; Koedam 1999, Koedam et al. 2005; Sommeijer et al. 2003). In contrast, in *M. beecheii*, in which all males are sons of the queen (Paxton et al. 2001), the production of queens is highest (14.6–19.8%, Moo-Valle et al. 2001; Darchen and Delage-Darchen 1975) (Fig. 4.8 and Table 4.2). Recently, the production of queens was evaluated in six colonies of *M. colimana*, a species in which workers apparently do not reproduce (Macías-Macias and Quezada-Euán 2015). It was found that queen production was on average 11.5%, a figure lower than in *M. beecheii*. It would be interesting to confirm these apparently discordant results in *M. colimana*, with molecular estimations of the number of queen matings and the maternity of males.

The model by Wenseleers et al. (2003) also predicts that when the relationship among sisters decreases ($R_f < 0.75$), the production of queens should increase. Such a situation is found in colonies headed by polyandric queens, or when multiple queens participate in reproduction (polygynia). An example of polygynic colonies is Brazilian *M. bicolor*, in which up to seven reproductive queens can be found in one colony (da Silva 1977; Nogueira-Neto 1997; Peters et al. 1999). Data obtained in *M. bicolor* has shown that queen production is high (15.6%; Ferreira et al. 2013). Estimates of mean effective maternity in *M. bicolor* are close to 1.1, meaning that workers in this species are more related to their sisters' sons ($r = 0.35$) than the queens'. In addition, the production of males, sons of the workers, was estimated in around 37% (Alves et al. 2012) confirming that in *M. bicolor*, the queens produce mostly female offspring (Koedam et al. 2007).

Interestingly, caste conflict models reveal that highly eusocial bees may be oppressed rather than altruistic. Wenseleers and Ratnieks (2004) likened the *Melipona* situation to a "tragedy of the commons", in which some individuals increase their exploitation of the common resource for their own benefit, but at the same time reduce the resources for the whole group. Nonetheless, as revealed from stingless bees, the group can impose different ways to control, usually through coercion and even elimination of selfish individuals (such as the killing of excess queens), and resolve conflict for the benefit of society.

References

Alves DA, Imperatriz-Fonseca VL, Francoy TM, Santos-Filho PS, Billen J, Wenseleers T (2011) Successful maintenance of a stingless bee population despite a severe genetic bottleneck. Conserv Genet 12:647–658

Alves DA, Imperatriz-Fonseca VL, Billen J, Wenseleers T (2012) Reproductive skew, male parentage and social parasitism in the facultatively polygyne eusocial bee *Melipona bicolor*. In: Anais do X Encontro sobre Abelhas, 2012. Riberão Preto, Brasil, p 111

Beye M, Hasselmann M, Fondrk MK, Page RE, Omholt SW (2003) The gene *csd* is the primary signal for sexual development in the honey bee and encodes a SR-type protein. Cell 114:419–429

Boleli IC, Paulino-Simões ZL, Gentile Bitondi MM (1999) Cell death in ovarioles causes permanent sterility in *Frieseomelitta varia* worker bees. J Morphol 242:271–282

Boomsma JJ (2007) Kin selection versus sexual selection: why the ends do not meet. Curr Biol 17:R673–R683

Borges AA, Ferreira-Caliman MJ, Nascimento FS, Campos LAO, Tavares MG (2012) Characterization of cuticular hydrocarbons of diploid and haploid males, workers and queens of the stingless bee *Melipona quadrifasciata*. Insect Soc 59:479–486

Bourke AFG, Ratnieks FLW (1999) Kin conflict over caste determination in social Hymenoptera. Behav Ecol Sociobiol 46:287–297

Cardoso-Júnior CAM, Fujimura PT, Santos-Júnior CD, Borges NA, Ueira-Vieira C, Hartfelder K, Goulart LR, Bonetti AM (2017) Epigenetic modifications and their relation to caste and sex determination and adult division of labor in the stingless bee Melipona scutellaris. Genet Mol Biol 40:61–68

Chinh TX, Gijs GJ, Meeuwsen FJAJ, Sommeijer MJ (2003) Patterns of male production in the stingless bee *Melipona favosa* (Apidae, Meliponini). Apidologie 34:161–170

Contel EPB, Kerr WE (1976) Origin of males in *Melipona subnitida* estimated from data of an isozymic polymorphic system. Genetica 46:271–279

Cruz-Landim C (2000) Ovarian development in Meliponine bees (Hymenoptera: Apidae): the effect of queen presence and food on worker ovary development and egg production. Genet Mol Biol 23:83–88

Dade HA (1985) Anatomy and dissection of the honeybee. International Bee Research Association, London

Darchen R, Delage-Darchen B (1975) Contribution a l'étude d'une abeille du Mexique *Melipona beecheii* B. (Hymenoptére: Apidae). Le dèterminisme des caste chez les Mélipones. Apidologie 6:295–339

Drumond PM, Oldroyd BP, Osborne K (2000) Worker reproduction in *Austroplebeia australis* Friese (Hymenoptera, Apidae, Meliponini). Insect Soc 47:333–336

Elias-Neto M, Nascimento A, Bonetti A, Nascimento F, Mateus S et al (2014) Heterochrony of cuticular differentiation in eusocial corbiculate bees. Apidologie 45:397–408

Faustino CD, Silva-Matos EV, Mateus S, Zucchi R (2002) First record of emergency queen rearing in stingless bees (Hymenoptera, Apinae, Meliponini). Insect Soc 49:111–113

Ferreira NT Jr, Blochtein B, Serrão JE (2013) Seasonal production and spatial distribution of *Melipona bicolor schencki* (Apidae; Meliponini) castes in brood combs in southern Brazil. Apidologie 44:176–187

Francini IB, Gross MC, Nunes-Silva CG, Carvalho-Zilse AG (2011) Cytogenetic analysis of the Amazon stingless bee *Melipona seminigra merrillae* reveals different chromosome number for the genus. Sci Agric 68:592–593

Francini IB, Nunes-Silva CG, Carvalho-Zilse AG (2012) Diploid male production of two Amazonian *Melipona* bees (Hymenoptera: Apidae). Psyche. Article ID 484618

Halcroft M, Haigh AM, Spooner-Hart R (2013) Ontogenic time and worker longevity in the Australian stingless bee, *Austroplebeia australis*. Insect Soc 60:259–264

Hartfelder K, Makert GR, Judice CC, Pereira GAG, Santana WC, Dalacqua R, Bitondi MMG (2006) Physiological and genetic mechanisms underlying caste development, reproduction and division of labor in stingless bees. Apidologie 37:144–163

Hasselmann M, Gempe T, Schiøtt M, Nunes-Silva CG, Otte M, Beye M (2008) Evidence for the evolutionary nascence of a novel sex determination pathway in honeybees. Nature 454:519–522

Jarau S, van Veen JW, Twele R, Reichle C, Herrera-Gonzales E, Aguilar I, Francke W, Ayasse M (2010) Workers make the queens in *Melipona* bees: identification of geraniol as a caste determining compound from labial glands of nurse bees. J Chem Ecol 36:565–569

Jarošík V, Honekel A (2007) Sexual differences in insect development time in relation to sexual size dimorphism. In: Fairbairn DJ, Blanckenhorn WU, Székely T (eds) Sex, size and gender roles. Oxford University Press, Oxford

Kerr WE (1948) Estudos sôbre o gênero Melipona. Annales Escola Superior Agricultura "Luiz de Queiroz" 5:181–276

Kerr WE (1950) Genetic determination of castes in the genus *Melipona*. Genetics 35:143–152

Kerr WE (1969) Some aspects of the evolution of social bees (Apidae). Evol Biol 3:119–175

Kerr WE, Akahira Y, Camargo CA (1975) Sex determination in bees. IV. Genetic control of juvenile hormone production in *Melipona quadrifasciata* (Apidae). Genetics 81:749–756

Koedam D (1999) Production of queens, workers and males in the stingless bee *Melipona favosa* (Apidae: Meliponinae): patterns in time and space. Neth J Zool 49:289–302

Koedam D, Contrera FAL, Fidalgo AdO, Imperatriz-Fonesca VL (2005) How queen and workers share in male production in the stingless bee *Melipona subnitida* (Apidae, Meliponini). Insect Soc 52:114–121

Koedam D, Cepeda Aponte OI, Imperatriz-Fonseca VL (2007) Egg laying and oophagy by reproductive workers in the polygynous stingless bee *Melipona bicolor* (Hymenoptea, Meliponini). Apidologie 38:55–66

Lopes DM, Pompolo SdG, Oliveira Campos LAd, Tavares MG (2008) Cytogenetic characterization of *Melipona rufiventris* Lepeletier 1836 and *Melipona mondury* Smith 1863 (Hymenoptera, Apidae) by C banding and fluorochromes staining. Genet Mol Biol 31:49–52

Lopes DM, Fernandes A, Praça-Fontes MM, Werneck HA, Resende HC, Campos LAO (2011) Cytogenetics of three *Melipona* species (Hymenoptera, Apidae, Meliponini). Sociobiology 57:1–10

Ma WJ, Kuijper B, de Boer JG, van de Zande L, Beukeboom LW, Wertheim B, Pannebakker BA (2013) Absence of complementary sex determination in the parasitoid wasp genus *Asobara* (Hymenoptera: Braconidae). PLoS One 8:e60459

Machado MFPS, Contel EPB, Kerr WE (1984) Proportion of males sons-of-the-queen and sons-of-workers in *Plebeia droryana* (Hymenoptera, Apidae) estimated from data of an MDH isozymic polymorphic system. Genetica 65:193–198

Macías-Macias JO, Quezada-Euán JJG (2015) Stingless bees in a temperate climate: oviposition behavior and duration of ontogenic development stages in *Melipona colimana* (Hymenoptera: Meliponini). J Apic Res 54:255–259

Michener CD (1974) The social behavior of the bees: a comparative study. Belknap Press, Harvard University, Cambridge

Moo-Valle H, Quezada-Euan JJG, Wenseleers T (2001) The effect of food reserves on the production of sexual offspring in the stingless bee *Melipona beecheii* (Apidae, Meliponini). Insect Soc 48:398–403

Moo-Valle H, Quezada-Euán JJG, Canto-Martín J, Gonzalez-Acereto JA (2004) Caste ontogeny and the distribution of reproductive cells on the combs of *Melipona beecheii* (Apidae: Meliponini). Apidologie 35:587–594

Morais MM, Nascimento FS, Pereira RA, Bego LR (2006) Colony internal conditions related to caste production in *Melipona compressipes fasciculata* (Apidae, Meliponini). Insect Soc 53:265–268

Nates-Parra G, Villa A, Vergara BC (1989) Ciclo de desarrollo de *Trigona* (*Tetragonisca angustula*) (development cycle of *Trigona*) Latreille 1811 (Hymenoptera, Trigonini). Acta Biol Colomb 1:91–98

Nijhout HF, Callier V (2015) Developmental mechanisms of body size and wing-body scaling in insects. Annu Rev Entomol 60:141–156

Nogueira-Neto P (1997) Vida e criação de abelhas indígenas sem ferrão. Editora Nogueirapis, São Paulo

Nogueira-Neto P (2003) Endocruzamentos em colônias de *Scaura longula* Lepeletier (Hymenoptera, Apidae). In: Melo GAR, Alves dos Santos I (Eds) Apoidea neotropica: homenagem aos 90 anos de Jesus Santiago Moure, pp 189-190. UNESC, Criciúma

Nunes TM, Mateus S, Favaris AP, Amaral MFZJ, von Zuben LG, Clososki GC, Bento JMS, Oldroyd BP, Silva R, Zucchi R, Silva DB, Lopes NP (2014) Queen signals in a stingless bee: suppression of worker ovary activation and spatial distribution of active compounds. Sci Rep 4(1):7449

Palmer KA, Oldroyd BP, Quezada-Euán JJG, Paxton RJ, May-Itzá WdJ (2002) Paternity frequency and maternity of males in some stingless bee species. Mol Ecol 11:2107–2113

Paxton RJ, Weiβschuh N, Engels W, Hartfelder K, Quezada-Euán JJG (1999) Not only single mating in stinglees bees. Naturwissenschaften 86:143–146
Paxton RJ, Ruhnke H, Shah M, Bego LR, Quezada-Euan JJG, Ratnieks FLW (2001) Social evolution in stingless bees: are the workers or is the queen in control of male production? In: II Seminario Mexicano sobre abejas sin aguijón. Universidad Autónoma de Yucatán, Mérida, México, pp 104–107
Peters JM, Queller DC, Imperatriz-Fonseca VL, Roubik DW, Strassmann JE (1999) Mate number, kin selection and social conflicts in stingless bees and honeybees. Proc R Soc Lond Ser B 266:379–384
Quezada-Euán JJG, López-Velasco A, Pérez-Balam J, Moo-Valle H, Velazquez-Madrazo A, Paxton RJ (2011) Body size differs in workers produced across time and is associated with variation in the quantity and composition of larval food in *Nannotrigona perilampoides* (Hymenoptera, Meliponini). Insect Soc 58:31–38
Quezada-Euán JJG, May-Itzá WdJ, Medina LA, Paxton RJ (2013) Diploid males at urban congregations of the stingless bee *Nannotrigona perilampoides* (Hymenoptera: Meliponini). In: Memorias VIII Congreso Mesoamericano de Abejas Nativas. CINAT, Costa Rica, pp 345–350
Ratnieks FLW (1988) Reproductive harmony via mutual policing by workers in eusocial Hymenoptera. Am Nat 132:217–236
Ratnieks FLW (2001) Heirs and spares: caste conflict and excess queen production in *Melipona* bees. Behav Ecol Sociobiol 50:467–473
Ribeiro MdF, Wenseleers T, Filho PdSS, Alves DdA (2006) Miniature queens in stingless bees: basic facts and evolutionary hypotheses. Apidologie 37:191–206
Rocha MP, Pompolo SG (1998) Karyotypes and heterochromatin variation (C-bands) in *Melipona* species (Hymenoptera, Apidae, Meliponini) chromosomes. Genet Mol Biol 21:41–45
Sakagami SF (1982) Stingless bees. In: Hermann HR (ed) Social insects, vol III. Academic Press, London, pp 361–423
Salmah S, Inoue T, Sakagami SF (1996) Incubation period and post-emergence pigmentation in the Sumatran stinglees bee *Trigona* (*Heterotrigona*) *itama* (Apidae: Meliponinae). Jpn J Entomol 64:401–411
Santos CF, Menezes C, Imperatriz-Fonseca VL, Arias MC (2013) A scientific note on diploid males in a reproductive event of a eusocial bee. Apidologie 44:519–521
Savard J, Tautz D, Richards S, Weinstock GM, Gibbs RA, Werren JH, Tettelin H, Lercher MJ (2006) Phylogenomic analysis reveals bees and wasps (Hymenoptera) at the base of the radiation of Holometabolous insects. Genome Res 16:1334–1338
da Silva DLN (1977) Estudos bionomicos em colônias mistas de Meliponinae. Boletim de Zoologia 2:7–106
Sommeijer MJ, Chinh TX, Meeuwsen FJAJ (1999) Behavioural data on the production of males by workers in the stingless bee *Melipona favosa* (Apidae, Meliponinae). Insect Soc 46:92–93
Sommeijer MJ, de Bruijn LLM, Meeuwsen FJAJ, Martens EP (2003) Natural patterns of caste and sex allocation in the stingless bees *Melipona favosa* and *M. trinitatis* related to worker behaviour. Insect Soc 50:38–44
Strassmann J (2001) The rarity of multiple mating by females in the social Hymenoptera. Insect Soc 48:1–13
Tavares M, Carvalho C, Soares F, Oliveira Campos L (2012) Genome size diversity in stingless bees (Hymenoptera: Apidae, Meliponini). Apidologie 43:731–736
Tavares MG, Lopes DM, Campos LAO (2017) An overview of cytogenetics of the tribe Meliponini (Hymenoptera: Apidae). Genetica 145:241–258
Teixeira LV (2012) Produção de rainhas em especies de abelhas sem ferrão com células de cria dispostas em cacho (Hymenoptera, Apidae, Meliponini). Tese Doctoral, Universidade Federal do Viçosa, Minas Gerais, Brasil
Terada Y (1974) Contribução ao estudo da regulação social em *Leurotrigona muelleri* e *Frieseomelitta varia*. Tese Mestrado, Universidade de São Paulo, Riberão Preto, Brasil

Tóth E, Strassmann EJ, Nogueira-Neto P, Imperatriz-Fonseca VL, Queller CD (2002) Male production in stingless bees: variable outcomes of queen–worker conflict. Mol Ecol 11:2661–2667

Tóth E, Strassmann EJ, Imperatriz-Fonseca VL, Queller CD (2003) Queens, not workers, produce the males in the stingless bee *Schwarziana quadripunctata quadripunctata*. Anim Behav 66:359–368

Velthuis HHW, Sommeijer MJ (1991) Roles of morphogenetic hormones in caste polymorphism in stingless bees. In: Gupta AP (ed) Morphogenetic hormones of arthropods. Rutgers University Press, New Brunswick, pp 346–383

Velthuis HHW, Cortopassi-Laurino M, Pereboom Z, Imperatriz-Fonseca VL (2003) Speciation, development, and the conservative egg of the stingless bee genus *Melipona*. Proc Exper Appl Entomol NEV Amster 14:53–57

Vollet-Neto A, dos Santos CF, Santiago LR, Alves DdA, Figueiredo JPd, Nanzer M, Arias MC, Imperatriz-Fonseca VL (2015) Diploid males of *Scaptotrigona depilis* are able to join reproductive aggregations (Apidae, Meliponini). J Hymenopt Res 45:125–130

Vollet-Neto A, Oliveira RC, Schillewaert S, Alves DA, Wenseleers T, Nascimento FS, Imperatriz-Fonseca VL, Ratnieks FLW (2017) Diploid male production results in queen death in the stingless bee *Scaptotrigona depilis*. J Chem Ecol 43:403–410

Wenseleers T, Ratnieks FLW (2004) Tragedy of the commons in *Melipona* bees. Proc R Soc Lon B 271(Suppl):S310–S312

Wenseleers T, Ratnieks FLW, Billen J (2003) Caste fate conflict in swarm-founding social Hymenoptera: an inclusive fitness analysis. J Evol Biol 16:647–658

Wenseleers T, Hart AG, Ratnieks FLW, Quezada-Euán JJG (2004) Queen execution and caste conflict in the stingless bee *Melipona beecheii*. Ethology 110:725–736

Wenseleers T, Ratnieks FLW, Ribeiro MF, Alves DA, Imperatriz-Fonseca VL (2005) Working-class royalty: bees beat the caste system. Biol Lett 1:125–128

Wille A (1983) Biology of the stingless bees. Annu Rev Entomol 28:41–64

Winston ML (1987) The biology of the honey bee. Harvard University Press, Cambridge

Zayed A (2009) Bee genetics and conservation. Apidologie 40:237–262

Zayed A, Packer L (2005) Complementary sex determination substantially increases extinction proneness of haplodiploid populations. Proc Natl Acad Sci USA 102:10742–10746

Chapter 5
Colony Function and Communication

Workers of *M. beecheii* performing different activities in the brood area.

How colonies work as a coherent unit is one fascinating aspect of insect societies. The mechanisms that these small organisms, with apparently limited capacity for reasoning and decision, use to produce sophisticated responses (highly structured nests, comb construction, food gathering, and colony defense), cooperating as a functional unit, have captured the scientific interest for centuries. Today, it is generally accepted that social coordination seems based on individual behavior, and that communication and information among colony members are key aspects to produce coordinated responses (Seeley 1995).

In social insects, there is no central control of the activities; no individual or group of individuals decide which tasks are performed, nor who, when, where, and

J. J. G. Quezada-Euán, *Stingless Bees of Mexico*,
https://doi.org/10.1007/978-3-319-77785-6_5

how they are performed (Seeley 1998). The large number of individuals conforming colonies would make this type of control inefficient, due to the time it may require for the information to flow up to one central control and back to the executers (Gordon 2016). Rather, it seems that workers react with individual responses to diverse stimuli, which in turn affect the behavior of nearby nestmates. Individual responses are somehow integrated to build group level organized and a coordinated behaviors (Page Jr. 2013). Organization relies to a great extent on the ability to learn and the capacity to communicate among colony members (Biesmeijer 1997). The latter, the capacity to communicate, is especially relevant as it allows passing on and receiving information among nestmates, from which coordinated actions result (Seeley 1995).

Surprisingly, not much is yet known about colony integration in Meliponini compared with the honey bees (Seeley 1995; Biesmeijer 1997). However, accumulating evidences are allowing a better understanding on how colonies work and the proximate mechanisms involved in communication. This chapter covers some general aspects of the organization of work in stingless bees, with descriptions of particular behaviors, like the process of oviposition and the mechanisms of defense. In the last two sections, I summarized information of chemical and physical communication in Meliponini.

5.1 Organization of Activities

Stingless bee societies are formed by three types of individual, males, queens (reproductive females), and workers (partially sterile females). The workers are by far the main body of colonies; in stingless bees their numbers range from a few hundred to over tens of thousands depending on the species (Sakagami 1982). Workers perform most activities, including the construction of different structures, feeding of the larvae and the queen, as well as defense of the nest, and collection of food and other materials.

Similar to the honey bee, the various tasks that stingless bee workers perform across their life can be roughly divided in activities inside and outside the colony. There is also an intermediate phase in which the workers perform guarding and orientation flights before they dedicate to external activities (Michener 1974). However, how the progression across tasks and task assignment are achieved is a complex phenomenon that is still not well understood. Like in the honey bee, the progression of worker stingless bees across different tasks is strongly related to age and is known as age polyethism (Wilson 1971). Interestingly, age polyethism is flexible, meaning that workers can progress or revert across different activities, depending on the needs of their colonies (Robinson and Huang 1998). Artificial manipulation of colonies has shown that if young workers are eliminated, foragers can revert to performing nest activities despite their older age (Sommeijer 1987).

Although stingless bee workers transit through a series of activities, these and the age at which they perform them depend on the status of individuals (threshold to different stimuli), colony needs, and changes in the environment (Biesmeijer 1997). The status of individuals depend on physiological changes related to age. Titers of morphogenetic hormones like ecdysteroids, and Juvenile Hormone (JH) vary with worker age and seem to affect individual sensibility to perform different tasks. In the honey bee, it has been shown that such hormones play basic roles in the transition of workers across different activities (Hartfelder et al. 2002; Robinson and Huang 1998; Page Jr. 2013). However, in stingless bees the interaction between the different hormones show important differences with the honey bees (Hartfelder et al. 2006). In some species, like *M. quadrifasciata*, the levels of JH and ecdysteroids show low hemolymph titers during the first days after emergence in both, workers and queens, which is similar to what occurs in the honey bee (Hartfelder et al. 2002). Nonetheless, in species like *M. scutellaris*, JH titers are highest in nurse bees decreasing markedly in foragers (Cardoso-Júnior et al. 2017a). A reverse situation is found in worker honey bees, in which JH titers constantly increase as workers age (Hartfelder et al. 2002; Page Jr. 2013). These findings suggest that stingless bees may have maintained the ancestral gonadotropic function for JH (Cardoso-Júnior et al. 2017a). Also in contrast with honey bees in which vitellogenin titers peak in young workers, stingless bees' vitellogenin patterns do not show marked relationship with worker age (Hartfelder et al. 2006).

There is also a genetic component of polyethism and some bees tend to perform certain activities more frequently than others; some may not even perform certain activities at all (Biesmeijer 1997). Recently, findings in *M. scutellaris* suggest a possible mechanism linking genotype component with larval diet and activity (Cardoso-Júnior et al. 2017b). It seems that differences in global methylation of the genome in the larval-pupal stages favor the activation of genes related to task allocation in the adult bee (Cardoso-Júnior et al. 2017b).

As explained before, flexibility in task performance is one important characteristic of polyethism in the highly eusocial bees. Due to this particularity, the term work division or division of labor, which suggests a rigid system of task performance, may be better changed to *task allocation* (Gordon 2016). The term task allocation includes elements of plasticity, suggesting that the assignment of individuals to different tasks could vary in response to the constantly changing stimuli in the environment and individual thresholds, determined physiologically and genetically (Gordon 2016).

Externally, workers of the highly eusocial bees are remarkably uniform (but see Grüter et al. 2012), in comparison with ants or termites, in which highly differentiated worker morphologies exist (Wilson 1971). Thus, in highly social bees, the term caste seems more appropriate for the two morphologically and physiologically different individuals of the female sex, the workers, and the queen (Michener 1974). The subdivision that exists among workers of the highly eusocial bees regarding allocation to different activities may be better referred to as subcaste.

In general, the activities performed by the workers in the highly social bees could be organized along two main principles (Ratnieks and Anderson 1999; Hart and Ratnieks 2001):

- *Task allocation*: Workers are assigned to major groups of tasks like nursing, defending, or foraging.

– *Task partition*: The main tasks are divided among groups of workers, usually of different age, in different subtasks. For instance, nectar collection is more efficiently performed by two groups of workers, foragers and receivers. Interestingly, task partitioning seems exclusive of highly eusocial species, being absent in the Bombinini and Euglossini (Hart and Ratnieks 2002).

Generally, the activities performed by stingless bee workers fall into five major groups (Sakagami 1982; Wille 1983):

(a) Cleaning and reparing the brood chamber
(b) Construction of cells and provisioning
(c) Nectar reception and processing
(d) Guard
(e) Nectar, pollen, and resin collection

Similar to honey bees, the first tasks occur in the nest and are performed by younger bees. As bees age, they start performing activities closer to the exterior. In *M. beecheii*, a detailed account on the activities of several workers, revealed a general pattern from "indoor to outdoor" performance as workers age. Nonetheless, a great deal of variation between individuals on the time and frequency dedicated to different tasks was also noted (Medina-Medina et al. 2014; Fig. 5.1). Young bees mainly

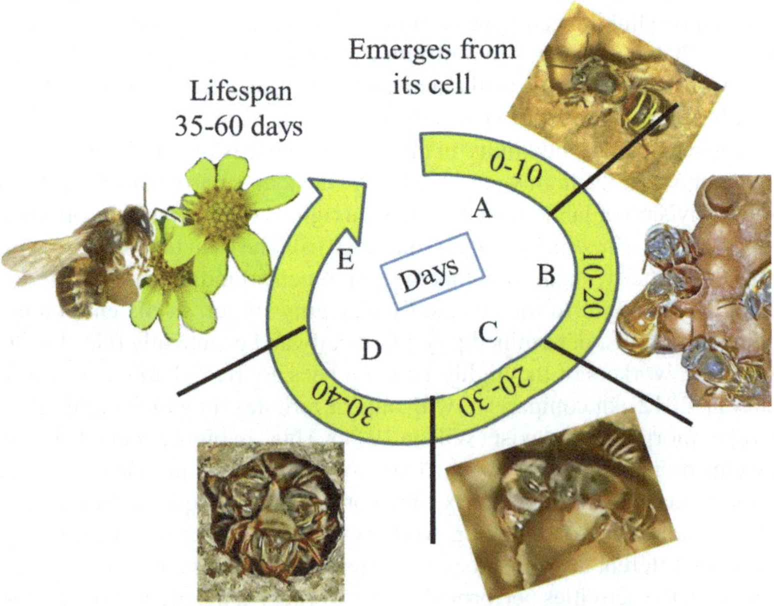

Fig. 5.1 A general sequence of activities during the life cycle of a *M. beecheii* worker: (**a**) Construction and cleaning; (**b**) larval food provisioning and production of trophic eggs; (**c**) nectar reception and processing; (**d**) guard, orientation flights; (**e**) foraging

engage in cleaning and construction in the brood area; they also have well-developed wax glands. Workers between 10 and 18 days produce trophic eggs, receive nectar from foragers, and spend a great deal of time in dehydrating it. Older bees (25–35 days) usually perform activities related with guarding and cleaning; they also start short-range flights presumably for orientation. After 35 days, workers start foraging.

The average life span calculated for workers of *M. beecheii* was 52 days, with a range between 35 and 60 days (Medina-Medina et al. 2014; Fig. 5.1).

Some distinctive tasks and behaviors only found in stingless bees (Sakagami 1982; Wille 1983) are:

1. The collection of large amounts of resin
2. Mass provisioning of larval food
3. Excretion inside the nest, which probably delays engaging in external activities
4. The production of trophic eggs to feed the queen

The type of activity affects worker longevity, possibly due to the physical wear involved. In Costa Rica, Biesmeijer and Toth (1998) found that *M. beecheii* nectar foragers are more active and live on average 3 days. In contrast, pollen foragers are only active 1–3 h daily, and can live up to 12 days.

An interesting feature of task allocation in eusocial bees is specialization in field bees (Kolmes 1985). In stingless bees, some individuals prefer collecting nectar over pollen and *vice versa* (Biesmeijer and Toth 1998). In *A. mellifera,* it has been documented that nectar foragers also specialize on some kinds of flowers (Seeley 1995). It seems that task specialization is beneficial for the colony, increasing individual efficiency and performance.

In *Melipona* specialization on the collection of certain resource type seems less common; most workers indistinctively engage in nectar, pollen, or resin foraging, compared with honey bees (Biesmeijer and Toth 1998). Perhaps the small number of workers per colony in *Melipona* can explain why specialization on particular resources may not be adaptive. Non-specialized workers may more ready shift to collecting different resources as they are available.

The causes of specialization are not well understood. In *A. mellifera* some physical characteristics of the workers could affect their preference for collecting nectar or pollen. Forager honey bees were compared morphologically, and although no substantial differences in body size were found, pollen collectors had significantly more olfactory *sensilla* plates on the antennae compared with nectar collectors (Riveros and Groenenberg (2010).

In *M. beecheii,* Quezada-Euán et al. (2015) compared the body size of bees bringing nectar, pollen, and resin to the colony. Similar to the honey bee, body size did not vary between bees collecting different resources. Although the level of specialization of each worker was not determined, these results suggest that in the highly eusocial honey bees and stingless bees, body size does not seem crucial in determining specialization (Quezada-Euán et al. 2015). It is possible that the elaborate mechanisms of communication for food collection in the highly eusocial honey bees and *Melipona* (Winston 1987; Nieh et al. 2003) could restrain large size differences between individuals. Large size differences could produce "distortion" in the interpretation of messages related to distance and/or direction to food sources (Waddington 1989; see section 5.6). It would be interesting to verify if similar principles apply in non-*Melipona* species with less sophisticated communication systems (Jarau 2009) and that present significant body size differences between workers (Quezada-Euán et al. 2011).

One activity in which Worker body size seems important is guarding. Indeed, in some meliponine species, a guard subcaste has been identified, larger compared to other workers in the colony (Grüter et al. 2012, 2017).

In addition to the possible effect of morphological traits on specialization, there also seems to be an influence of experience and learning. In *Pb. emerina* workers specialize in the collection of resins only after a period of handling these materials which occurs after the second week of life (dos Santos et al. 2010).

One important aspect in stingless bees is that males engage in some activities (nectar dehydration, wax production) at least sporadically (Van Veen et al. 1997; Velthuis et al. 2005). In this regard males of stingless bees are more similar to workers (morphologically and behaviorally), compared to drones in the honey bee (Hartfelder et al. 2006; Medina et al. 2016).

The evidence to date suggests that work allocation shares some common features between stingless bees and honey bees but also important contrasts. It is possible that similarities in the general mechanisms (genetic and physiologic) behind age polyethism exist in both groups of bees (Robinson and Huang 1998; Hartfelder et al. 2006). However, it is important to note that substantial differences exist in the mechanisms controlling task allocation between both types of bee. Much work is still needed to better understand the organization of work and its proximate causes in the stingless bees.

5.2 Task Partitioning

For efficiency, one major task can be divided into subtasks, usually performed by workers of different age (Hart and Ratnieks 2002).

One activity in which partitioning has been studied in detail is the collection of food. In *Apis* nectar gatherers transfer nectar loads to receivers inside the colony, who store it in the combs; foragers do not discharge the nectar they bring into the cells (Hart and Ratnieks 2001). This type of partitioning increases efficiency and also provides feedback between field and hive bees during nectar collection (Seeley 1985).

Nectar transference between nectar collectors and receivers has been documented for various species of stingless bees: *M. favosa* (Sommeijer 1984), *M.panamica* (Nieh and Roubik 1998), and *Ttgn. carbonaria* (Nieh et al. 1999/2000). In a study in Yucatan, Hart and Ratnieks (2002) evaluated the rate of food transference in four species of stingless bee with different systems of recruitment, from primitive (*N. perilampoides*) to more developed (*M. beecheii*) (Nieh et al. 1999/2000;

Schmidt et al. 2008). It was found that all species exhibited task partitioning during nectar collection. However, there were substantial differences in the rates of transference between collectors and receivers. In *M. beecheii* nectar collectors transferred loads to a larger number of receivers (up to 12). In *Fr. nigra, Pb. frontalis,* and *Scp. pectoralis,* with presumably less developed communication systems, nectar was transferred to less receivers ($\overline{x}$ = 1.8). Interestingly, the lowest number of transferences was observed in *N. perilampoides* ($\overline{x}$ = 1.5), which has a comparatively more primitive communication system. These results indicate that although task partitioning during food collection seems a common feature in stingless bees, there are differences in the intensity of interactions among workers, seemingly related to the sophistication of communication of food resources (Hart and Ratnieks 2002).

Surprisingly, *M. beecheii* foragers transferred food to a much larger number of receivers ($\overline{x}$ = 5.7), even compared with *A. mellifera,* in which average transference from foragers to receivers is between 1.9 and 2.7 times (Kirchner and Lindauer 1994; Hart and Ratnieks 2001). Such a high rate of nectar transference in *Melipona* could be related to the more sophisticated communication system in this genus (Lindauer and Kerr 1960; Nieh and Roubik 1995, 1998; Nieh 1998), possibly involving some mechanism of food recruiting in addition to nectar processing alone. By sharing a load with multiple receivers, foragers may stimulate recruiting of more food collectors. On the other hand, a large number of food discharges could indicate that receivers may also be selective and refuse the nectar carried by some foragers, or that the demand for nectar has decreased. Nevertheless, in the honey bee nectar collection is largely determined by the number of receivers, rather than demand, that seems constant (Seeley 1985).

Another example of task partitioning in stingless bees is the management of waste. In most stingless bees, waste dumps are temporarily built in some areas of the nest. In *M. beecheii*, some workers engage in gathering waste pellets, Interestingly, they usually transfer such pellets (93% of cases) to other workers, who take them out of the nest (Medina-Medina et al. 2014).

The results on food collection and waste management in stingless bees, show that task partitioning seems to be exclusive to highly eusocial bees, albeit with various degrees of complexity across species. Interestingly, it seems that task partition has only evolved in species that found colonies through swarming, and is absent in species whose colonies are founded by individual females (Hart and Ratnieks 2002).

5.3 Food Provisioning and Oviposition Process

One of the most studied behaviors in stingless bees is the process of food provisioning and oviposition. The process of food provisioning of brood cells and the subsequent oviposition by the queen is a highly ritualized conduct showing diverse degrees of interaction and antagonism between queen and workers (Fig. 5.2.).

Fig. 5.2 Phases of the process of provisioning and oviposition (POP) in *M. beecheii*: (**a**) Final phase of cell construction with a "collared" cell. (**b**) The queen approaches the cell and stands on one side while the workers take turns to discharge larval food; the queen frequently taps them on the thorax with her front legs. (**c**) The queen inspects the cell occasionally, possibly ingesting some food. (**d**) After consuming the trophic egg or eggs laid by workers, the queen lays her egg. (**e**) The queen moves away. (**f**) One worker climbs on top and rotates while capping the cell with the cerumen collar

Sakagami and Zucchi (1966) studied the complex sequence of behaviors known as provisioning and oviposition process (POP). They divided the process in four general phases which also take place in *Apis*, but in a different order:

1. Cell construction (C)
2. Oviposition (O)
3. Provisioning of the larval food (P)
4. Cell closing (CC)

In honey bees the completion of the four phases takes roughly 8 days, because of the prolonged feeding of the larvae in open cells. The order, thus, is C-O-P-CC. In stingless bees, the whole process may only take 2 h and the sequence is C-P-O-CC. A distinctive characteristic of the process in stingless bees is the ritualized interaction between queen and workers between C and P.

Although the basic phases of the POP are similar across different stingless bee species (Sakagami 1982; Drumond et al. 2000), there is a great deal of variation in the duration of the different phases, aggression between queen and workers, number of cells oviposited, etc. Such interspecific variation has been subject of various phylogenetic analyses (Zucchi 1993; Drumond et al. 2000).

One possible explanation for the complex queen-worker interaction during the POP is that through ritualized behaviors, the queen reaffirms her reproductive dominance over the workers. This may be so because in Meliponini the queen's chemical control of the workers seems generally less developed compared to the honey bee (Wille 1983; Zucchi 1993). Nonetheless, recent evidence indicates that chemical control of worker reproduction may occur in some species (*Friesella schrottkyi*) by means of queen cuticular compounds (Nunes et al. 2014a). Although stingless bees differ from honey bees in which the queen's signal is mostly produced in the mandibular glands (Nunes et al. 2014a), these findings indicate that physical and chemical control of worker reproduction by the queen may be found at different degrees in stingless bees, and that both strategies may not be excluding.

The general process of the POP in *M. beecheii* is presented in Fig. 5.2. In this species the whole process from construction, to cell capping, takes around 14.9 min. Cell capping alone may take up to 8 min (van Veen 2000; Avila et al. 2005). The highest frequency of POPs in *M. beecheii* occur at night. During the day, the frequency of POPs is one approximately every 2 h, but in the evening they may occur every hour (van Veen 2000).

During the POP, in most species of stingless bee, the workers produce a type of egg as food for the queen. These eggs, known as trophic eggs, lack nuclei and are infertile (Koedam et al. 1996). Trophic eggs are also found in other insects in which they are used to feed the offspring (Crespi 1992; Perry and Roitberg 2006). Stingless bees are in this sense a rare exception in the insect world, because trophic eggs are used to feed the queen. Thus, trophic eggs seem to play an important, but little understood, part in the interactions between workers and queen during the POP (Sakagami 1982).

A characteristic of trophic eggs is an apparently incipient stage of development, compared with fertile eggs. The chorion is thinner and more fragile. It is believed that under the influence of the queen, trophic eggs are rapidly released from the worker's ovaries, and thus, they cannot complete their development (Cruz-Landim 2000). In contrast, workers also produce fertile functional eggs and these have a thick chorion with a reticular pattern, similar to that of the queen's (Koedam et al.

1996; Velthuis 1997). Chemically, there are differences in the surface of trophic and functional eggs too. In *Melipona* the queen's eggs are rich in hydrocarbons C21–C29 which may help in repelling water, in combination with a reticular surface. In contrast, trophic eggs with little disturbance can easily sink into the larval food (Jungnickel et al. 2001).

Trophic eggs are found in almost all neotropical stingless bees, but the site where they are deposited during the POP varies across species (Sakagami 1982).

Trophic eggs in most *Melipona* species, are laid on top of the larval food or attached to the wall inside the cell (Fig. 5.3); in other stingless bees, eggs are more frequently laid on the rim of the cell (Sakagami 1982; Wille 1983).

In several species of *Melipona*, trophic eggs are smaller than the queen's (Aparecido-Pereira et al. 2006). However, in *M. beecheii* trophic eggs are relatively large and of similar size to the queen's (Fig. 5.3). Curiously, eggs produced by the queen of different *Melipona* can have different shapes, but the size tends to remain constant across species, in spite of differences in body size of the adult workers (Velthuis et al. 2003).

Fig. 5.3 Worker (O, trophic) and queen (R) eggs in *M. beecheii*. The image underneath shows the position of both types of egg on the larval food

Fig. 5.4 Successive mode of cell provisioning and oviposition in *M. beecheii*. A few cells are built simultaneously but only one is provisioned and oviposited after each POP (third from left to right)

An important difference in the POP of different species of stingless bees is the number of cells that are built, provisioned, and oviposited by the queen (Roubik 1989). In accordance to the number of cells and their location during the oviposition, different POP patterns have been identified:

1. Successive: In this pattern, the queen lays an egg in one cell which is sealed before the process starts in another cell. A few cells can be provisioned with larval food simultaneously (*Melipona, Frieseomelitta*). (Figs. 5.2 and 5.4).
2. Synchronic: A number of cells, up to a couple dozen, are simultaneously provisioned; when they are ready, the queen lays eggs in rapid succession (*Nannotrigona, Plebeia*). A few cells may be simultaneously capped by the workers, not one by one, as in the successive pattern.
3. Semi-synchronic: On the same comb there may be a few cells that are oviposited successively, while others are oviposited in synchrony, like in *Scaptotrigona* (Fig. 5.5).
4. Composite: The queen can lay eggs in two different combs, while the cells on one comb are provisioned and oviposited in succession, in other combs the provision and oviposition may occur in synchrony (*Lestrimelitta*) (Fig. 5.6).

Fig. 5.5 Semi-synchronic cell provisioning in *S. pectoralis*. On the same comb, a group of cells have been provisioned and are ready to be oviposited, while another group of cells are simultaneously built. The queen lays eggs consecutively in the cells that are ready

Fig. 5.6 Composite cell provisioning in *L. niitkib*. On each of two combs, more than a dozen cells are being built and provisioned simultaneously. The queen may lay eggs in the cells on one comb synchronically and successively, before moving to the other

5.4 Defense

In accordance to Hermann (1984), social insects face four principal classes of natural enemies: arthropod predators, vertebrate predators, insect parasitoids, and parasites and pathogens. One distinctive feature of highly eusocial insects is the build up of large reserves of food, representing a valuable source of energy and protein for other animals. Not surprisingly, together with the increase of reserves, the potential attraction of various types of robbers and predators increased as well (Breed et al.

2012). To avoid the potential risk of plunder, social insects have developed a series of strategies to protect their nests.

Because stingless bees can accumulate large amounts of honey and pollen, it is puzzling that this group of bees lost the functionality of the sting, one weapon commonly used by other social Hymenoptera for the protection of their nests. Nonetheless, these insects have evolved a series of seemingly equally efficient strategies to defend their nests (Kerr and Lello 1962; Grüter et al. 2012; Nunes et al. 2014a), which will be revised in this section.

It is not clear why stingless bees lost the functionality of their sting. However, it seems that the sting in bees is generally under-relaxed selection, allowing a wide range of morphologies and even gross reduction. Indeed, sting function has been lost in members of three more bee families apart from the Apidae (stingless bees); these are the Stenotritidae, Andrenidae, and Megachilidae (Packer 2003).

Some of the first stingless bees were small, and it is likely that they followed a defense strategy based more on crypsis and retreat (Wille 1979). Modern minute stingless bees still follow this type of strategy, they are timid (Kerr and Lello 1962; Wille 1979). In the course of time, the value of the sting as a defense mechanism in shy and inconspicuous species, may have been limited.

In *A. mellifera*, it is believed that the sting and venom evolved to defend the nest against large predators, mainly mammals (Winston 1987). Interestingly, the sting of the workers of *A. mellifera* has barbs and a ganglion associated, which keeps it buried in the skin while venom is pumped (Dade 1985). Another evidence of the use of the sting of *A. mellifera* against large predators is the presence of hyaluronidase and phospholipase in the venom. These two enzymes dilute the connective material of the skin making it easy for the other materials to disperse. Melittin and other compounds are responsible for the swelling and pain (Choo et al. 2010; Daneels et al. 2015). However, the production of venom is costly, as well as the massive attacks involved in defending the nest against large predators.

Presently army ants, cleptobiotic bees (*Lestrimelitta*), and frequently conspecific colonies are the main predators of stingless bees (Gloag et al. 2008; Grüter et al. 2016). It is likely that the sting and venom resulted gradually less useful against such attackers (Gloag et al. 2008; Grüter et al. 2016). Against other insects a tactic to immobilize the enemy or prevent invasion of the nest would be more advantageous (Shackleton et al. 2015). Attacking also represents high costs for the colony and a better strategy in species with low populations as many stingless bees, would be withdrawal and camouflaging. However, presently, there seem to be two main strategies of defense in stingless bees; one is crypsis and retreat, but in other species-aggressive responses are the norm. In a few outstanding cases (*Oxytrigona*), defense also involves the use of chemical weapons (Wille 1979). Nonetheless, mass attacks seem to occur with more frequency in species with populous colonies.

As mentioned before, army ants and other meliponines are probably the main predators of stingless bee colonies today. Different strategies have evolved to defend against these attackers. In the case of army ants, stingless bees make use of resin to block their entrance or repel the attacks.

A major threat to stingless bees is pillage by colonies of the same or different species. Like other eusocial insects, stingless bees engage in robbing if the opportunity arises, and some species are particularly prone to it (Grüter et al. 2016). However, there are two genera of stingless bees that have specialized in stealing food and materials from other nests, a behavior known as cleptobiosis (Breed et al. 2012). Bees of the neotropical *Lestrimelitta* with approximately two dozen species (Camargo and Pedro 2007) and the African *Cleptotrigona,* with one species (Eardley 2004), are obligate cleptobionts. Workers of both species have lost the corbiculae and plumose hairs involved in the collection of pollen, and obtain food and building materials exclusively from robbing other stingless bees (Fig. 5.9). *Lestrimelitta* and *Cleptotrigona* are not phylogenetically related, indicating that cleptobiosis evolved independently in two different continents (Eardley 2004).

The attacks of cleptobiotic *Lestrimelitta* seem to be of particular importance in having shaped the defense strategies of stingless bees, most notably the evolution of morphologically distinct worker soldiers in some species. Worker polymorphism is rare in flying social insects (Grüter et al. 2017), although it is common in ants and termites (Hölldobler and Wilson 1990). It is argued that the evolution of morphological subcastes would be difficult in wasps and bees because of the limitations posed by flight (Waddington 1989; Quezada-Euán et al. 2013). Therefore, the pressure that cleptobionts exert on stingless bee colonies must be considerable for the evolution of such differentiated groups of workers. Apparently, a soldier subcaste has evolved more frequently in species that are preferred targets of *Lestrimelitta* (Grüter et al. 2012, 2017). Nonetheless, soldiers have been found only in species that defend aggressively by fighting cleptobionts. Species that are preferred hosts of *Lestrimelitta* but that do not fight (Sakagami et al. 1993) may not have guard subcastes, but this has not been evaluated.

Biting is a frequent strategy used by stingless bee workers to defend their nests. The mandibles are mainly used in the collection and handling of food and nest materials. However, the size and sharpness of the mandibles seem to have evolved in relation to the defensive capacity and aggressive behavior of different species (Shackleton et al. 2015). Biting in stingless bees frequently involve a suicidal behavior. During the defense, workers bite a target and do not dislodge their mandibles, even when the head is sectioned from the body (Shackleton et al. 2015) (Fig. 5.7).

In species from Yucatan, the mandibles of the workers have remarkably different size and shape (Fig. 5.8). In two of the most defensive species, *Scp. pectoralis* and *Pt. bilineata*, a marked development in sharpness and relative size of mandibular teeth is evident. It is noticeable that large mandibles with sharp teeth are also present in *M. beecheii* (Fig. 5.8), compared with Brazilian species of *Melipona,* which lack sharp teeth (Shackleton et al. 2015).

The mandibles of *L. niitkib* only have a couple small teeth, but a large blunt surface and reduced curvature on the inner margin are evident (Fig. 5.8). This type of mandible is probably designed to cause severe injuries when biting. Cleptobionts use force to raid some species and their mandibles may be specially designed to

Fig. 5.7 Worker of *Scp. pectoralis* biting. Biting is probably the most important defense strategy in stingless bees, and has a suicidal component

Fig. 5.8 External surface of the right mandible of workers of various species of stingless bees in Yucatan. The teeth are located on the left margin and the curvature on the superior margin of the mandible, respectively

overcome their hosts' aggressions. However, robber bees also use chemicals during raids, especially citral produced in their mandibular glands. Whether the shape of their mandibles is related to the use of chemical substances is unknown (Sakagami et al. 1993; Quezada-Euán et al. 2013). Aggression is not frequent in species with small workers, like *N. perilampoides* and *Plebeia*; instead they retreat when attacked. During the attacks of cleptobiotic bees, workers of these species cluster in circles with their heads inwards, probably to protect against the bites of robbers

Fig. 5.9 Above: Intraspecific mass attack between colonies of *L. niitkib*. Below: Workers of *Fr. nigra* clustering with their heads inwards during an attack by *L. niitkib*

(Fig. 5.9). Perhaps this apparently defeatist behavior is important to avoid unnecessary deaths against a stronger enemy (Sakagami et al. 1993).

The intensity of the aggressive response of Yucatecan stingless bees falls into two main categories, intense or timid (Table 5.1). Intraspecific variation in the aggressive response is also evident (pers. Obs.). In *M. beecheii,* some colonies are aggressive and some are tame, but it is not known if differences in the aggressive response are linked to environmental variables or genetic components. In *A. mellifera* it is known that environmental (colony size, amount of food reserves, environ-

Table 5.1 Defensive strategies of some species of stingless bees from Yucatán

Species	Intensity of defense	Worker population[a]	Strategy against mammals	Strategy against insects
Melipona beecheii	Medium	800–1500	Bite, pheromone	Block nest entrance, suicidal bite
Scaptotrigona pectoralis	Intense	2000–4000	Massive attack, bite, recruit more attackers	Suicidal bite
Nannotrigona perilampoides	None	700–1200	Retreat	Resin to immobilize intruders, retreat
Plebeia	None	600–1000	Retreat	Resin to immobilize intruders, retreat
Frieseomelitta nigra	None	500–1000	Retreat	Block entrance, resin to immobilize intruders, retreat
Trigona fuscipennis	Intense	5000–10,000	Massive attack, caustic bite, recruit more attackers	Suicidal bite, resins
Partamona bilineata	Intense	2000–4000	Massive attack, bite	Suicidal bite, resins
Trigona fulviventris	None	500–1000	Retreat	Block entrance, resins
Cephalotrigona zexmeniae	None	500–1200	Retreat	Block entrance, resins
Lestrimelitta niitkib	None	3000–5000	Retreat	*Suicidal bite*

[a]From Quezada-Euán and González-Acereto 2002

mental temperature, and humidity), but also genetic factors affect the intensity of the aggressive response of colonies (Winston 1987).

Guarding and defense are among the final activities performed by workers (Sommeijer 1984). The risk involved in guarding and external activities is higher compared with nest activities. In this regard, the loss of workers of an advanced age would be less detrimental to colonies than losing young ones. In fact, older workers have a comparatively shorter life span and have already made a significant contribution to colony fitness (Tofilski 2002).

Stingless bees are avid collectors of resins and use them in the construction, but also during the defense of their colonies. Indeed, resin use can be considered a primary line of defense, in probably all stingless bee species (Leonhardt and Blüthgen 2009). Some species, like *Fr. nigra*, maintain large reserves of resin near the entrance and if an intruder breaks in, guards immobilize it, while others deposit drops of resin on legs and wings to bury it. This mechanism called mummification has been reported in various species of stingless bees, and is considered a defense strategy against large intruders that cannot be easily removed from the colony. Resin also protects the colony from bacteria and other microorganisms that may grow on corpses (Greco et al. 2010).

Stingless bees are frequently in contact with resin, enriching their cuticular chemical profile (Leonhardt et al. 2015). It has been suggested that resins compounds on the cuticle could be used as protection against microorganisms (Simone-

Finstrom and Spivak 2010). Nonetheless, experimental manipulation of Asian stingless bees in which resin compounds were removed from the cuticle, were not more susceptible to fungus infections compared with bees kept intact (Leonhardt et al. 2015). However, resins were effective repellents against ants. Workers with large amounts of resin-derived compounds on their bodies were significantly less attractive to those predators (Leonhardt et al. 2015). Perhaps this may explain why resin collection increases when ant attacks become more frequent (Leonhardt and Blüthgen 2009). Resins can also be used as chemical mediators in inter- and intra-specific coexistence. It has been discovered that chemical compounds in the resins moderate aggression in coexisting Asian species (Leonhardt et al. 2010a).

Although resins may not be involved in the defense of stingless bees against disease, they may be important in the protection of a valuable resource, their food reserves. Stingless bees can store large food reserves which under tropical conditions may be easily spoiled (Roubik 1989). Some resin compounds incorporated in the cerumen of food pots could regulate the growth of microorganisms that may spoil pollen or nectar (Roubik 1989). On the other hand, some bacteria and yeasts may be beneficial for preserving the stored food. Perhaps stingless bees select different resins in accordance to antimicrobial properties, explaining why they are choosy when collecting them. Workers learn to collect resin from specific plants, and even slight variations of the plant's chemical profiles can deter collection (Leonhardt et al. 2010b).

Bees, like other insects, are protected against disease by humoral and cellular components of their immune system. Surprisingly, evidence of a reduced individual immune response has been found in social insects, compared with solitary species (López-Uribe et al. 2016). In the honey bee, there is evidence that fewer genes related with the immune response are generally active, compared with solitary bees (Evans et al. 2006). It is argued that social insects may rely more on collective mechanisms, such as hygienic behavior, than on individual immune function to reduce the risk of disease (López-Uribe et al. 2016).

Social bees have also developed an association with certain microbes (microbionts) to protect against pathogens and to preserve food (Morais et al. 2013; Leonhardt and Kaltenpoth 2014; Kwong et al. 2017). Among the important bacteria living in stingless bees colonies are lactobacillus (*Bacillus*) important for the digestion of pollen. The lactobacilli soften the hard pollen exine and may produce antibiotics or lactic acid to protect food reserves from the invasion by other microorganisms (Menezes et al. 2013).

Some lactobacilli also inhabit the bee's digestive system, and my be important probiotics for the prevention of disease (Vásquez et al. 2012). These microbes can be found in the gut just a few hours after the bee emerges (Vásquez et al. 2012). This suggests that *Lactobacillus* firms of highly social species seem transmitted both, horizontally (among individuals in the same generation), and vertically (between generations) (McFrederick et al. 2013). The association between corbiculate bees and gut microbiota seems an old one. At least five core gut bacterial lineages may have been acquired early in the evolution of eusocial corbiculate bees, possibly 80 million years ago (Kwong et al. 2017). Interestingly, lactobacilli species seem common across corbiculate taxa, with only a few firms restricted to particular hosts (McFrederick et al. 2013). For instance, *L. kunkeei* can be found in the digestive tract of *M. beecheii* and in the honey bee (Vásquez et al. 2012). In contrast, Leonhardt and Kaltenpoth (2014) identified some *Lactobacillus* firms only in Australian stingless bees.

Some bacteria living in the colonies of stingless bees may protect against pathogens. For instance, *Streptomyces*, found in *Trigona* colonies from Brazil, produces antibiotics with inhibitory effect on the pathogens causing American and European foulbrood in the honey bee (Menezes et al. 2013). However, direct effects of this and other bacterium in disease protection have not yet been found in stingless bees.

Some stingless bees have reached impressive symbiosis with their microbiota. For instance, the larvae of Brazilian *Scp. depilis* feed on a fungus (genus *Monascus*) that grows on the liquid food and its elimination triggers larval mortality (Menezes et al. 2015). Similarly, *Bacillus meliponotrophicus* found in *Trigona* and *Melipona* may be involved in some unknown process vital for the colonies. The application of antibiotics/streptomycin to kill the bacterium resulted in their collapse (Machado 1971; Morais et al. 2013).

Microorganisms in the pollen and honey of stingless bees are important for fermentation and predigestion, but may also control spoiling bacteria and yeast (Morais et al. 2013). For instance, dehydration of pollen reserves in *Ptilotrigona* mediated by the yeast *Candida* protects them from decay (Camargo et al. 1992). It is known that some stingless bees (*Trigona* and *Partamona*) actively collect fungal spores, but the use in colonies has not been studied (Morais et al. 2013).

Although a large variety of microorganisms are associated with stingless bees, many aspects of their ecology and effect on the well-being of individuals, and colonies, have just started to be explored and understood (Morais et al. 2013).

As mentioned before, social bees seem to largely depend on collective behavioral mechanisms to fight disease. One of these mechanisms is hygienic behavior. The ability to detect and remove dead or contaminated brood and adults from nests has been known for over 50 years in the honey bee (Rothenbuhler 1964). In *A. mellifera*, hygienic behavior is performed by workers between 15 and 20 days old, before they start foraging (Spivak and Downey 1998). Hygienic bees use olfactory signals to detect diseased or dead brood, which they uncap and remove from the colony. Although hygienic behavior has a genetic component in the honey bee, the frequency of colonies expressing it is low, in the range of 10%. (Lapidge et al. 2002; Bigio et al. 2013).

Given its economic importance, hygienic behavior had been extensively studied in the honey bee (Rothenbuhler 1964; Spivak and Downey 1998). However, it was not known if the rarity of disease in stingless bees may be due by similar mechanisms. Medina-Medina et al. (2009) and Nunes-Silva et al. (2009) investigated if Meliponini may also use hygienic behavior for colony immunity. Medina-Medina et al. (2009) worked with two species of stingless bee from Yucatan (*M. beecheii* and *Scp. pectoralis*), while Nunes-Silva et al. (2009) investigated the Brazilian *Pb. remota*. In both studies a protocol similar to that used in *A. mellifera* to assess hygienic behavior (Spivak and Downey 1998) was applied. Sections of comb with dead pupae were introduced in test colonies and the number and time in which dead individuals were removed were registered. Interestingly, all three species of stingless bees exhibited hygienic behavior, but there were inter- and intraspecific differences in the number of pupae removed after 48 h, and the time to remove them completely from the combs. The most hygienic species were *Pb. remota* and *Scp. pectoralis*, with 96% and 66% dead pupae removed in 48 h, respectively. In comparison, *M. beecheii* colonies only removed between 30 and 40% dead pupae in the same period of time. In addition, it took *M. beecheii* 9 days to remove all dead pupae, while *Scp. pectoralis* accomplished that in just 3 days. It was also evident that intraspecific differences in hygienic behavior existed, suggesting this trait may have a genetic component in stingless bees. Recently, Toufailia et al. (2016) studied hygienic behavior in Brazilian *M. scutellaris*, *Scp. depilis*, and *Ttr. angustula*. The authors also reported high levels of hygienic behavior after 48 h in all three species (>60%). A significant negative correlation between freeze-killed brood removal and the frequency of deformed *Scp. depilis* was found, implying a link between hygienic behavior and disease presence in stingless bee colonies.

Interestingly, although hygienic behavior is present in honey bees and stingless bees, marked diffences exist between both taxa. The levels of pupa removal of stingless bees is generally above 60% (in some species >90%) while in the honey bee it is usually <50% (Medina-Medina et al. 2009; Nunes-Silva et al. 2009; Toufailia et al. 2016). Moreover, hygienic behavior has been recorded in all colonies of stingless bees studied, while in honey bees, the frequency of colonies exhibiting hygienic behavior is low (Bigio et al. 2013).

In addition, stingless bees remove infected cells completely, not just cappings, as occurs in the honey bee (Medina-Medina et al. 2009). The elimination of cocoons and waste in brood cells, reduces potential sources of infection to adults, brood, and food stores (Medina-Medina et al. 2014). In contrast, in the honey bee, workers can eliminate infected individuals, but potentially infected waste is left in the colony. In Meliponini, a combination of a high level of hygienic behavior (high rate of brood removal) and complete elimination of potential sources of infection seem highly effective mechanisms to face and cope with disease. Low rates of infection and spread of disease may result from the stingless bee system of pathogen management (Toufailia et al. 2016; Díaz et al. 2017).

5.5 Communication by Chemical Means

In the highly eusocial insects information is extensively passed on among individuals using chemical signals (Wilson 1990). In the stingless bees, the identification of specific chemical signals as well as evidence on their involvement in communication are still comparatively limited. Nonetheless, the amount of information on the chemical world of stingless bees has significantly increased in the last two decades. Thanks to new methods of analysis and the possibility of artificially producing many of the identified compounds greatly improve the possibility of new and exciting discoveries on the chemical ecology of these insects (Couvillon and Ratnieks 2008; Jarau 2009; Nunes et al. 2009a).

A distinctive feature of stingless bees is the impressive number and diversity of chemical compounds on individuals and nests (Leonhardt et al. 2015). Some of them, probably the majority, only serve a structural purpose. Nonetheless, those conveying information or semiochemicals can have important intra- or interspecific effects (Wyatt 2003). Those semiochemicals mediating intraspecific communication are known as pheromones (Free 1987). Pheromones are compounds produced in glands and secreted externally, serving as signals that affect the behavior, development, and/or physiology of conspecifics (Free 1987). Those semiochemicals acting at interspecific level are known as allelochemicals or allomones (Free 1987; Jarau 2009).

Apart from their type of action, semiochemicals can work as signals or cues (Barth et al. 2008; Jarau 2009). A semiochemical is considered a signal if it evolved to convey specific information. In contrast, a cue is incidentally used to transmit information, but did not evolve specifically for that purpose (Barth et al. 2008).

One aspect of the life of stingless bee colonies, in which chemicals are frequently involved, is the collection of food. From the food source itself, to the transmission of information on location and direction, chemical signals are involved in the process. Floral fragrances are important sources of chemicals that attract potential visitors. Fragrances serve the plant guiding pollinators looking for potential food (Raguso 2001). Indeed, flowers have evolved fragrance to attract pollinators evoking discriminatory behavior (Proctor et al. 1996; Chittka and Thomson 2001). Bees use fragrances to find flowers (and rewards), and they are capable of associating them with specific food sources (Chittka and Thomson 2001; Raguso 2001). The smell of food is of particular relevance to species with relative simple mechanisms of communication, like *Plebeia.* In these species, foragers carry the smell of flowers which seems the almost exclusive mechanism used by nestmates to find such sources (Jarau 2009). The smell of food can be important in species with more sophisticated mechanisms of food exploitation too.

A higher level of chemical communication involves the use of scent marks left by the workers when visiting food sources. Species of *Tetragonisca*, *Scaptotrigona*, *Nannotrigona*, and some *Melipona* use marks to find (or reject) flowers that have been previously visited by nestmates. Experiments using artificial feeders show that sources previously visited can be more attractive compared to non- visited ones (Villa and Weiss 1990). Interestingly, scent marks may equally attract nestmates and non-nest mates indicating that such cues are not colony specific (Nieh 2004; Nieh et al. 2004; Jarau 2009).

The use of Scent marks may depend on reward accessibility. In *Tr. fulviventris*, if nectar is easily obtained, with little energy invested in the process, scent marks are less intensely used. However, if food is difficult to obtain, bees more frequently leave scent marks on them. Curiously, scent marks seem to have a repellent effect in some situations, and flowers previously visited can become less attractive (Goulson et al. 2001). Evidently, scent marks can have different effects depending on the context in which food is presented. Scent marks are also involved in interspecific interactions; artificial feeders first visited by *M. beecheii* resulted less attractive to *Tr. corvina* foragers (Boogert et al. 2006).

Although the effects of scent marks had been known, their origin had been debatable. Initial works in *Melipona* suggested the anal glands as the source of scent marks (Kerr and Rocha 1988; Aguilar and Sommeijer 2001). However, there was no clear correlation between gland depositions and attraction to food sources (Aguilar and Sommeijer 2001). Moreover, feeders intensively marked were not visited at a similar rate (Nieh 1998; Hrncir et al. 2004a), suggesting that anal glands were not the origin of scent marks in *Melipona*.

Another candidate source of scent marks are the tendon glands inside the femur and tibia. These glands open at the base of the pretarsus on the distal portion of the legs. When the external duct of the tendon gland was obstructed in *M. seminigra* workers visiting a feeder, significantly less visits occurred, compared to feeders visted by workers with intact glands (Hrncir et al. 2004a). In addition, when extracts of the tendon glands were experimentally applied to feeders, the number of visits increased. Such results suggest tendon glands as a more plausible origin of scent marks. Chemical analysis of tendon gland extracts revealed that frequent compounds are saturated hydrocarbons ($\geq$10%, pentacosane and heptacosane), and their corresponding alkenes (Jarau et al. 2004).

Although scent marks are important in food location, the effect can be considered a cue rather than a signal. The scent is left as the bee walks on the food, and is therefore left passively. Foragers do not seem to leave scent marks on food sources on purpose.

Another group of chemicals used in food location can be effectively considered signals. They evolved specifically to guide nestmates to food sources (Schorkopf et al. 2007; Barth et al. 2008). These signals are collectively known as trail pheromones and have been documented in various genera of stingless bees (*Trigona*, *Scaptotrigona*, *Geotrigona*, *Cephalotrigona*, and *Oxytrigona*). Trail pheromones represent a more efficient method for locating food sources compared to scent marks. They can guide nestmates across distance but also altitude. The indication of altitude is an important feature of food location in stingless bees in rain forests, which is absent in *A. mellifera* (Jarau 2009). With the help of the glossa, successful foragers leave pheromone droplets at regular intervals building a path from the food source to the nest, that can be followed by nestmates. Droplets are usually deposited more frequently in proximity to the food source (Schorkopf et al. 2007). It has been shown that trail pheromones are produced in the labial glands of the head, and not the mandibular glands, as was originally believed (Schorkopf et al. 2007, 2009).

Most species of stingless bee use either scent marks or trail pheromones as guides to food sources. Nonetheless, there also seem to be some forms of food communication not mediated by such chemical methods. A study in *Pt. orizabaensis* showed that workers of this species can rapidly recruit nestmates to food sources, but no evidence was found for the use of marks from the legs nor trail pheromones produced in the labial glands (Flaig et al. 2016). Chemical signals (marks or trails) are useful to colony members for food location. However, there is also the potential risk of such signals being used by conspecific and heterospecific competitors, a form of "espionage " defined as eavesdropping (Nieh et al. 2004). The risk of eavesdropping may represent a strong selective force upon mechanisms of food collection based on chemical cues. Perhaps, "hidden" food communication in some species (as *Pt. orizabaensis*; Flaig et al. 2016) may be the evolutionary response to avoid the risk of eavesdropping by competitors (Lichtenberg et al. 2011). There may be still some interesting forms of food communication in stingless bees waiting to be discovered.

The mandibular glands of stingless bees may not be involved in recruitment to food sources, but they participate in other forms of chemical communication. Many compounds found in the mandibular glands are highly volatile, disseminating rapidly. When threatened, workers of many species spread their mandibles, possibly releasing alarm pheromones. One alarm pheromone frequently found in the mandibular glands of stingless bees is 2-heptanol (Keeping et al. 1982; Smith and Roubik 1983; Johnson et al. 1985; Engels et al. 1987; Cruz-Lopez et al. 2007). This compound is also found in honey bee workers (Free 1987). Nonetheless, in some species, like *M. beecheii,* 2-heptanol has not been detected in the mandibular glands of workers (Cruz-Lopez et al. 2005). Interestingly, the mandibular secretions of males may trigger an aggressive response of *Scaptotrigona* and *Partamona* workers. Males of these species seem to represent a first front in colony defense producing alarm signals that alert workers (Schorkopf 2016). An extreme case of the use of the mandibular glands in defense, is in species of the genus *Oxytrigona.* The mandibular glands of *Oxytrigona* workers produce caustic chemicals (mainly formic acid) that burn the skin when biting (Roubik et al. 1987).

Compounds produced in the mandibular glands of stingless bee gynes and males seem to play an important role during sexual attraction (Engels et al. 1990). However, no particular sexual attractant has yet been identified in mandibular bouquets of stingless bees.

Another important aspect of chemical communication in stingless bees is nestmate recognition. Stingless bees, like other highly eusocial insects need rapid and efficient identification of nestmates from potential intruders (Nunes et al. 2008). Most evidences from highly eusocial insects, including bees, suggest that the information for nestmate recognition is encoded in the hydrocarbons covering the cuticle (Vander Meer and Morel 1998; Lenoir et al. 2001; van Zweden and D'Ettorre 2010).

Cuticular hydrocarbons are hydrophobic compounds protecting the insect from water loss (Gibbs 1995). There are differences in the volatility of cuticular compounds which depend on the length of the molecule, the presence of unsaturated bonds and methylation. Unsaturated hydrocarbons can be highly volatile, and are good candidates for recognition cues (Gibbs and Pomonis 1995). Indeed, in the stingless bees alkenes and alkadienes seem mainly responsible for nestmate recognition (Jungnickel et al. 2004; Buchwald and Breed 2005; Pianaro et al. 2007; Nunes et al. 2008; Nascimento and Nascimento 2012; Septanil et al. 2012). One indication of the importance of unsaturated hydrorcarbons in stingless bee recognition is a highly diversified production of alkene isomers in this taxon (Martin et al. 2017). Cuticular hydrocarbons seem predominantly of genetic origin, but can also be acquired from the environment (Breed et al. 1985; Page Jr et al. 1991; Leonhardt et al. 2015; Gutiérrez et al. 2016). In the case of stingless bees, cerumen has been proposed as a candidate for external recognition hydrocarbons (Nunes et al. 2011), although this hypothesis has not been confirmed (Jones et al. 2012). Similarly, resin-derived terpenoids substantially enrich the cuticular profile of stingless bees, but their role in nestmate recognition is still unclear (Leonhardt et al. 2010a, b; Leonhardt et al. 2015).

It seems that the mixture of cuticular hydrocarbons provides an individual fingerprint. Guard bees are in charge of discriminating nestmates from potential intruders. However, individual fingerprints change continuously, and guards must constantly update their internal recognition pattern (Couvillon and Ratnieks 2008; Nunes et al. 2008; Nascimento and Nascimento 2012). Interestingly, although individual fingerprints show variation, nestmate fingerprints are more similar to the guards' than non-nestmates (Couvillon and Ratnieks 2008; Nascimento and Nascimento 2012). Guard bees compare the chemical fingerprint of individuals entering the nest with a pattern in their brain of guards, if these do not match aggression is elicited against non-nestmates (Nash and Boomsma 2008; van Zweden and D'Ettorre 2010; Nunes et al. 2011). It is suggested that in stingless bees a system based on undesirably absent compounds is used for the recognition of intruders (Couvillon and Ratnieks 2008).

Nestmate recognition is important for the protection of food and materials in the nest. However, specialized intruders have evolved chemical disguise or insignificance to avoid recognition (Lenoir et al. 2001; Martin et al. 2007; 2010; van Zweden and D'Etorre 2010). *Lestrimelitta* workers are specialized cleptobionts which during nest raids release large amounts of citral (lemon-smelling pheromone) produced in the mandibular glands (Sakagami et al. 1993). Citral may weaken the defensive response of hosts by disrupting chemical recognition (Blum et al. 1970). However, no conclusive evidence for citral as a masking allomone has been found, and its use by *Lestrimelitta* seems to vary depending on host species, and the context in which the attacks are conducted (Sakagami et al. 1993). In some hosts, citral releases aggression, but in others like *Fr. varia,* compounds from the labial glands of *Lestrimelitta* could possibly have a repellent effect (von Zuben et al. 2016; Table 5.2)

Interestingly, *Lestrimelitta* is host selective (Sakagami et al. 1993; Quezada-Euán and González-Acereto 2002). It is not clear how these obligate cleptobionts select their hosts. It has been suggested that the quality of reserves and aggressive response of the host, and perhaps genetic factors, may be involved in host selection. It is also possible that specialization may be part of this system, and scouts gaining easier access to some species could learn by association (Sakagami et al. 1993; Jarau 2009).

Lestrimelitta mass attacks would require some sort of nestmates recruitment (Fig. 5.9). It is possible that *Lestrimelitta* scouts previously enter nests to obtain some sort of information to select their prey (Sakagami et al. 1993).

Similar to other social cleptobionts *Lestrimelitta* scouts may use chemical deception or insignificance to avoid recognition (Quezada-Euán et al. 2013). To evaluate this possibility, the similarity of cuticular profiles was compared between five potential host species and *L. niitkib* from the Yucatan Peninsula. The study

Table 5.2 Some candidate semiochemicals detected on stingless bees with their origin and function

Function	Type	Origin	Candidate Semiochemicals	Species	Reference
Recruitment to food sources	Fragances	Floral	Volatile compounds	All	Jarau (2009)
	Fingerprint	Tendon glands	Alkanes: Pentacosane (C25), heptacosane (C27) and their alkenes	*M. seminigra*	Jarau et al. (2004)
	Trail pheromones	Labial gland	Hexyl decanoate Octyl octanoate	*Tr. recursa* *Tr. spinipes*	Jarau et al. (2006) Schorkopf et al. (2007)
	Raiding pheromones	Mandibular gland	Citral Dicetones	*Lestrimelitta* *Oxytrigona*	Blum et al. (1970) Bian et al. (1984)
Recognition	Nestmate recognition	Cuticle ¿Mandible?	Long chain saturated and unsaturated hydrocarbons (C23-C31)	*Tr. fulviventris* *Scp. bipunctata* *Fr. varia - L. limao* *M. asilvai*	Buchwald and Breed (2005) (Jungnickel et al. 2004) (Nunes et al. 2008) Nascimento and Nascimento (2012)
			Long chain unsaturated hydrocarbons (C23-C33)	*N. perilampoides* *M. beecheii,* *Pb. frontalis,* *Scp. pectoralis,* *Fr. nigra, L. niitkib*	(Quezada-Euán et al. 2013)
	Identification and changes by caste and gender	Mandibular	2- heptanone, nonanal, 2 nonanol, neral, geranial in foragers absent in nest bees	*Tr. gribodoi*	Keeping et al. (1982)
			2-heptanol, 2-tridecanone, 2-pentadecanone increase with age	*Scp. postica*	Francke et al. (1983)
		Cuticular	Queen: alkenes and alkadienes (27:1 to 31:1 carbons) Gyne: n-butylpalmitate	*Fr. varia*	Nunes et al. (2009a)

(continued)

Table 5.2 (continued)

Function	Type	Origin	Candidate Semiochemicals	Species	Reference
			Saturated (C23-C27) young workers, alkenes more frequent in older workers	Worker *Sch. quadripunctata*	Nunes et al. (2009b)
			Geranylgeranyl acetate	Worker *N. testaceicornis*	Pianaro et al. (2009)
			(Z)9-Nonacosene	Male *N. testaceicornis*	
			Tetradecanal	*Pb. droryana* Worker	
			Linoleic and linolenic acids and S-2-Nonanol	*Pb. droryana* Male	
			Mixture of saturated and unsaturated hydrocarbons	Diploid males *M. quadrifasciata*	Borges et al. (2012)
			Mixture of hydrocarbons	Age differences in workers of *M. quadrifasciata*	Cruz-Landim et al. (2012)
			C27 frequent in nest workers, while Z-27:1 more frequent in foragers	*Scp. postica*	(Poiani et al. 2014)
	Indication of queen presence and regulation of ovarian activation	Head	Queen: alkenes (tricosene C23:1)	*Scp. postica*	Engels et al. (1987)
		Integumental glands of the abdomen and/or Dufour gland	Mixture of cuticular hydrocarbons (predominance of C25)	*Frs. schrottkyi*	Nunes et al. (2014a)
Defense	Attack releasers and caustic compounds	Mandibular	2-heptanol	*M. fasciata* *M. interrupta*	Smith and Roubik (1983)
			2-heptanol 2-nonanol	*Tr. silvestriana*	Johnson et al. 1985

			Formic acid 2-heptanone, tetradecyl acetate	*Oxytrigona*	Roubik et al. (1987) Cruz-Lopez et al. (2007)
			2-heptanol, 2-nonanol	*Fr. silvestrii* *Fr. varia*	Patricio et al. (2004)
			Geraniol, Farnesyl acetate (absence of 2-heptanol)	*M. beecheii*	Cruz-López et al. (2005)
			2-heptanol, nonanal, benzaldehyde, 2-tridecanone	*Tr. spinipes* *Scp. depilis*	Schorkopf et al. (2009)
		Resins	Terpenoids tame aggression Terpenoids repel predators	*Ttga. collina* *Ttgn. carbonaria* *Au. australis*	Leonhardt et al. (2010) Leonhardt et al. (2015)
			Pheromones produced by males mediate in defensive response	Workers and males: *Scp. depilis,* *Scp. bipunctata,* *Pt. cupira*	Schorkopf (2016)
		Mandibular gland	Citral, 9-nonacosene (release aggression in *Fr. varia* and *M. flavolineata)*	*L. limao*	von Zuben et al. (2016) Nunes et al. (2014b)
		Labial gland	Hexadecyl-and hexadecenyl-acetate (repel *Fr. varia)*		von Zuben et al. (2016)

(continued)

Table 5.2 (continued)

Function	Type	Origin	Candidate Semiochemicals	Species	Reference
Reproduction	Pheromones in sexual attraction	Head	Gyne: secondary alcohols 7 to 13 Carbons	*Scp. postica*	Engels et al. (1987)
		Abdomen Mandibular	Gyne:alkanes C23 to C27 Heptanol, nonanol, tridecanol	*Scp.postica*	Flach et al. (2006)
			Males do not react to 2-heptanol and 2-nonanol	*N. testaceicornis Pb. droryana*	Pianaro et al. (2009)
			Males react to 2-nonanol of gynes	*Scp. mexicana*	Verdugo-Dardon et al. (2011)
			Isopropyl-hexanoate present in gynes attracts males	*Ttr. angustula*	Fierro et al. (2011)
			Cuticular hydrocarbons do not attract heterospecific males	*Plebeia, Trigona, Nannotrigona, Scaptotrigona*	dos Santos et al. (2015)

involved assessing the speed to recognize *L. niitkib* workers by guards of species that are frequently raided, and others that are not raided. Interestingly, *L. niitkib* cuticular profile was not insignificant, but remarkably similar to some of its preferred hosts. Alkenes C27:1 and C29:1 were predominant in two frequently raided species (*N. perilampoides* and *Pb. frontalis*) and *L. niitkib.* The workers of chemically similar species also took longer to react to *L. niitkib* compared with dissimilar ones, suggesting that cleptobionts may pass unnoticed by guards. It was noted that cuticular similarities may arise as a result of the phylogenetic closeness of *Lestrimelitta* with *Nannotrigona* and *Plebeia* (Rasmussen and Cameron 2010). The results of this study indicate that chemical deception may be used in combination with other forms of host attack by these obligate cleptobionts (Quezada-Euán et al. 2013).

Apart as clues in nestmate recognition, cuticular hydrocarbons possibly indicate the status of colony members (Cruz-Landim et al. 2012). Differences between the cuticular profile of gynes and physogastric queens have been reported in various species (Nunes et al. 2009a, b). Curiously, the cuticular profiles of workers seem to be more similar to those of males than queens (Borges et al. 2012). Although cuticular compounds generally indicate the presence of the queen to members of her colony (Engels 1987), a notable finding, is that the cuticular hydrocarbons of the queen in *Frs. schrotkyii* seem also responsible of the ovarian suppression of workers, a process somehow similar to that in *A. mellifera* (Nunes et al. 2014a).

In spite of the well-known role of chemical cues and pheromones in the attraction of mates in other insects, evidence is still scarce in stingless bees (Billen and Morgan 1998; Grüter and Keller 2016). In *Scp. postica,* secondary alcohols (Engels et al. 1987) and hydrocarbons of the queen cuticle (Flach et al. 2006), seem to be involved in the attraction of mates. Flach et al. (2006) found that some chemicals in the fragrance of an orchid flower are remarkably similar to the cuticular profile of *Scp. postica* queens, and deceive males of this species to pollinate the plant. In *Scp. mexicana,* males were capable of identifying virgin and physogastric queens by their chemical profiles. The relative amounts of 2-nonanol and other alcohols of the queen's cuticle seem to act as discriminating cues (Verdugo-Dardón et al. 2011). Similarly, isopropyl-hexanoate, a compound found on the abdomens of gynes of *Ttr. angustula,* elicited a strong electroantennogram response in the males of this species, suggesting its possible role as a sexual attractant (Fierro et al. 2011).

A summary of candidate semiochemicals and their suggested contribution in different aspects of the communication in stingless bees is presented in Table 5.2.

5.6 Communication by Physical Means

The best studied context in which bees frequently use physical cues is in the communication during food exploitation. In the honey bee, the use of dances to indicate distance and direction to the food as well as in recruitment is well known (Seeley 1985, 1995). The stingless bees also use physical mechanisms to communicate (vibrations, runs, sounds), but the information conveyed in many of these signals is not well understood (Nieh et al. 2003; Hrncir et al. 2004b; Schmidt et al. 2008).

When foragers return to their nests after finding a source of food, they become agitated, run rapidly around the nest, and frequently bump into other workers (Hrncir et al. 2004a). These collisions or bumps are collectively known as jostling, and have been interpreted as a way through which foragers communicate their success in finding food, or may be used to recruit other foragers (Hrncir et al. 2004a; Barth et al. 2008). In *M. quadrifasciata,* the number of workers engaging in syrup collection increases in relation with the number of individuals being jostled by scouts (Hrncir et al. 2000).

In addition to jostling, successful foragers can also use the thoracic muscles to vibrate the substrate, producing sounds and air currents. Workers probably use the subgenual organ of their legs or the Johnston's organ of the antennae to perceive such vibrations (Hrncir et al. 2006; Barth et al. 2008). It has been detected that vibrations encoding information related with food finding are usually in the range of 300–600 Hz, and that foragers frequently perform these vibrations when transferring food to nest bees (Hrncir et al. 2006). In *M. seminigra,* foragers returning from the field frequently shake some apparently inactive workers in the nests, but it is not clear if this stimulates them to start foraging (Hrncir et al. 2004b).

Mechanical cues seem to indicate a food resource, but whether they encrypt some other information related to distance and orientation (as occurs in *A. mellifera*), has not been conclusively demonstrated in stingless bees. In some *Melipona,* no correlation has been found between the number and intensity of the vibrations performed by successful foragers with the distance or orientation to the food source (Hrncir et al. 2000; Hrncir et al. 2004b). In contrast, studies in other species have detected significant correlation between the intensity and frequency of the vibrations performed by foragers with the distance to food (Nieh 1998; Nieh and Roubik 1998). In *M. seminigra,* there is a great deal of variation in the pulsations of different foragers coming from the same food source, raising doubts about their validity as reliable indicators of distance (Hrncir et al. 2004b). It has been suggested that the contrasting results found in *Melipona* may relate to differences in the ability to communicate spatial dimensions among species (Nieh 2004).

A clear evidence for the transmission of information by means of vibrations is related to the quality of food. In various species, it has been consistently found that the duration and frequency of thorax vibrations increase when the sugar content of the food increases (*M. costaricensis:* Aguilar and Briceño 2002; *M. mandacaia, M. bicolor:* Nieh et al. 2003; *M. seminigra:* Hrncir et al. 2004b; *N. testaceicornis:* Schmidt et al. 2008).

Bees live most of their lives in obscurity. However, they use vision in external activities. Workers use ultraviolet light to find nectar guides on flowers and can detect polarized light (Winston 1987; Chittka and Thomson 2001). It is documented that stingless bees are capable of associating color with different food sources (Villa and Weiss 1990). Workers of *M. seminigra* can use optic flow (moving images) in route to the food source to obtain an indication of distance. However, whether this information is transmitted to other workers and how it is done remain unclear (Hrncir et al. 2003).

The spectra of light may be differently perceived by stingless bee species. In a comparative study, *M. mondury* preferred UV-reflecting over UV-absorbing bee-blue-green objects, whereas *M. quadrifasciata* showed an opposite preference. This result suggests that differential visual adaptations may have evolved to avoid inter-specific competition during food collection in stingless bees (Koethe et al. 2016).

Stingless bees learn rapidly and are capable of associating time and space with food availability (Breed et al. 2002). Workers can learn the location and the time when food is available and can remember these traits for over long periods of time. Learning and association have only been found in eusocial Hymenoptera (Breed et al. 2002). When foraging, some visual cues, mainly the presence of other bees on the food, could be used as additional guides or repellents to such sources (Sommerlandt et al. 2014).

Evidently, stingless bees are capable of communicating food location and other important features of their environment to other colony members. However, the cues or signals they use, how they are transmitted, and the information they may convey, are still not completely revealed.

References

Aguilar I, Briceño D (2002) Sounds in *Melipona costaricensis* (Apidae: Meliponini): effect of the sugar concentration and nectar source distance. Apidologie 33:375–388

Aguilar I, Sommeijer M (2001) The deposition of anal excretions by *Melipona favosa* foragers (Apidae: Meliponinae): behavioural observations concerning the location of food sources. Apidologie 32:37–48

Aparecido-Pereira R, Morais MM, Gioli LD, Santos NF, Rossi MA, Bego LR (2006) Comparative morphology of reproductive and trophic eggs in *Melipona* bees (Apidae, Meliponini). Braz J Morphol Sci 23:349–354

Avila B, Moo-Valle H, Valladares P, Camposeco F, Quezada-Euán JJG (2005) Descripción del proceso de aprovisionamiento y oviposición en colonias de *Melipona beecheii* (Apidae: Meliponini). Reporte de investigación, opción apicultura. FMVZ-Universidad Autónoma de Yucatán, Maestría en Producción Animal Tropical

Barth FG, Hrncir M, Jarau S (2008) Signals and cues in the recruitment behavior of stingless bees (Meliponini). J Comp Physiol A 194:313–327

Bian Z, Fales HM, Blum MS, Jones TH, Rinderer TE, Howard DF (1984) Chemistry of cephalic secretion of fire bee *Trigona* (*Oxytrigona*) *tataira*. J Chem Ecol 10:451–461

Biesmeijer JC (1997) The organisation of foraging in stingless bees of the genus *Melipona*; an individual approach. Ph.D. thesis, Utrecht University, Utrecht, 263 pp

Biesmeijer JC, Toth E (1998) Individual foraging, activity level and longevity in the stingless bee *Melipona beecheii* in Costa Rica (Hymenoptera, Apidae, Meliponinae). Insect Soc 45:427–443

Bigio G, Schürch R, Ratnieks FLW (2013) Hygienic behaviour in honey bees (Hymenoptera: Apidae): effects of brood, food, and time of the year. J Econ Entomol 106:2280–2285

Billen J, Morgan ED (1998) Pheromone communication in social insects: sources and secretions. In: Vander Meer RK, Breed MD, Espelie KE, Winston ML (eds) Pheromone communication in social insects: ants, wasps, bees and termites. Westview Press, Boulder, pp 3–33

Blum MS, Crewe RM, Kerr WE, Keith LH, Garrison AW, Walker MM (1970) Citral in stingless bees: isolation and functions in trail laying and robbing. J Insect Physiol 16:1637–1648

Borges AA, Ferreira-Caliman MJ, Nascimento FS, Campos LAO, Tavares MG (2012) Characterization of cuticular hydrocarbons of diploid and haploid males, workers and queens of the stingless bee *Melipona quadrifasciata*. Insect Soc 59:479–486

Boogert NJ, Hofstede FE, Aguilar Monge I (2006) The use of food source scent marks by the stingless bee *Trigona corvina* (Hymenoptera: Apidae): the importance of the depositor's identity. Apidologie 37:366–375

Breed MD, Butler L, Stiller TM (1985) Kin discrimination by worker honey bees in genetically mixed groups. Proc Natl Acad Sci USA 82:3058–3061

Breed MD, Stocker EM, Baumgartner LK, Vargas E (2002) Time-place learning and the ecology of recruitment in a stingless bee, *Tr. amalthea* (Hymenoptera, Apidae). Apidologie 33:251–258

Breed MD, Cook C, Krasnec MO (2012) Cleptobiosis in social insects. Psyche. Article ID 484765

Buchwald R, Breed MD (2005) Nestmate recognition cues in the stingless bee *Trigona fulviventris*. Anim Behav 70:1331–1337

Camargo JMF, Garcia MVB, Junior ERQ, Castrillon A (1992) Notas previas sobre a bionomia de *Ptilotrigona lurida* (Hymenoptera, Apidae, Meliponinae): associação de leveduras em pólen estocado. Boletim do Museu Paraense Emílio Goeldi 8:391–395

Camargo JMF, Pedro SRM (2007) Meliponini Lepeletier, 1836. In: Moure JS (ed) Catalogue of the bees (Hymenoptera, Apoidea) in the Neotropical region. Sociedade Brasileira de Entomologia, Curitiba, pp 272–578

Cardoso-Júnior CAM, Pereira Silva R, Araújo Borges N, de Carvalho WJ, Walter SL, Paulino Simões ZL, Bitondi MMG, Ueira Vieira C, Bonetti AM, Hartfelder K (2017a) Methyl farnesoate epoxidase (*mfe*) gene expression and juvenile hormone titers in the life cycle of a highly eusocial stingless bee, *Melipona scutellaris*. J Insect Physiol 101:185–194

Cardoso-Júnior CAM, Fujimura PT, Santos-Júnior CD, Borges NA, Ueira-Vieira C, Hartfelder K, Goulart LR, Bonetti AM (2017b) Epigenetic modifications and their relation to caste and sex determination and adult division of labor in the stingless bee *Melipona scutellaris*. Genet Mol Biol 40(1):61–68

Chittka L, Thomson JD (2001) Cognitive ecology of pollination, animal behavior and floral evolution. Cambridge University Press, Cambridge

Choo YM, Lee KS, Yoon HJ, Kim BY, Sohn MR et al (2010) Dual strategy of bee venom serine protease: prophenoloxidase-activating factor in arthropods and fibrin(ogen)olytic enzyme in mammals. PLoS One 5:e10393

Couvillon MJ, Ratnieks FLW (2008) Odour transfer in stingless bee marmelada (*Frieseomelitta varia*) demonstrates that entrance guards use an 'undesirable-absent' recognition system. Behav Ecol Sociobiol 62:1099–1105

Crespi BJ (1992) Cannibalism and trophic eggs in subsocial and eusocial insects. In: Elgar MA, Crespi BJ (eds) Cannibalism, ecology and evolution among diverse taxa. Oxford University Press, Oxford, pp 176–213

Cruz-Landim C (2000) Ovarian development in Meliponine bees (Hymenoptera: Apidae): the effect of queen presence and food on worker ovary development and egg production. Genet Mol Biol 23:83–88

Cruz-Landim C, Ferreira-Caliman MJ, Gracioli-Vitti LF, Zucchi R (2012) Correlation between mandibular gland secretion and cuticular hydrocarbons in the stingless bee *Melipona quadrifasciata*. Genet Mol Res 11:966–977

Cruz-Lopez L, Malo EA, Morgan ED, Rincon M, Guzman M, Rojas JC (2005) Mandibular gland secretion of *Melipona beecheii*: chemistry and behavior. J Chem Ecol 31:1621–1632

Cruz-Lopez L, Aguilar S, Malo EA, Rincon M, Guzman M, Rojas JC (2007) Electroantennogram and behavioral responses of workers of the stingless bee *Oxytrigona mediorufa* to mandibular gland volatiles. Entomol Exp Appl 123:43–47

Dade HA (1985) Anatomy and dissection of the honeybee. International Bee Research Association, London

Daneels EL, Van Vaerenberg M, Debyser G, Devreese B, Graaf DC d (2015) Honeybee venom proteome profile of queens and winter bees as determined by a mass spectrometric approach. Toxins 7:4468–4483

Díaz S, de Souza Urbano S, Caesar L, Blochtein B, Sattler A, Zuge V, Haag KL (2017) Report on the microbiota of *Melipona quadrifasciata* affected by a recurrent disease. J Invertebr Pathol 143:35–39

dos Santos CG, Blochtein B, Megiolaro FL, Imperatriz-Fonseca VL (2010) Age Polyethism in *Plebeia emerina* (Friese) (Hymenoptera: Apidae) colonies related to propolis handling. Neotrop Entomol 39:691–696

dos Santos CF, Ferreira-Caliman MJ, Nascimento FS (2015) An alien in the group: eusocial male bees sharing nonspecific reproductive aggregations. J Insect Sci 15:157

Drumond PM, Zucchi R, Oldroyd BP (2000) Description of the cell provisioning and oviposition process of seven species of *Plebeia* Schwarz (Apidae, Meliponini), with notes on their phylogeny and taxonomy. Insect Soc 47:99–112

Eardley CD (2004) Taxonomic revision of the African stingless bees (Apoidea: Apidae: Apinae: Meliponini). Afr Plant Protect 10:63–96

Engels W (1987) Pheromones and reproduction in Brazilian stingless bees. Memorias Instituto Oswaldo Cruz, Rio de Janeiro 82(Suppl III):35–45

Engels E, Engels W, Schroder W, Francke W (1987) Intranidal worker reactions to volatile compounds identified from cephalic secretions in the stingless bee, *Scaptotrigona postica* (Hymenoptera, Meliponinae). J Chem Ecol 13:371–386

Engels W, Engels E, Lübke G, Schröder W, Francke W (1990) Volatile cephalic secretions of drones, queens and workers in relation to reproduction in the stingless bee, *Scaptotrigona postica*. Entomologia Generalis 15:91–101

Evans JD, Aronstein K, Chen YP, Hetru C, Imler JL, Jiang H, Kanost M et al (2006) Immune pathways and defence mechanisms in honeybees *Apis mellifera*. Insect Mol Biol 15:645–656

Fierro MM, Cruz-López L, Sánchez D, Villanueva-Gutiérrez R, Vandame R (2011) Queen volatiles as a modulator of *Tetragonisca angustula* drone behavior. J Chem Ecol 37:1255–1262

Flach A, Marsaioli AJ, Singer RB, Amaral M d CE, Menezes C, Kerr WE, Batista-Pereira LG, Correa AG (2006) Pollination by sexual mimicry in *Mormolyca ringens*: a floral chemistry that remarkably matches the pheromones of virgin queens of *Scaptotrigona* sp. J Chem Ecol 32:59–70

Flaig IC, Aguilar I, Schmitt T, Jarau S (2016) An unusual recruitment strategy in a mass-recruiting stingless bee, *Partamona orizabaensis*. J Comp Physiol A 202:679–690

Francke W, Schroder W, Engels E, Engels W (1983) Variation in cephalic volatile substances in relation to worker age and behavior in the stingless bee, *Scaptotrigona postica*. Z Naturforsch 38c:1066–1068

Free JB (1987) Pheromones of social bees. Cornell University Press, Ithaca

Gibbs A (1995) Physical properties of insect curticular hydrocarbons: model mixtures and lipid interactions. Comp Biochem Physiol 112B:667–672

Gibbs A, Pomonis JG (1995) Physical properties of insect cuticular hydrocarbons: the effects of chain length, methyl-branching and unsaturation. Comp Biochem Physiol 112B:243–249

Gloag R, Heard T, Beekman M, Oldroyd B (2008) Nest defence in a stingless bee: what causes fighting swarms in *Trigona carbonaria* (Hymenoptera, Meliponini)? Insect Soc 55:387–391

Gordon DM (2016) From division of labor to the collective behavior of social insects. Behav Ecol Sociobiol 70:1101–1108

Goulson D, Chapman JW, Hughes WOH (2001) Discrimination of unrewarding flowers by bees; direct detection of rewards and use of repellent scent marks. J Insect Behav 14:669–678

Greco MK, Hoffmann D, Dollin A, Duncan M, Spooner-Hart R, Neumann P (2010) The alternative pharaoh approach: stingless bees mummify beetle parasites alive. Naturwissenschaften 97:319–323

Grüter C, Menezes C, Imperatriz-Fonseca VL, Ratnieks FLW (2012) A morphologically specialized soldier caste improves colony defence in a Neotropical eusocial bee. Proc Natl Acad Sci USA 109:1182–1186

Grüter C, Keller L (2016) Inter-caste communication in social insects. Curr Opin Neurobiol 38:6–11

Grüter C, von Zuben LG, Segers FHID, Cunningham JP (2016) Warfare in stingless bees. Insect Soc 63:223–236

Grüter C, Segers FHID, Menezes C, Vollet-Neto A, Falcón T, von Zuben L, Bitondi MMG, Nascimento FS, Almeida EAB (2017) Repeated evolution of soldier sub-castes suggests parasitism drives social complexity in stingless bees. Nat Commun 8:4

Gutiérrez E, Ruiz D, Solís T, May-Itzá W d J, Moo-Valle H, Quezada Euán JJG (2016) Does larval food affect cuticular profiles and recognition in eusocial bees? A test on *Scaptotrigona* gynes (Hymenoptera: Meliponini). Behav Ecol Sociobiol 70:781–789

Hart AG, Ratnieks FLW (2001) Why do honey-bee (*Apis mellifera*) foragers transfer nectar to several receivers? Information improvement through multiple sampling in a biological system. Behav Ecol Sociobiol 49:244–250

Hart AG, Ratnieks FLW (2002) Task partitioned nectar transfer in stingless bees (Meliponini): work organisation in a phylogenetic context. Ecol Entomol 27:163–168

Hartfelder K, Bitondi MMG, Santana WC, Simões ZLP (2002) Ecdysteroid titers and reproduction in queens and workers of the honey bee and of a stingless bee: loss of ecdysteroid function at increasing levels of sociality? J Insect Physiol 32:211–216

Hartfelder K, Makert GR, Judice CC, Pereira GAG, Santana WC, Dallacqua R, Bitondi MMG (2006) Physiological and genetic mechanisms underlying caste development, reproduction and division of labor in stingless bees. Apidologie 37:144–163

Hermann HR (1984) Defensive mechanisms: general considerations. In: Hermann HR (ed) Defensive mechanisms in social insects. Praeger, New York, pp 1–31

Hölldobler B, Wilson EO (1990) The ants. Harvard University Press, Cambridge

Hrncir M, Jarau S, Zucchi R, Barth FG (2000) Recruitment behavior in stingless bees, *Melipona scutellaris* and *M. quadrifasciata*. II. Possible mechanisms of communication. Apidologie 31:93–113

Hrncir M, Jarau S, Zucchi R, Barth FG (2003) A stingless bee (*Melipona seminigra*) uses optic flow to estimate flight distances. J Comp Physiol 189:761–768

Hrncir M, Jarau S, Zucchi R, Barth FG (2004a) On the origin and properties of scent marks deposited at the food source by a stingless bee, *Melipona seminigra*. Apidologie 35:3–13

Hrncir M, Jarau S, Zucchi R, Barth FG (2004b) Thorax vibrations of a stingless bee (*Melipona seminigra*). II. Dependence on sugar concentration. J Comp Physiol A 190:549–560

Hrncir M, Schmidt VM, Schorkopf DLP, Jarau S, Zucchi R, Barth FG (2006) Vibrating the food receivers: a direct way of signal transmission in stingless bees (*Melipona seminigra*). J Comp Physiol A 192:879–887

Jarau S (2009) Chemical communication during food exploitation in stingless bees. In: Jarau S, Hrncir M (eds) Food exploitation by social insects: ecological, behavioral, and theoretical approaches. CRC, Boca Raton, pp 223–250

Jarau S, Hrncir M, Ayasse M, Schulz C, Francke W, Zucchi R, Barth FG (2004) A stingless bee (*Melipona seminigra*) marks food sources with a pheromone from its claw retractor tendons. J Chem Ecol 30:793–804

Jarau S, Schulz CM, Hrncir M, Francke W, Zucchi R, Barth FG, Ayasse M (2006) Hexyl decanoate, the first trail pheromone compound identified in a stingless bee, *Trigona recursa*. J Chem Ecol 32:1555–1564

Johnson LK, Haynes LW, Carlson MA, Fortnum HA, Gorgas DL (1985) Alarm substances of the stingless bee, *Trigona silvestriana*. J Chem Ecol 11:409–416

Jones SM, van Zweden JS, Grüter C, Menezes C, Alves DA, Nunes-Silva P, Czaczkes T, Imperatriz-Fonseca VL, Ratnieks FLW (2012) The role of wax and resin in the nestmate recognition system of a stingless bee, *Tetragonisca angustula*. Behav Ecol Sociobiol 66:1–12

Jungnickel H, Velthuis HHW, Imperatriz-Fonseca VL, Morgan ED (2001) Chemical properties allow stingless bees to place their eggs upright on liquid larval food. Physiol Entomol 26:300–305

Jungnickel H, da Costa AJS, Tentschert J, Flávia E, Patricio LRA, Imperatriz-Fonseca VL, Drijfhout F, Morgan ED (2004) Chemical basis for inter-colonial aggression in the stingless bee *Scaptotrigona bipunctata* (Hymenoptera: Apidae). J Insect Physiol 50:761–766

Keeping MG, Crewe RM, Field BI (1982) Mandibular secretions of the old world stingless bee, *Trigona gribodoi* Magrettii: isolation, identification, and compositional changes with age. J Apic Res 21:65–73

Kerr WE, Lello E (1962) Sting glands in stingless bees—a vestigial character. J NY Entomol Soc 70:190–214

Kerr WE, Rocha R (1988) Communicação em *Melipona rufiventris* e *Melipona compressipes*. Ciência e Cultura 40:1200–1202

Kirchner WH, Lindauer M (1994) The causes of the tremble dance. Behav Ecol Sociobiol 35:303–308

Koedam D, Velthausz PH, van de Krift T, Dohmen MR, Sommeijer MJ (1996) Morphology of reproductive and trophic eggs and their controlled release by workers in *Trigona* (*Tetragonisca*) *angustula* Illiger (Apidae, Meliponinae). Physiol Entomol 21:289–296

Koethe S, Bossems J, Dyer AG, Lunau K (2016) Colour is more than hue: preferences for compiled colour traits in the stingless bees *Melipona mondury* and *M. quadrifasciata*. J Comp Physiol A 202:615–627

Kolmes SA (1985) An information-theory analysis of task specialization among worker honey bees performing hive duties. Anim Behav 33:181–187

Kwong WK, Medina LA, Koch H, Sing KW, Yu Soh EJ, Ascher JS, Jaffé R, Moran NA (2017) Dynamic microbiome evolution in social bees. Sci Adv 3:e1600513

Lapidge K, Oldroyd B, Spivak M (2002) Seven suggestive quantitative trait loci influence hygienic behavior of honeybees. Naturwissenschaften 89:565–568

Lenoir A, D'Ettorre P, Errard C (2001) Chemical ecology and social parasitism in ants. Annu Rev Entomol 46:573–599

Leonhardt SD, Blüthgen N (2009) A sticky affair: resin collection by Bornean stingless bees. Biotropica 41:730–736

Leonhardt SD, Kaltenpoth M (2014) Microbial communities of three sympatric Australian stingless bee species. PLoS One 9:e105718

Leonhardt SD, Jung LM, Schmitt T, Blüthgen N (2010a) Terpenoids tame aggressors: role of chemicals in stingless bee communal nesting. Behav Ecol Sociobiol 64:1415–1423

Leonhardt SD, Zeilhofer S, Schmitt T (2010b) Stingless bees use terpenes as olfactory cues to find resin sources. Chem Senses 35:603–611

Leonhardt SD, Wallace HM, Blüthgen N, Wenzel F (2015) Potential role of environmentally derived cuticular compounds in stingless bees. Chemoecology 25:159–167

Lichtenberg EM, Hrncir M, Turatti IC, Nieh JC (2011) Olfactory eavesdropping between two competing stingless bee species. Behav Ecol Sociobiol 65:763–774

Lindauer M, Kerr WE (1960) Communication between the workers of stingless bees. Bee World 41:29–41–65–71

López-Uribe MM, Sconiers WB, Frank SD, Dunn RR, Tarpy DR (2016) Reduced cellular immune response in social insect lineages. Biol Lett 12:20150984

Machado JO (1971) Simbiose entre as abelhas sociais brasileiras (Meliponinae, Apidae) e uma espécie de bactéria. Ciência e Cultura 23:625–633

Martin SJ, Jenner EA, Drijfhout FP (2007) Chemical deterrent enables a social parasitic ant to invade multiple hosts. Proc R Soc B 274:2717–2721

Martin SJ, Carruthers JM, Williams PH, Drijfhout FP (2010) Host specific social parasites (*Psithyrus*) indicate chemical recognition system in bumblebees. J Chem Ecol 36:855–863

Martin SJ, Shemilt S, da S Lima CB, de Carvalho CAL (2017) Are isomeric alkenes used in species recognition among neo-tropical stingless bees (*Melipona* spp). J Chem Ecol 43:1066–1072

McFrederick SQ, Cannone JJ, Gutell RR, Kellner K, Plowes RM, Mueller UG (2013) Specificity between lactobacilli and hymenopteran hosts is the exception rather than the rule. Appl Environ Microbiol 79:1803–1812

McFrederick SQ, Wcislo WT, Taylor DR, Ishak HD, Dowd SE et al (2012) Environment or kin: whence do bees obtain acidophilic bacteria? Mol Ecol Notes 21:1754–1768

Medina-Medina LA, Hart AG, Ratnieks FLW (2009) Hygienic behavior in the stingless bees *Melipona beecheii* and *Scaptotrigona pectoralis* (Hymenoptera: Meliponini). Genet Mol Res 8:571–576

Medina-Medina LA, Hart AG, Ratnieks FLW (2014) Waste management in the stingless bee *Melipona beecheii* Bennett (Hymenoptera: Apidae). Sociobiology 61:428–434

Medina RG, Fairbairn DJ, Bustillos A, Moo-Valle H, Medina S, Quezada-Euán JJG (2016) Variable patterns of intraspecific sexual size dimorphism and allometry in three species of eusocial corbiculate bees. Insect Soc 63:493–500

Menezes C, Vollet-Neto A, León Contrera FA, Venturieri GC, Imperatriz-Fonseca VL (2013) The role of useful microorganisms to stingless bees and stingless beekeeping. In: Vit P, Pedro SRM, Roubik DW (eds) Pot honey: a legacy of stingless bees. Springer, New York, pp 153–172

Menezes C, Vollet-Neto A, Marsaioli AJ, Zampieri D, Fontoura IC, Luchessi AD, Imperatriz-Fonseca VL (2015) A Brazilian social bee must cultivate fungus to survive. Curr Biol 25:2851–2855

Michener CD (1974) The social behavior of the bees: a comparative study. Belknap Press, Harvard University, Cambridge

Morais PB, Calaça PSST, Rosa CA (2013) Microorganisms associated with stingless bees. In: Vit P, Pedro SRM, Roubik DW (eds) Pot honey: a legacy of stingless bees. Springer, New York, pp 173–186

Nascimento DL, Nascimento FS (2012) Acceptance threshold hypothesis is supported by chemical similarity of cuticular hydrocarbons in a stingless bee, *Melipona asilvai*. J Chem Ecol 38:1432–1440

Nash DR, Boomsma JJ (2008) Communication between hosts and social parasites. In: D'Ettorre P, Hughes DP (eds) Sociobiology of communication: an interdisciplinary perspective. Oxford University Press, Oxford, pp 55–80

Nieh JC (1998) The role of a scent beacon in the communication of food location by the stingless bee, *Melipona panamica*. Behav Ecol Sociobiol 43:47–58

Nieh JC (2004) Recruitment communication in stingless bees (Hymenoptera, Apidae, Meliponini). Apidologie 35:159–182

Nieh JC, Roubik DW (1995) A stingless bee (*Melipona panamica*) indicates food location without using a scent trail. Behav Ecol Sociobiol 37:63–70

Nieh JC, Roubik DW (1998) Potential mechanisms for the communication of height and distance by a stingless bee, *Melipona panamica*. Behav Ecol Sociobiol 43:387–399

Nieh JC, Tautz J, Spaethe J, Bartareau T (1999) The communication of food location by a primitive stingless bee, *Trigona carbonaria*. Zoology 102:238–246

Nieh JC, Contrera FAL, Rangel J, Imperatriz-Fonseca VL (2003) Effect of food location and quality on recruitment sounds and success in two stingless bees, *Melipona mandacaia* and *Melipona bicolor*. Behav Ecol Sociobiol 55:87–94

Nieh JC, Barreto LS, Contrera FAL, Imperatriz-Fonseca VL (2004) Olfactory eavesdropping by a competitively foraging stingless bee, *Trigona spinipes*. Proc R Soc Lond B 271:1633–1640

Nunes TM, Nascimento FS, Turatti IC, Lopes NP, Zucchi R (2008) Nestmate recognition in a stingless bee: does the similarity of chemical cues determine guard acceptance? Anim Behav 75:1165–1171

Nunes TM, Turatti IC, Lopes NP, Zucchi R (2009a) Chemical signals in the stingless bee, *Frieseomelitta varia*, indicate caste, gender, age, and reproductive status. J Chem Ecol 35:1172–1180

Nunes TM, Turatti IC, Mateus S, Nascimento FS, Lopes NP, Zucchi R (2009b) Cuticular hydrocarbons in the stingless bee *Schwarziana quadripunctata* (Hymenoptera, Apidae, Meliponini): differences between colonies, castes and age. Genet Mol Res 8:589–595

Nunes TM, Mateus S, Turatti IC, Morgan E, Zucchi R (2011) Nestmate recognition in the stingless bee *Frieseomelitta varia* (Hymenoptera, Apidae, Meliponini): sources of chemical signals. Anim Behav 81:463–467

Nunes TM, Mateus S, Favaris AP, Amaral MFZJ, von Zuben LG, Clososki GC, Bento JMS, Oldroyd BP, Silva R, Zucchi R, Silva DB, Lopes NP (2014a) Queen signals in a stingless bee: suppression of worker ovary activation and spatial distribution of active compounds. Sci Rep 4:7449

Nunes TM, von Zuben LG, Costa L, Venturieri GC (2014b) Defensive repertoire of the stingless bee *Melipona flavolineata* Friese (Hymenoptera: Apidae). Sociobiology 61:541–546

Nunes-Silva PN, Imperatriz-Fonseca VL, Gonçalves LS (2009) Hygienic behavior of the stingless bee *Plebeia remota* (Holmberg, 1903) (Apidae, Meliponini). Genet Mol Res 8:649–654

Packer L (2003) Comparative morphology of the skeletal parts of the sting apparatus of bees (Hymenoptera: Apoidea). Zool J Linnean Soc 138:1–38

Page RE Jr (2013) The spirit of the hive: the mechanisms of social evolution. Harvard University Press, Cambridge

Page RE Jr, Metcalf RA, Erickson EH Jr, Lampman RL (1991) Extractable hydrocarbons and kin recognition in the honey bee (*Apis mellifera* L.). J Chem Ecol 17:745–756

Patricio EFLRA, Cruz-López L, Morgan ED (2002) Electroantennography in the study of two stingless bee species (Hymenoptera: Meliponini). Braz J Biol 64:827–831

Perry JC, Roitberg BD (2006) Trophic egg laying: hypotheses and tests. Oikos 112:706–714

Pianaro A, Flach A, Patricio EFLRA, Nogueira-Neto P, Marsaioli AJ (2007) Chemical changes associated with the invasion of a *Melipona scutellaris* colony by *Melipona rufiventris* workers. J Chem Ecol 33:971–984

Pianaro A, Menezes C, Kerr WE, Singer RB, Patricio EFLRA, Marsaioli AJ (2009) Stingless bees: chemical differences and potential functions in *Nannotrigona testaceicornis* and *Plebeia droryana* males and workers. J Chem Ecol 35:1117–1128

Poiani SB, Morgan ED, Drijfhout FP, Cruz-Landim Cd (2014) Separation of *Scaptotrigona postica* workers into defined task groups by the chemical profile on their epicuticle wax layer. J Chem Ecol 40:331–340

Proctor M, Yeo P, Lack A (1996) The natural history of pollination. Timber Press, Portland

Quezada-Euán JJG, González-Acereto JA (2002) Notes on the nest habits and host range of cleptobiotic *Lestrimelitta niitkib* (Ayala 1999) (Hymenoptera: Meliponini) from the Yucatán peninsula, México. Acta Zool Mex 86:245–249

Quezada-Euán JJG, López-Velasco A, Pérez-Balam J, Moo-Valle H, Velazquez-Madrazo A, Paxton RJ (2011) Body size differs in workers produced across time and is associated with variation in the quantity and composition of larval food in *Nannotrigona perilampoides* (Hymenoptera, Meliponini). Insect Soc 58:31–38

Quezada-Euán JJG, Ramírez J, Eltz T, Pokorny T, Medina R, Monsreal R (2013) Does sensory deception matter in eusocial obligate food robber systems? A study of *Lestrimelitta* and stingless bee hosts. Anim Behav 85:817–823

Quezada-Euán JJG, May-Itzá WdJ, Montejo E, Moo-Valle H (2015) Isometric worker size variation in relation to individual foraging preference and seasonal colony growth in stingless bees. Insect Soc 62:73–80

Raguso RA (2001) Floral scent, olfaction, and scent-driven foraging behavior. In: Chittka L, Thomson JD (eds) Cognitive ecology of pollination: animal behavior and floral evolution. Cambridge University Press, Cambridge, pp 83–105

Rasmussen C, Cameron SA (2010) Global stingless bee phylogeny supports ancient divergence, vicariance, and long distance dispersal. Biol J Linn Soc 99:206–232

Ratnieks FLW, Anderson C (1999) Task partitioning in insect societies. II. Use of queueing delay information in recruitment. Am Nat 154:536–548

Riveros AJ, Groenenberg W (2010) Sensory allometry, foraging task specialization and resource exploitation in honeybees. Behav Ecol Sociobiol 64:955–966

Robinson GE, Huang ZY (1998) Colony integration in honey bees: genetic, endocrine and social control of division of labor. Apidologie 29:159–170

Roubik DW (1989) Ecology and natural history of tropical bees. Cambridge University Press, Cambridge

Roubik DW, Smith BH, Carlson RG (1987) Formic acid in caustic cephalic secretions of stingless bee *Oxytrigona* (Hymenoptera: Apidae). J Chem Ecol 13:1079–1086

Rothenbuhler WC (1964) Behavior genetics of nest cleaning in honey bees. IV. Responses of F1 and backcross generations to disease-killed brood. Am Zool 12:578–583

Sakagami SF (1982) Stingless bees. In: Hermann HR (ed) Social insects, vol III. Academic Press, London, pp 361–423

Sakagami SF, Zucchi R (1966) Estudo comparativo do comportamento de varias especies de abelhas sem ferrão, com especial referencia oa processo de aprovisionamento e postura das celulas

(Comparative study of various stingless bees behaviour with special provitioning reference process and cell posture). Ciencia e Cultura 18:283–296
Sakagami SF, Roubik DW, Zucchi R (1993) Ethology of the robber stingless bee, *Lestrimelitta limao* (Hymenoptera: Apidae). Sociobiology 21:237–277
Shackleton K, Toufailia A, Balfour NJ, Nasciento FS, Alves DA, Ratnieks FLW (2015) Appetite for self-destruction: suicidal biting as a nest defense strategy in *Trigona* stingless bees. Behav Ecol Sociobiol 69:273–281
Schmidt VM, Hrncir M, Schorkopf DLP, Mateus S, Zucchi R, Barth FG (2008) Food profitability affects intranidal recruitment behaviour in the stingless bee *Nannotrigona testaceicornis*. Apidologie 39:260–272
Schorkopf DLP, Jarau S, Francke W, Twele R, Zucchi R, Hrncir M, Schmidt VM, Ayasse M, Barth FG (2007) Spitting out information: *Trigona* bees deposit saliva to signal resource locations. Proc R Soc B 274:895–898
Schorkopf DLP, Hrncir M, Mateus S, Zucchi R, Schmidt VM, Barth FG (2009) Mandibular gland secretions of Meliponine worker bees: further evidence for their role in interspecific and intraspecific defence and aggression and against their role in food source signalling. J Exp Biol 212:1153–1162
Schorkopf DLP (2016) Male Meliponine bees (*Scaptotrigona* aff. *depilis*) produce alarm pheromones to which workers respond with fight and males with flight. J Comp Physiol A 202:667–678
Seeley TD (1985) Honeybee ecology, a study of adaptation in social life. Princeton University Press, Princeton
Seeley TD (1995) The wisdom of the hive, the social physiology of honeybee colonies. Harvard University Press, Cambridge
Seeley TD (1998) Thoughts on information and integration in honey bee colonies. Apidologie 29:67–80
Septanil MPB, Mateus S, Turatti IT, Nunes TM (2012) Mixed colonies of two species of congeneric stingless bees (Hymenoptera: Apinae, Meliponini) display environmentally-acquired and endogenously-produced recognition signals. Physiol Entomol 37:72–80
Simone-Finstrom M, Spivak M (2010) Propolis and bee health: the natural history and significance of resin use by honey bees. Apidologie 41:295–311
Smith BH, Roubik DW (1983) Mandibular glands of stingless bees (Hymenoptera: Apidae): chemical analysis of their contents and biological function in two species of *Melipona*. J Chem Ecol 9:1465–1472
Sommeijer MJ (1984) Distribution of labour among workers of *Melipona favosa* F: age polyethism and worker oviposition. Insect Soc 31:171–184
Sommeijer MJ (1987) Age-polyethism in stingless bees and evidence of flexible individual ontogenetic sequences. In: Eder J, Rembold H (eds) Chemistry and biology of social insects. Peperny, München, pp 129–130
Sommerlandt FMJ, Huber W, Spaethe J (2014) Social information in the stingless bee, *Trigona corvina* Cockerell (Hymenoptera: Apidae): the use of visual and olfactory cues at the food site. Sociobiology 61:401–406
Spivak M, Downey DL (1998) Field assays for hygienic behavior in honey bees (Hymenoptera: Apidae). J Econ Entomol 91:64–70
Tofilski A (2002) Influence of age polyethism on longevity of workers in social insects. Behav Ecol Sociobiol 51:234–237
Toufailia HA, Alves DA, Bento JMS, Marchini LC, Ratnieks FLW (2016) Hygienic behaviour in Brazilian stingless bees. Biol Open 5:1712–1718
Vander Meer RK, Morel L (1998) Nestmate recognition in ants. In: Vander Meer RK (ed) Pheromone communication in social insects: ants, wasps, bees and termites. Westview Press, Boulder, pp 79–103
van Veen JW (2000) Cell provisioning and oviposition in *Melipona beecheii* (Apidae, Meliponinae), with a note on caste determination. Apidologie 31:411–419

van Veen JW, Sommeijer MJ, Meeuwsen F (1997) Behaviour of drones in *Melipona* (Apidae, Meliponinae). Insect Soc 44:435–447

van Zweden JS, D'Ettorre P (2010) Nestmate recognition in social insects and the role of hydrocarbons. In: Blomquist GJ, Bagnères AG (eds) Insect hydrocarbons: biology, biochemistry and chemical ecology. Cambridge University Press, Cambridge, pp 222–243

Vásquez A, Forsgren E, Fries I, Paxton RJ, Flaberg E, Szekely L, Olofsson TC (2012) Symbionts as major modulators of insect health: lactic acid bacteria and honeybees. PLoS One 7:e33188

Velthuis HHW (1997) The biology of stingless bees. Dept. of Ethology, Utrecht University, Utrecht

Velthuis HHW, Cortopassi-Laurino M, Pereboom Z, Imperatriz-Fonzeca VL (2003) The conservative egg of the genus *Melipona* and its consequences for speciation. In: Melo GAR, Alves-dos-Santos I (eds) Apoidea Neotropica: Homenagem aos 90 Anos de Jesus Santiago Moure. Editora UNESC, Criciúma, pp 171–176

Velthuis HHW, Koedam D, Imperatriz-Fonseca VL (2005) The males of *Melipona* and other stingless bees, and their mothers. Apidologie 36:169–185

von Zuben LG, Schorkopf DLP, Elias LG, Vaz ALL, Favaris AP, Clososki GC, Bento JMS, Nunes TM (2016) Interspecific chemical communication in raids of the robber bee *Lestrimelitta limao*. Insect Soc 63:339–347

Verdugo-Dardón M, Cruz-López L, Malo EA, Rojas JC, Guzmán-Díaz M (2011) Olfactory attraction of *Scaptotrigona mexicana* drones to their virgin queen volatiles. Apidologie 42:543–550

Villa JD, Weiss MR (1990) Observations on the use of visual and olfactory cues by *Trigona* spp foragers. Apidologie 21:541–545

Waddington KD (1989) Implications of variation in worker body size for the honey bee recruitment system. J Insect Behav 2:91–103

Wille A (1979) Phylogeny and relationships among the genera and subgenera of the stingless bees (Meliponinae) of the world. Rev Biol Trop 27:241–277

Wille A (1983) Biology of the stingless bees. Annu Rev Entomol 28:41–64

Wilson EO (1971) The insect societies. Belknap Press of Harvard University Press, Cambridge

Wilson EO (1990) Success and dominance in ecosystems: the case of the social insects. In: Kinne O (ed) Excellence in ecology. Book 2. Ecology Institute, Oldendorf/Luhe

Winston ML (1987) The biology of the honey bee. Harvard University Press, Cambridge

Wyatt TD (2003) Pheromones and animal behaviour. Communication by smell and taste. Cambridge University Press, Cambridge

Zucchi R (1993) Ritualized dominance, evolution of queen-worker interactions and related aspects in stingless bees. (Hym., Apidae). In: Sakagami SF, Inoue T, Yamane S (eds) Evolution of insect societies. Hakuhinsha, Tokyo, pp 207–249

Chapter 6
Reproduction

In collaboration with Teresita Solís

Mating in *M. beecheii*

There are two ways in which social insects can reproduce: one is through the production of sexual offspring by queens, males, and (in some species) workers, and the other is through colony fission (Ruttner 1985; Koeniger 1986). This means that, in social insects, effective reproduction depends not only on individual performance, but also to a large extent on the functioning of the colony. Both aspects are tightly linked; survival of the colony is in the best interests of all colony members to ensure high output of reproductives and daughter colonies (Crozier and Pamilo 2003).

Sexual offspring in stingless bees can be produced by queens and males, but also workers. As explained in Chap. 4, the workers of many species are capable of pro-

J. J. G. Quezada-Euán, *Stingless Bees of Mexico*,
https://doi.org/10.1007/978-3-319-77785-6_6

ducing male offspring, though whether the queen or workers dominate in male production varies considerably between species, and also between colonies within a species (Paxton et al. 2001; Tóth et al. 2002). Because the males of stingless bees live a solitary life when reaching sexual maturity, they seem a relatively "cheap" way of genetic dispersal, although male production generally occurs only when colonies have accumulated sufficiently large food reserves (Moo-Valle et al. 2001). In *Melipona*, gynes also invade colonies (Sommeijer et al. 2003), which may represent another type of "cheap" genetic dispersal, via females. Nonetheless, the importance of gyne invasion of non-related colonies in natural populations remains unknown. The same applies to workers; although workers can potentially parasitize other colonies, no evidence of this has been found.

In addition to individual reproduction, propagation in social insects through the female line, involves the production of workers that depart with the queen in a swarm (Wilson 1971). Notably, reproduction of colonies in the highly eusocial bees (meliponines and honey bees) invariably occurs through colony fission (splitting) (Michener 1974; Hölldobler and Wilson 1990). The mother colony divides into two (and exceptionally more) colonies, each with a group of workers and generally one queen. Thus, the investment in colony reproduction through swarming is large compared to the investment in the production of individual sexual offspring (Fig. 6.1).

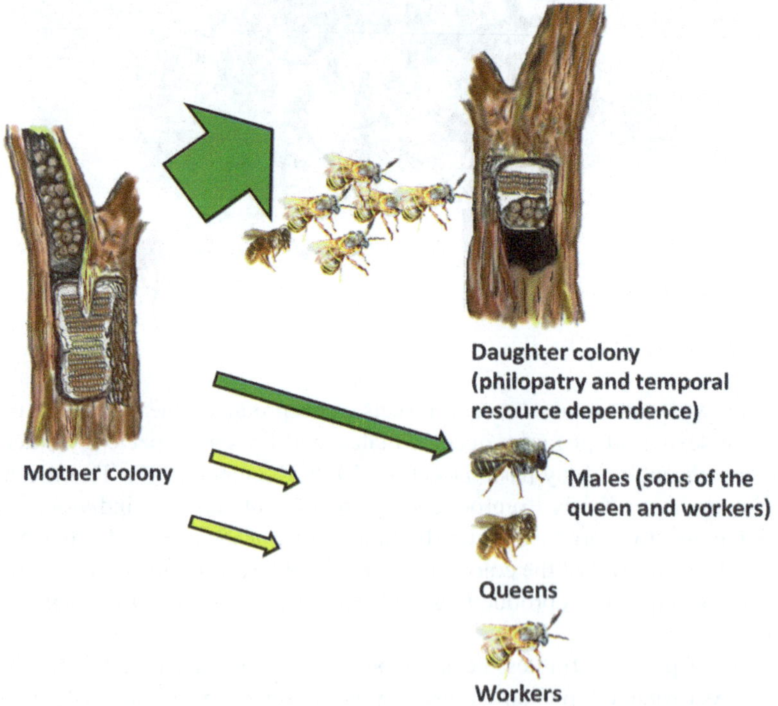

Fig. 6.1 Reproductive investment and potential routes of genetic dispersal in Meliponini. The thickness of the arrow represents the magnitude of resource investment, and its length the potential distance dispersal. Female dispersal via queens and workers has not been well documented in non-managed populations

In addition, colony reproduction seems especially demanding in the case of stingless bees because daughter colonies are not completely independent after departing. There is a long-lasting bond between mother and daughter colony until the latter is self-sufficient (Engels and Imperatriz-Fonseca 1990). This also implies that in stingless bees dispersal via females involves comparatively shorter distances (philopatry), than via males.

Although highly eusocial bees share some common features in the reproduction of colonies and individuals, there are also marked contrasts in their reproductive biology which are presented in Table 6.1.

Table 6.1 Comparison of some reproductive traits between stingless bees and honey bees

Individual traits	Stingless bees	Honey bees
Sexual dimorphism	Males larger than gynes in *Melipona* Males smaller than gynes in non-*Melipona*	Males larger than gynes in *Apis mellifera* but smaller in other *Apis* species
Virgin queen	At emergence, ovaries well developed in non-*Melipona* gynes and underdeveloped in *Melipona* Ovarioles have increased in size rather than number (usually $n = 4$) Queens tolerate each other Workers eliminate excess queens	At emergence ovaries well developed Ovarioles have increased in number rather than length ($n > 100$) Queens do not tolerate each other Queens eliminate other queens
Males	Sexually immature at emergence Solitary lifestyle when sexual maturity is reached	Sexually immature at emergence Live in colonies when sexual maturity is reached
Reproductive workers	Frequent in the presence of a queen (queen's chemical control of ovarian development rare). No worker policing has been documented	Rare (queen's chemical control of ovarian development). Worker policing exists
Queen mating	Species predominantly monandrous In non-*Melipona*, males congregate nearby colonies searching for virgin queens Males of *Melipona* possibly congregate far away from colonies and mating seems to occur in these areas	All species are polyandrous Males congregate far away from colonies
Male mating	Monogynous—males die after mating	Monogynous—males die after mating
Mated queen	Physogastric—unable to fly	Non-physogastric—capable of flying
Contact workers-queen	Workers have little contact with the queen—the queen is frequently aggressive towards workers Ritualized oviposition	Workers frequently groom the queen; there is a queen retinue—no evident physical aggression of the queen Non-ritualized oviposition
Queen feeding	Mainly trophic eggs produced by the workers	Royal jelly produced in glands in the head of workers
Colony traits		
Composition of swarm	Non-mated gynes and workers	Mother queen and workers

(continued)

Table 6.1 (continued)

Individual traits	Stingless bees	Honey bees
After swarming	Non existent	Frequent in some populations
Contact between mother and daughter colony	Daughter colony located nearby Long-lasting contact with mother colony	Daughter colony moves far away No further contact between mother and daughter colonies
Preparation of the new nest	Anticipated	No preparation before swarming
Reproductive strategy	Few offspring colonies with high survival rates ("K" strategy)	Tropical ecotypes: more offspring colonies with relatively low survival rates ("R" strategy)

6.1 Individual Reproduction

The production and behavior of sexual offspringin stingless bees show important differences with the honey bees (Table 6.1).

When completing their development, queens of stingless bees can emerge rapidly from their cell, as in *Melipona,* or prolonged, sometimes needing the help of workers, like in most other genera. Moreover, the queens of genera different to *Melipona* seem to take longer to develop (pers. Obs.), in comparison with the queens of *Melipona* (Moo-Valle et al. 2004). Possibly, these differences in duration of development can explain why *Melipona* gynes emerge with underdeveloped ovaries, compared with other genera of stingless bees (Wille 1983).

In the meliponines, the behavior of workers towards queens varies across species. In most non-*Melipona* species, gynes at emergence are generally not attacked by workers (Gutiérrez et al. 2016). In contrast, in *Melipona*, workers are frequently aggressive to gynes; indeed, most excess queens are rapidly killed in the colonies (Wenseleers et al. 2004; Jarau et al. 2009; Kärcher et al. 2013). On the other hand, stingless bee queens are rather indifferent to each other. There is no aggression among gynes or with the mother queen (Kärcher et al. 2013). In comparison, honey bee queens are not tolerant to other queens; they fight each other to death (Winston 1987).

After the queen emerges she can be kept alive for some time. Some non-*Melipona* species build special "compartments" where they maintain gynes, these could be empty food pots or spaces in the involucrum. Such lodgings are known as royal chambers and may serve to keep gynes as insurance in case of the loss of the mother queen. Imperatriz-Fonseca and Zucchi (1995) report that in some species, the gyne participates in the construction of her own chamber and can even control the access of workers (e.g., in *Frieseomelitta, Plebeia*). In other species like *Cel. longicornis*, a few workers form a shield around gynes to protect them against attacks by other workers (Imperatriz-Fonseca and Zucchi 1995). Because swarming rates in stingless bees are low, most queens are eventually sacrificed.

Melipona species are outstanding for the excess number of queens produced (see Chap. 4). However, most queens are killed (Wenseleers et al. 2004; Kärcher et al. 2013). In *M. beecheii,* for instance, gynes live on average only 27 h (Wenseleers

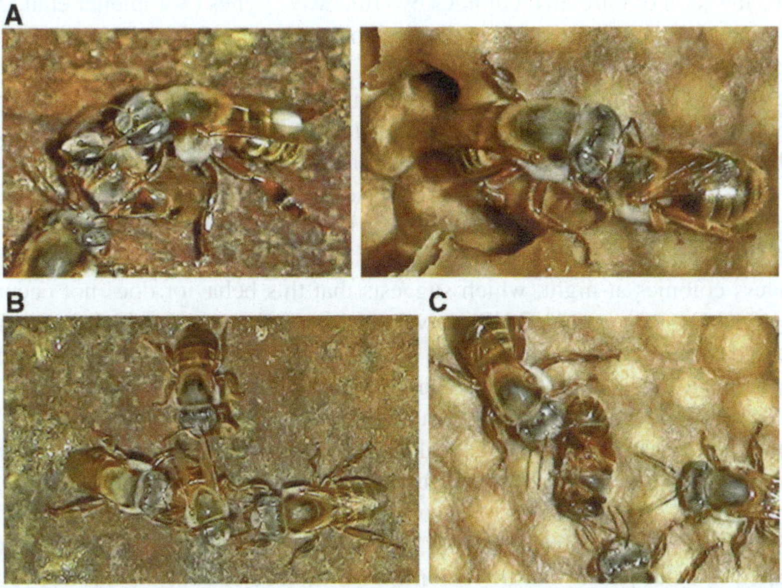

Fig. 6.2 The process of gyne sacrifice in *M. beecheii*. (**a**) Two aspects of decapitation by a single worker. (**b**) Pulling extremities by various workers. (**c**) A gyne faking death to avoid worker aggression

et al. 2004). In this species, two forms of queen sacrifice have been recorded (Fig. 6.2). Approximately 25% of sacrifices are made by a single worker. She climbs on the gyne's dorsum and, inserting the mandibles in the space between the head and the thorax, quickly proceeds to decapitate her. In a second and more frequent behavior, a group of workers hold the queen and pull her from the legs, head, and abdomen, trying to dismember her (Fig. 6.2). This last type of killing can last up to 1 h. In many cases, gynes try to hide from the workers, and can even fake death to avoid elimination (Fig. 6.2, Wenseleers et al. 2004). It seems that less "fit" queens fake death more frequently, as those engaging in this type of behavior are eventually killed (van Veen et al. 1999). If queens avoid immediate killing, they can reach a stage when start performing a behavior called abdominal inflation. Abdominal inflation start in *M. favosa* and *M. beecheii* some 12–15 h after gyne emergence and becomes more frequent with time (Koedam et al. 1995; van Veen et al. 1999). In *M. beecheii* all queens eventually accepted in colonies performed abdominal inflation. In some way queens with an inflated abdomen could be more attractive to workers (van Veen et al. 1999). Gynes in *Scaptotrigona* inflate the abdomen and use it to hit the workers that try to bite them. Perhaps this behavior is used to defend from worker aggression and is also part of a process of queen domination (Gutiérrez et al. 2016). Abdominal inflation could also be related with pheromone production (van Veen et al. 1999), but this has not been demonstrated.

The excess production of gynes in *Melipona* and aggressive harassment by their worker sisters, are possibly the origin of a phenomenon observed in meliponaries, i.e., the invasion of unrelated colonies by "runaway" gynes (Sommeijer et al. 2003). This behavior was first observed in the Caribbean species *M. favosa,* but no evidence of reproductive takeover could be confirmed. The authors interpreted the aggression of the workers as another way by which colonies (and workers) could increase reproductive fitness via dispersal of their gyne sisters, if these were to become queens in invaded colonies (Sommeijer et al. 2003) (Fig. 6.1). The rate of gyne colony usurp was quantified by means of molecular analyses in *M. scutellaris* from Brazil. Gynes of this species exhibited surprisingly high rates of colony invasion estimated in ~25% (Wenseleers et al. 2011). Interestingly, gynes seem to preferentially take over queenless colonies at night, which suggests that this behavior does not occur randomly (van Oystaeyen et al. 2013). Given the high frequency at which queen usurping could occur, it is puzzling why colonies do not seem to have developed mechanisms to reject alien queens (Kärcher et al. 2013; van Oystaeyen et al. 2013).

In *M. beecheii* we have observed (unpublished data) that many gynes exit their colonies and seek refuge in areas like empty boxes or under pieces of wood. However, most of these gynes do not survive long, probably because they need to be fed by workers, although it has been suggested that they can probably feed on flowers (Sommeijer and de Bruijn 2003). Marking large number of *M. beecheii* gynes has failed to detect them in other colonies, but molecular analysis has not yet been conducted to confirm if gyne colony takeover could be common in this species. Although gyne usurping seems frequent in meliponarios, there is still no evidence of its occurrence in non-managed populations (Wenseleers et al. 2011) and, thus, whether it represents a significant mode of female gene dispersal in natural populations of stingless bees.

Female reproductive parasitism in stingless bees can potentially occur via workers too. Workers, similar to gynes, can abandon their natal nest and invade other colonies, where they can produce sons (Beekman and Oldroyd 2008). To date no such phenomenon has been documented in any species of stingless bee. However, worker parasitism across generations has been detected. In *M. scutellaris* workers carry on producing sons, even after their mother queen has been superseded. In this sense, daughters of the old queen parasite workers of the new generation (their nieces), with which they are less genetically related (Alves et al. 2009).

Probably because of the large number of gynes produced in *Melipona*, colonies seem highly selective in the process of queen replacement. In a study to evaluate the process of queen acceptance in *M. beecheii*, it was found that colonies can remain queenless for up to 3 weeks and can kill as many as 20 queens in this period (Flores-Grande et al. unpublished data). Similar results have been obtained with other species, like Brazilian *M. compressipes* (Kerr 1987). Even though in *Melipona* there is strong selection on queens, the role of factors such as genotype, aggression, and chemical profiles, among others, is still not well understood (Kerr 1987; Gutiérrez et al. 2016). The speed of queen acceptance has important applications in management because, if the queen is lost, the colony may remain queenless for several weeks, which can affect growth and possibly compromise its survival.

After selection, the accepted gyne experiences notorious changes in behavior before her mating flight. In this period, the gyne increases her activity, jostling and performing considerable antennal contact and trophallaxis or food exchange with workers (van

Veen 1999). Aggression towards workers, like pushing or jostling, could be considered part of a gradual process of domination. In fact, physogastric queens frequently push and hit the workers during the POP (Sakagami 1982; Jarau et al. 2009).

Regarding queen mating, Flores-Grande et al. (2003) found that gynes of *M. beecheii* made a single mating flight. On the day of the mating flight, the gyne exits the colony and makes a short "orientation" flight. She performs a first flight facing the entrance of her colony in a zigzag pattern, possibly registering landmarks to help in finding the colony. Slowly, the gyne flies away from the colony until she rapidly disappears into the forest. The age at which gynes of *M. beecheii* undertook their mating flights was 4.6 days in average, and the duration of the flight was only 10.8 min. In studies with *M. beecheii* in Costa Rica, the duration of mating flights was over 100 min (van Veen and Sommeijer 2000). It is possible that such differences in mating flights for the same species relate to the availability of males in both areas, and the possibility to find them nearby. Mating flights took place more frequently at midday, between 11:30 and 12:50 h. The average temperature registered for the mating flights in Yucatan was 31.1 °C, the environmental humidity was 38.2%, and wind speed was 0.58 m/s. The average age at which queens started egg laying was 10.4 days after emergence, and 5.8 days after their mating flight (Flores-Grande et al. 2003).

As a norm, stingless bees queens are monandrous, they mate with only one male (Peters et al. 1999; Paxton et al. 1999; Palmer et al. 2002). After mating, the genital capsule of the male detaches from his body and remains inserted in the queen's reproductive tract for a couple of days, forming what is known as the mating plug (Melo et al. 2001) (Fig. 6.3). It is thought that the male capsule causes distension of

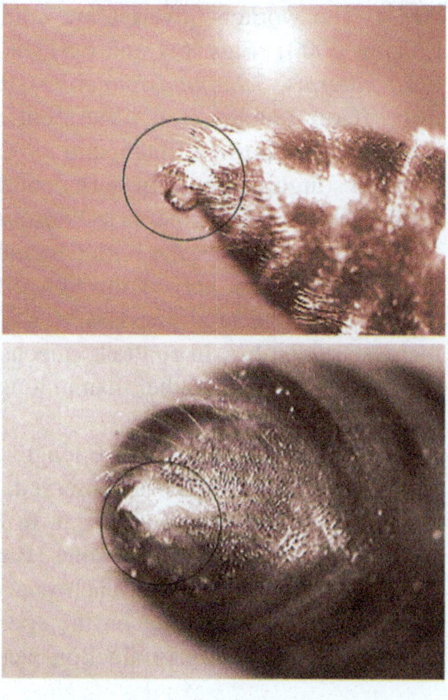

Fig. 6.3 Mating plug formed by the genital capsule of the male in the reproductive tract of a queen of *M. beecheii* (photo by Fredy Camposeco)

the *bursa copulatrix* of the queen, and produces a mechanical signal. This signal may be related to the activation of the *corpora allata,* and the production of JH that triggers ovary growth. In a few *M. quadrifasciata* queens, removing the male genital capsule in the first 3 days after mating resulted in reduced activation of the queen's ovaries (Melo et al. 2001).

In contrast with males of honey bees and bumble bees, the mating plug of stingless bee males has no glandular products. Thus, it is believed that there is no male chemical stimulation that may induce ovarian development in stingless bee queens, as known in the other corbiculate taxa (Colonello and Harfelder 2005).

After the mating flight, the queen reduces food soliciting, and her activity within the nest markedly decreases, as does aggression towards workers (van Veen 1999). At the same time the abdomen experiences a remarkable increase in size, known as physogastria. Physogastric queens (also seen in termites and some ants) experience a dramatic hypertrophy of the ovaries. In the process, the fat bodies in the abdomen also experience considerable growth. The fat bodies are the main source of vitellogenin, a major lipoprotein released into the hemolymph and taken up by the ovarioles for the production of eggs (Bordereau 1982; Paes de Oliveira and Cruz-Landim 2003; Paes de Oliveira et al. 2012).

Physogastria produces extreme distension of the abdomen, causing the intersegmental membranes of the metasoma to expand (Bordereau 1982). The abdominal terga and sterna become well separated in stingless bee queens, which does not occur in honey bee or bumble bee queens (Fig. 6.4). As a result of the physogastric state, stingless bee queens can double their weight and become incapable of flying.

The females, (workers and queens) of stingless bees, normally have four ovarioles in each of their two ovaries (Cruz-Landim et al. 1998; Martins and Serrâo 2004). The workers' ovaries are comparatively shorter than the queens' (Cruz-Landim 2000) (Figs. 6.4 and 6.5). In a few species (*Tr. spinipes, Pb. remota, N. testaceicornis, Sch. quadripunctata*), the queens can have up to eight ovarioles per ovary (Cruz-Landim et al. 1998), but in *Lestrimelitta,* the number of ovarioles per ovary can be up to 15 (Sakagami 1982). Nonetheless, the number of ovarioles per ovary in stingless bees is in stark contrast to the more than 100 present in honey bee queens (Engels and Imperatriz-Fonseca 1990; Martins and Serrâo 2004).

It is suggested that the high demand for egg production in queens of the highly eusocial bees was solved through different evolutionary routes. In Meliponini, queens experienced an increase in the relative length of the ovaries more than an increase in number. In contrast, queens of *Apis* have experienced an increase in the number of ovarioles, rather than in longitude (Engels and Imperatriz-Fonseca 1990; Martins and Serrâo 2004).

Intraspecific variation in the length (and less frequently number) of ovarioles in stingless bee queens has been observed. This variation may be related to differences in nutrition during gyne development. In *Tr. spinipes* larvae fed with larger amounts of food had more and longer ovarioles (Buschini and Campos 1995; Lisboa et al. 2005).

In stingless bee queens, each ovary usually produces between one and a few dozen per day, depending on the species (Engels and Imperatriz-Fonseca 1990). The rate of queen egg production per day has been recorded in some species: in

Fig. 6.4 Physogastric queen of *M. beecheii* (above left), showing the distended abdomen with clearly visible abdominal membranes and separated plates compared with *A. mellifera*, in which queens are non-physogastric. In the image below, the ovaries of worker, virgin (with the four ovarioles separated), and physogastric queen of *M. beecheii,* are shown

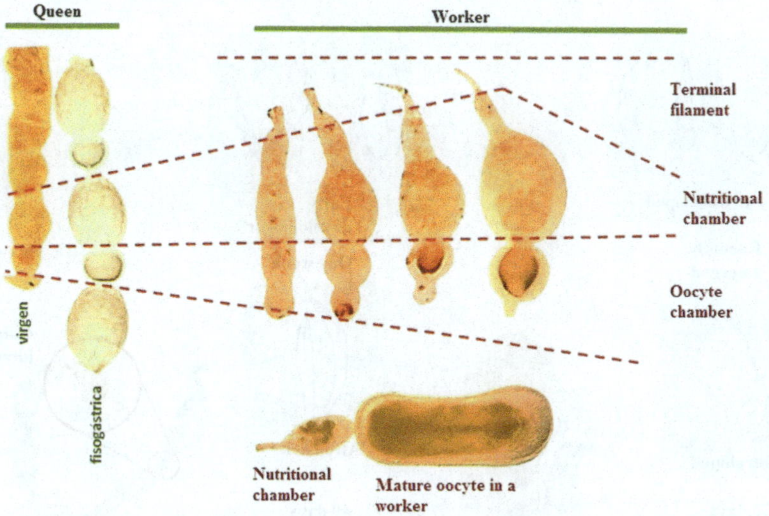

Fig. 6.5 Oogenesis in a worker and the queen of *M. beecheii*, with the parts of the ovarian follicle: the oocyte chamber and the nutritional chamber. The image below shows an egg nearly ready to be laid; the nutritional chamber is atrophied

M. compressipes it is between 25 and 30 eggs, in *M. quadrifasciata* between 10 and 20 (Kerr 1949), in *P. remota* between 60 and 180 (Van Benthem et al. 1995), in *M. beecheii* between 25 and 35, and in *L. niitkinb,* between 100 and 200 eggs per day (pers. Obs.). The rate of egg production of queens seem to depend on colony food conditions (Moo-Valle et al. 2001).

The production of eggs in the ovaries of queens (and workers) starts in the ovarian functional unit, the follicle. The ovarian follicle is formed by two sections, the nutritional chamber and the oocyte chamber (Fig. 6.5). The nutritional chamber is the section that supplies nutrients to the oocyte by means of special cells called trophocytes. The trophocytes provide vitellogenin and other proteins needed for the development of the oocyte (Cruz-Landim 2009). In the process of oogenesis, as nutrients are absorbed by the oocyte chamber, the nutritional chamber experiences gradual atrophy. The final stages of egg formation involve the production of the vitelline membrane as well as the chorion, forming the egg shell (Cruz-Landim 2009). The chorion of the bee's egg has two layers, a proteinic endochorion and the exochorion mainly made of carbohydrate (Cruz-Landim 2009).

The ovaries of the queen have several follicles undergoing maturation at the same time; in contrast, the ovaries of workers only have one maturing follicle, and exceptionally two, in each ovary (Cruz-Landim 2000; Figs. 6.4 and 6.5).

The rate of ovarian development among workers of the same colony generally depends on age. Young workers less than 1 week old usually have no developed ovaries, but after this period, they start producing trophic or reproductively functional eggs. The production of reproductive eggs is more frequent in workers before they start foraging (van Veen 1999). In older workers (foragers), the ovaries usually become atrophied (Fig. 6.6). Functional worker eggs are not fertilized, but develop

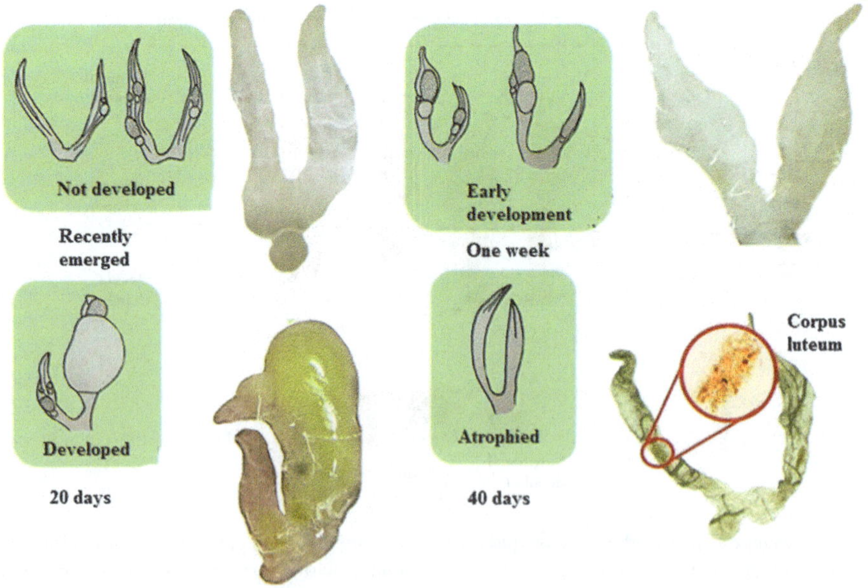

Fig. 6.6 Comparative development of the ovaries in workers of different age in *M. beecheii*

into males by parthenogenesis, similar to those of the queen (Koedam et al. 1999). The characteristics of trophic and functional eggs in stingless bees are explained in more detail in Sect. 5.3.

An outstanding fact of mated queens of social insects is the extremely long viability of the sperm they store, which may span several years (Baer et al. 2016). It is believed that the viability of sperm is maintained thanks to the stable conditions of the spermatheca. The spermatheca is a spherical sac at the base of the queen's oviducts where sperm is stored after mating (Cruz-Landim et al. 2003). To date it is not quite understood how the viability of the sperm is maintained in the spermatheca. Honey bee queens may actively produce proteins to enhance sperm viability during storage (den Boer et al. 2009). Interestingly, compounds that keep the sperm alive may not be produced in the spermatheca; they could be absorbed from the hemolymph, and secreted around the stored sperm (Cruz-Landim et al. 2003). The spermatheca is connected to the reproductive tract by a small tube in the transition between the *bursa copulatrix* and the midportion of the oviduct (Fig. 6.4).

Hymenopteran queens have control over sperm use by means of the spermatheca, and can switch between producing fertilized and non-fertilized eggs. In the honey bee, failure of the queen to fertilize worker-destined eggs seems to be a signal that workers use to replace her (Baer et al. 2016). In stingless bees, the longevity of mated queens and the causes leading to supersedure are scarcely documented. In queens of *M. compressipes* and *M. scutellaris*, a maximum longevity of 7 years has been estimated (Kerr 1996; Kerr et al. 1996), although a more conservative figure of 22 months could be more realistic (Carvalho-Zilse and Kerr 2004). Similarly, in *M. beecheii* queens more often seem to live an average of 2 years (van Veen 1999).

The males of stingless bees can be produced by the queen or the workers. As explained in Chap. 4, the relative numbers of sons of the queen and workers vary across species. In species in which males are mostly produced by the workers, the cells containing males tend to form clusters in some sections of the combs (Koedam et al. 1999). In the case of species like *M. beecheii* (Paxton et al. 2001) and *M. colimana,* in which the queens produce most of the males, there is no clustering of male cells (Moo-Valle et al. 2004; Macías-Macías and Quezada-Euán 2015).

Male production in colonies of stingless bees is frequently associated with the amount of food reserves (Moo-Valle et al. 2001; Chinh et al. 2003). Colonies tend to produce more males when they have large food reserves. In *Melipona,* when food reserves dwindle, male production suffers more than queen production, and can drop to almost zero (Moo-Valle et al. 2001). Generally, males represent an additional cost because their production would be at the expense of more valuable workers (Velthuis et al. 2005). Nevertheless, it is interesting noting that a marked seasonality in the production of males is not evident in stingless bees, in contrast to *A. mellifera*. For instance, in *M. beecheii* it is possible to find males in colonies all year round (Moo-Valle et al. 2001). In stingless bees asynchrony in the production of males and gynes seems a common feature of colonies. However, asynchronous production of males and gynes in a population, will lead to the simultaneous occurrence of both sexes from different colonies when mating occurs (Moo-Valle et al. 2001; Chinh et al. 2003; de de Bruijn and Sommeijer 2006).

In highly eusocial bees, gene dispersal via males could be cheaper compared with colony fission, in which many workers depart with the queen in a swarm (Roubik 1990). Overall male production in stingless bees (as in other eusocial Hymenoptera) is on average larger than gyne production (the numerical sex ratio is male biased), even in species with excess production of gynes, like in *Melipona.* In *M. beecheii* for instance, annual male production is twice that of gyne production (Moo-Valle et al. 2001). In other stingless bees, male bias could be even larger because fewer queens are produced. For instance, in the case of *Sch. quadripunctata,* the production of males to gynes is in the range of 200:1 (Velthuis et al. 2005). At the moment of mating, the rate of males to gynes can increase even more, because many queens (particularly in *Melipona*) can be sacrificed before they mate. The production of males to gynes at the time of mating can be easily in the range of 1000:1 (Engels and Imperatriz-Fonseca 1990). Under these conditions, in *Melipona* the probability of mating for individual males is in the range of 0.01–0.002 (Velthuis et al. 2005).

Evidently, excessive Male-biased production in stingless bees represents stark competition for females. However, the males of highly eusocial bees do not compete among themselves using strength, displacement, or fighting (Engels and Imperatriz-Fonseca 1990). This means that selection by females seems scarce or nonexistent; rather, males seem to engage in a scramble competition for queens (Paxton 2005). Nonetheless, some male physical characteristics could be advantageous in scramble competition for reproduction. In *M. beecheii* for instance, male fresh weight can vary between 25 and 65 mg, representing a weight difference of nearly threefold among individuals in the same population (Bustillos 2016; Fig. 6.7).

Although the benefit of large size has not been shown in stingless bees, there is evidence from other Apidae, that larger males may have mating advantages (Schlüns et al. 2003; Couvillon et al. 2010). Likewise, it may be possible that larger, or heavier stingless bee males, can more readily chase or get hold of gynes.

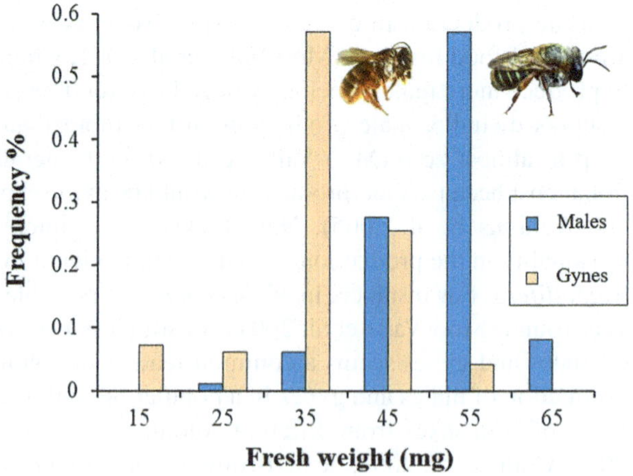

Fig. 6.7 Distribution of male and gyne fresh weight (mg) at emergence in *M. beecheii* colonies from Yucatan, Mexico (Bustillos 2016)

On the other hand, small males may also have reproductive advantages. For instance, small males of *Scp.* aff. *depilis* persist for longer periods in male congregations. It is speculated that persistence may increase their opportunities for mating, but no correlation was found between body size and sperm viability in males persisting in a congregation (Koffler et al. 2016). Nonetheless, it seems that stingless bee males could be selected for endurance (Roubik 1990). It could also be possible that small males persist longer in a mating congregation because small size restricts dispersal over long distances; note, though, that it is not known if larger males tend to disperse over longer distances. Although males of different sizes can follow contrasting reproductive strategies in other bee species (Alcock 1997), it remains to be shown, if the striking differences in body size found in male stingless bees, may have different adaptive value.

The resources of a colony may impact male traits. Males *of M. beecheii* reared in colonies with large pollen reserves are larger and produce more spermatozoa, compared with males reared in colonies with lesser reserves (Pech-May et al. 2012; Fig. 6.8).

In stingless bees, the male reproductive tract is formed by the structures for the production and storage of sperm, plus the copulatory organ or genital capsule (Fig. 6.9). As in other bees, spermatogenesis in male stingless bees takes place during the pupal stage and, when individuals emerge from their cells, the process is completed (Dallacqua and Cruz-Landim 2003). This means that males emerge with a fixed number of spermatozoa and no additional production occurs in adulthood. However, males emerge sexually immature; the spermatozoa in their testes must migrate to the seminal vesicles before mating and insemination can effectively

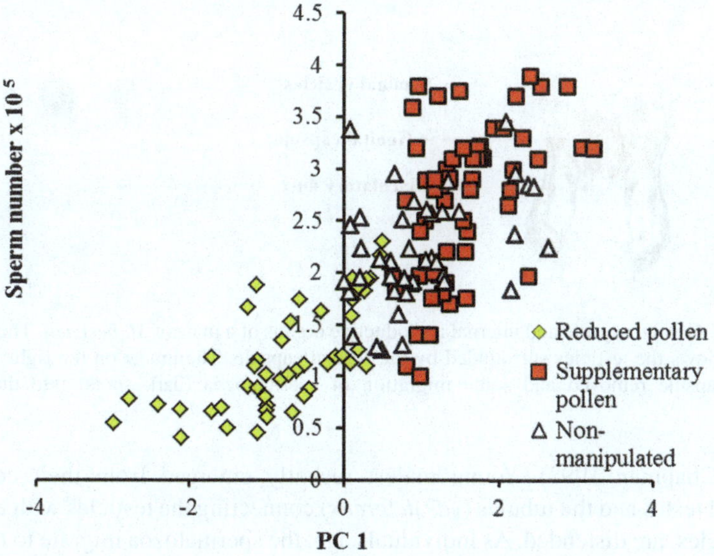

Fig. 6.8 Sperm production in three groups of *M. beecheii* males reared in colonies with reduced pollen reserves, excess pollen, and colonies non-manipulated (natural). PC1 is a multivariate measure of body size obtained from three morphometric traits (Pech-May et al. 2012)

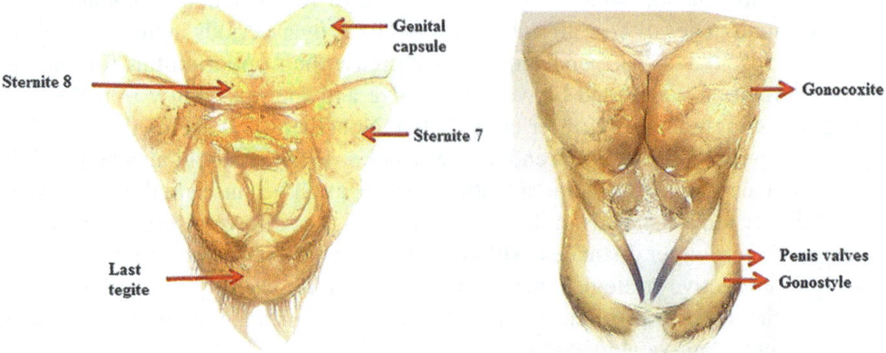

Fig. 6.9 The genitalia (left) and genital capsule (right), dissected from a male of *M. beecheii*

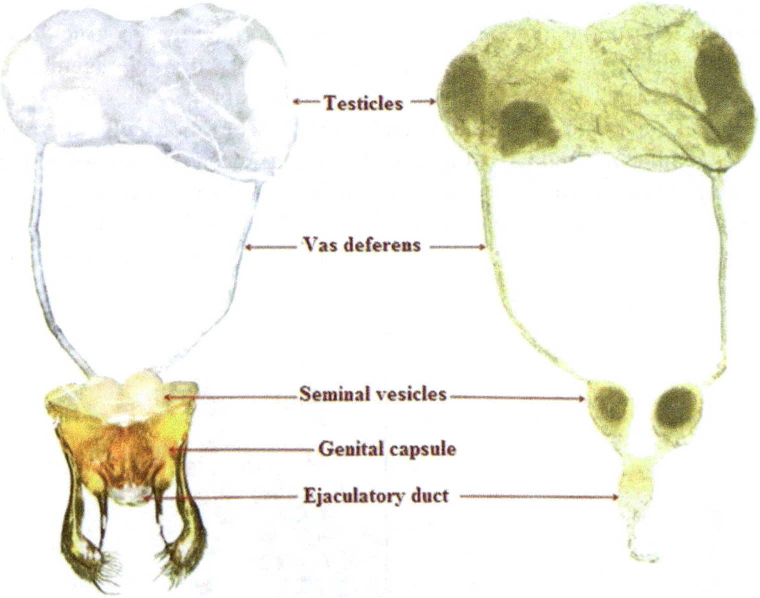

Fig. 6.10 Genital capsule and internal reproductive organs of a male of *M. beecheii*. The image on the left shows the testicles surrounded by the scrotal capsule. The image on the right shows the genital capsule removed and some migration of spermatozoa (dark spots) into the seminal vesicles

occur (Chapman 1998). Young males, recently emerged from their cell, have enlarged testes and the tubules (*vas deferens*) connecting the testicles with the seminal vesicles, are distended. As individuals age, the spermatozoa migrate to the seminal vesicles, and the testes gradually become atrophied in sexually mature individuals (Brito et al. 2010) (Figs. 6.10 and 6.11). The testes of stingless bees have four seminiferous tubules each (Ferreira et al. 2004) (Fig. 6.11).

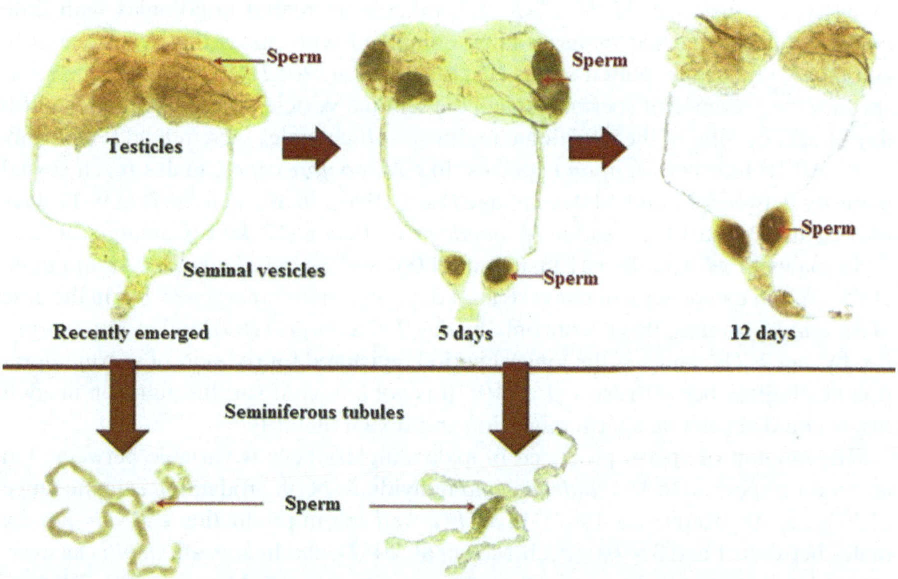

Fig. 6.11 The process of male sexual maturation male sexual maturity in *M. beecheii*. The image to the left shows a recently emerged male with no presence of spermatozoa (dark areas) in the seminal vesicles. The image in the center shows the start of the migration of the spermatozoa, which is almost finished in the male of the image to the right. In the image below, sperm concentrates in the confluence of the four seminiferous tubules of each testicle in five days old males, when migration starts to the seminal vesicles

In contrast with other Apidae, the male reproductive tract of stingless bees has simple genital ducts and lacks accessory glands (Ferreira et al. 2004; Colonello and Hertfelder 2005). In other bees, male accessory glands produce secretions that serve as nutrients for the spermatozoa in their movement along the female tract (den Boer et al. 2009). These secretions can also modify the post-copulatory behavior of the female (Dallacqua and Cruz-Landim 2003; Ferreira et al. 2004). Because of the lack of accessory glands in stingless bees, it is believed that substances helping the survival of the spermatozoa are found in other sections of the male reproductive tract (Dallacqua and Cruz-Landim 2003; Brito et al. 2010). In addition to anatomical simplification, the reproductive system of male stingless bees has comparatively thicker *vas deferens* and the ejaculatory duct is short. This type of elementary reproductive apparatus (compared with males of other taxa), lacking accessory glands is exclusive to the male Meliponini, and has been classified as type IV (Ferreira et al. 2004) (Fig. 6.11).

When stingless bee males emerge, they remain in their nest for a period of time before reaching sexual maturity, after which they abandon the nest and start an independent life, never returning to their natal colony (Roubik 1990). The time males spend in their colonies is variable, and is related with the speed at which they reach sexual maturity. In *M. beecheii* males can be sexually mature at 18 days, but there is considerable variation, ranging between 6 and 27 days (Camposeco 2005). Intra-colony factors could be related with the speed at which males reach sexual maturity. For instance, *M. beecheii* males that were reared in colonies with little pollen reserves, reached maturity later, compared with males that were reared in colonies with surplus pollen stores (Pech-May et al. 2012). However, in the same species, the presence of spermatozoa in the seminal vesicles was evident on the fifth day of age, in spite of the nutritional regime at which males were reared (Pech-May et al. 2012). In other *Melipona* species, like *M. compressipes*, males reach sexual maturity between 11 and 14 days of age (Kerr 1996), in *M. scutellaris* at 9–11 days (Abreu and Penna 1992), and in *M. quadrifasciata* at 8–12 days (Camargo 1972).

In males of *M. mondury* (Lima et al. 2006) and *Sc. xanthotricha* (Araújo et al. 2005), sperm can be seen in the vesicles 4 days after their emergence, but in the case of *M. quadrifasciata,* this occurs only by day 7 (Camargo 1984), and in *Frs. schrottkyi,* by day 9. The latter, is the longest period registered for the start of sperm migration in stingless bees (Brito et al. 2010). It is not known if surplus nutrition in adult males could accelerate sperm migration and sexual maturity.

The amount of sperm produced in male stingless bees is variable between, but also within species. In *M. quadrifasciata* individual sperm production is in the range of $1–1.2 \times 10^6$ (Kerr et al. 1962). In *M. beecheii* sperm production can vary among males between 1 and 3×10^6 (Pech-May et al. 2012), and in *Scp.* aff. *depilis* an average of 1.4×10^6 spermatozoa per male has been registered (Meneses et al. 2014).

Before reaching sexual maturity, males are mostly inactive in the nest (van Veen et al. 1997). Nevertheless, in contrast to honey bee males, they can engage is some activities (nectar dehydration, wax production, collection of pollen, and propolis), and can be indirectly involved in others, such as colony defense (Cortopassi-Laurino 1979; Boongird and Michener 2010; Schorkopf 2016). Nogueira-Neto (1997) considers that the activities performed by stingless bee males are related more to their preparation to leave the nest, rather than a real contribution to the colony. In this regard, it is males older than 20 days which normally engage in nectar dehydration (van Veen et al. 1997; van Veen 1999). In *M. beecheii*, van Veen (1999) observed that, as males age, they tend to spend more time in the area of food pots, which seems in agreement with the suggestion by Nogueira-Neto (1997).

The possibility of resin and pollen collection by males of some stingless bee species is surprising. It has been proposed that this behavior may occur more frequently in species in which males have hind tibiae comparatively similar to the workers'. In such species males are called isopodal, to identify them from species in which males and workers have very different hind legs, which are called anisopodal (Boongird and Michener 2010). Whether males of isopodal species are recruited to collect food, or if they do it independently, is still unknown.

When reaching sexual maturity, stingless bee males leave their natal colony (van Veen 1999), after which they survive by feeding on flowers (Kerr et al. 1962; Pers. Obs.). Eventually, males can join one aggregation, or perhaps move between several of these male groups (Fig. 6.12). In *Ttr. dorsalis,* males that were individually marked in an aggregation, remained in it for only a short time (Roubik 1990), which suggests that they may search for a particularly suitable one, or move constantly among different aggregations (dos Santos et al. 2016). Moving among aggregations could increase their opportunities for mating.

Fig. 6.12 Aggregation of *N. perilampoides* males. The behavior of the males at aggregations alternates between flying in a cloud that moves around colonies (left), and resting nearby (image to the right)

Two forms of male aggregations have been proposed in stingless bees. In species other than *Melipona*, males actively search for virgin females, and they congregate nearby colonies with young gynes (Fig. 6.12). In contrast, in *Melipona* it is believed that it is the gynes that search for male aggregations, which seem to be located far away from colonies (Velthuis et al. 2005). Nonetheless, *Melipona* male groups have also been reported close to meliponarios (Sommeijer et al. 2004; dos Santos et al. 2014), and individual males can be seen perching nearby colonies (Pers. Obs.).

In the case of non-*Melipona* species, when a gyne is accepted in a colony, males start to congregate nearby a few days after (Pers. Obs.). How males detect the presence of gynes in such colonies is still a matter of speculation. It has been suggested that males at long distance perceive some kind of queen pheromone, or that foragers carry queen smell, indirectly guiding males to their colony (Roubik 1990).

Male aggregations are formed by hundreds or thousands of males which originate from a large number of colonies, some located at considerable distances (dos Santos et al. 2016). Molecular analyses of males at aggregations have revealed the number of colonies from which they originate. In the case of *Scp. mexicana* males at aggregations originate from 21 to 55 colonies (Kraus et al. 2008; Mueller et al. 2012), up to 70 in *Ttr. angustula* (dos Santos et al. 2016), and up to 132 in the Asian *Ttga. collina* (Cameron et al. 2004). It is important to note that only a few males seem to originate from colonies nearby; as a rule the majority of males come from distant colonies (Mueller et al. 2012; dos Santos et al. 2016). Thus, male stingless bees seem capable of moving over long distances and, some studies have directly assessed this. Carvalho-Zilse and Kerr (2004) marked males of *M. scutellaris* and released them at different distances from their colony. Their results showed that some males can travel up to 1 km, although males released 100–400 m returned with more frequency to the meliponary. Using molecular markers, dos Santos et al. (2016) estimated the distance that *Ttr. angustula* males can disperse to be 600 m on average, but some dispersed over 2 km. It is important to note that such distances were calculated between two points in single events. It is possible that males travel considerably longer distances during their life span.

The behavior of males moving away from their natal colony and the fact that male aggregations are formed by males of multiple origin, seem effective adaptations to reduce the possibility of inbreeding. The wandering behavior of males is crucial in stingless bees, particularly considering that queens are predominantly monandrous, and that colonies are philopatric. The long dispersal of stingless bee males seems a key aspect to reduce inbreeding and effectively increase panmixia (Kraus et al. 2008). Recent findings of only subtle genetic differentiation among populations of some common species, like *Ttr. angustula* (Santiago et al. 2016) and *Pt. bilineata* (Landaverde-González et al. 2017), indicate the importance of male-mediated gene flow in stingless bees and the risk of losing genetic diversity if male dispersal is affected.

In the case of *Melipona,* male aggregations have been rarely seen. It has been suggested that males do not aggregate nearby colonies because although *Melipona* colonies constantly produce queens, most are killed. In this regard, it would not be adaptive for males to search for gynes near a colony (Sommeijer et al. 2004). Detailed reports of male aggregations have been provided on *M. favosa* and *M. subnitida* (Sommeijer and de Bruijn 1995; Sommeijer et al. 2004; dos Santos et al. 2014). The males of these species have been seen perching in groups, which contrast to the flying aggregations of males in *A. mellifera,* and non-*Melipona* stingless bees. Such perching sites have been found usually at 1.5 m height, where the number of males can be between 300 and 400. Similar to the aggregations of other stingless bees, the first males are seen early in the morning and the last just before dusk, but they do not overnight at the site (Sommeijer and de Bruijn 1995; Sommeijer et al. 2004). It is important to note that although the presence of queens has been reported in *Melipona* male aggregations, no attempts to mate have been observed (Sommeijer et al. 2004; dos Santos et al. 2014). It could be argued that these perching aggregations are resting males. In fact, observations in confinement chambers suggest that mating seems to take place in flight in *Melipona*, as in the other eusocial bees (pers. Obs.).

If stingless bee males and queens have some way to communicate their presence to each other is still under speculation (Engels 1987). It has been suggested that males of *M. favosa* release some kind of pheromone or chemical signal that attracts queens (Sommeijer et al. 2004). Interestingly, in *Scp. mexicana*, Galindo-López and Kraus (2009) showed that males were attracted to groups of other males. Curiously, in the field, males of that species were not attracted to colonies with virgin queens. The authors proposed that some type of chemical attraction among males could occur in this species. Interestingly, males of different species can attend the same aggregation. Interspecific similarities in the cuticular compounds and possible confusion caused by this type of recognition system have been dismissed as explanation for mixed aggregations of males (dos Santos et al. 2015). The explanation of mixed male aggregations is still unclear.

Interestingly, workers of *M. favosa* and *M. quadrifasciata* may also be found at male aggregations. It has been proposed that workers may attract males to these locations by means of pheromones or by bringing resin or other materials to these sites from their nests (Roubik 1990; Sommeijer et al. 2004; dos Santos et al. 2014).

As males are produced constantly in stingless bee colonies (Moo-Valle et al. 2001; de Bruijn and Sommeijer 2006), male aggregations can be active during most part of the year.

6.2 Colony Reproduction and Dynamics

The formation of new nests in many highly eusocial insects involves colony fission (reproductive division or swarming), although how this occurs varies across taxa (Table 6.1). In the case of stingless bees, colony fission occurs at low frequencies, and involves a long period of preparation. After swarming, a bond between mother and daughter nests can last months before they separate completely (Engels and Imperatriz-Fonseca 1990). In contrast, in honey bees, swarming occurs more frequently and rapidly, and does not involve further contact between mother and daughter colonies.

Swarming has rarely been observed in stingless bees; there are only few accounts on how it occurs (Engels and Imperatriz-Fonseca 1990). A prerequisite for swarming is large worker populations and food reserves. In colonies with these conditions, workers start searching for a potential new nest site nearby, sometimes in the same tree or substrate as the mother colony. This may explain the frequent aggregation of nests observed in many stingless bee species. In the honey bee, searching for a new site where the swarm will establish, only occurs after the bees have left the mother colony, and this can be located a kilometers away from the maternal nest (Seeley 1985).

Once a potential nest site has been selected, workers start conditioning the new cavity and build structures using cerumen and resin taken from the mother colony (Fig. 6.13). Within a few days, food pots are built and bees start moving pollen and honey from the mother colony too. Interestingly, Nogueira-Neto (1997) noted that the food carried by supplying bees is frequently a mixture of pollen and honey. Once stores have been developed in the new nest, swarming takes place. A virgin or a few virgin queens exit the mother colony with a group of workers that move into the new nest. Stingless bee swarming is different compared with honey bees, in which the mother queen leaves the nest before a virgin queen emerges (Winston 1987). Because the physogastric stingless bee queen cannot fly, it is the gynes that relocate with the swarms. At the new nest site, a young queen is selected and, after a few days, she performs her nuptial flight. The workers continue moving food and building materials from the mother colony until the new nest is self-sufficient (Sakagami 1982; Engels and Imperatriz-Fonseca 1990; Fig. 6.13).

The conservative system of swarming in stingless bees results in low rates of colony reproduction. The mother colony would need sufficient time (probably years) to accumulate the reserves needed to support her survival, and that of a daughter colony. The reproductive strategy of stingless bee colonies seems to be the production of few offspring (Slaa 2006) (Fig. 6.1). However, new stingless bee colonies seem to have high rates of survival. In a study in Costa Rica, the survival of

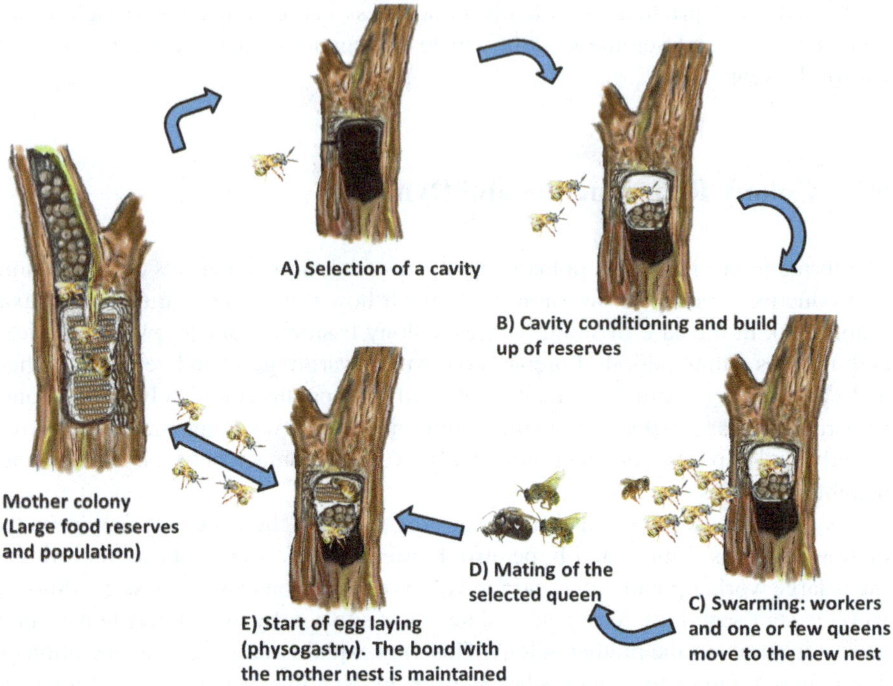

Fig. 6.13 The process of swarming (colony fission) in Meliponini. The letters and arrows indicate the progression of events

new stingless bee colonies was close to 70% at the end of the first year. In comparison, nearly all new Africanized honey bee colonies established in the same area died (Slaa 2006). In this regard, stingless bees seem to fit the description for K-strategists, characterized by monogamy, slow growth, and production of a few progenies, each with a high probability of survival. In comparison, r-strategists, like Africanized honey bees, have high fecundity and relatively little investment in individual progeny (Mac Arthur and Wilson 1967; Rinderer and Hellmich II 1991), but also experience high rates of mortality.

However, the strategy of colony foundation involving philopatry can be disadvantageous when habitats are altered. Fewer nesting sites or food resources can increase competition, but also the length of time necessary for colonies to reproduce (Engels and Imperatriz-Fonseca 1990; Brown and Albrecht 2001). The relatively limited dispersal of stingless bee colonies could also explain why populations of some species exhibit marked genetic structure for maternal rather than biparentally inherited genetic markers (Quezada-Euán et al. 2012: Santiago et al. 2016).

Meliponini colonies may take years for swarming (Roubik 1989), depending on the amount of reserves and building materials accumulated. In each annual cycle, colonies may gain reserves and increase size during the flowering season, but lose some during dearth, depending on the severity of shortage (Roubik 1982; Quezada-

Euán et al. 2015). Colonies may probably experience alternative episodes of increasing and losing size, until they reach a status capable of sustaining mother and daughter colonies (Quezada-Euán et al. 2015). Large stingless bee species, like *Melipona,* could be more sensitive to habitat alteration because they need comparatively more food and spacious cavities to reproduce (Brown and Albrecht 2001; Macías-Macías et al. 2014). Various species of *Melipona* have been considered under threat and risk of extinction (Kerr 2002); *M. beecheii* is one of them, especially in the Mayan region of southern Mexico (Quezada-Euán et al. 2001; Villanueva-Gutiérrez et al. 2013).

In the Yucatan, Santos (2006) surveyed the diversity of colonies in a deforested northern part of the Peninsula. No nests of *M. yucatanica* or *M. beecheii* were found. In contrast, small meliponine nests of *Fr. nigra* and *N. perilampoides* were common. This may be because these species are less demanding in the size of cavities or food, they need for successful nesting (Samejima et al. 2004).

Interestingly, the longevity of small stingless bee species in pristine forests can be shorter when compared to deforested regions, or urban areas (Slaa 2006). For instance, the annual survival of nests of *Ttr. angustula* in a forest was estimated at 0.74, but in deforested areas it increased to 0.92. Moreover, the life span of colonies of this species has been estimated to be 12.5 years in deforested areas but only 3.8 years in forests (Slaa 2006).

The estimated natural density of stingless bee colonies varies between a few to several nests per ha (Hubbell and Johnson 1977; Roubik 1983; Eltz et al. 2002; Antonini and Martins 2003; Roubik 2006). Such contrasting findings in the density of nests seem a result of two main factors, food resources and suitable spaces for nesting (Eltz et al. 2002). In Costa Rica, Hubbell and Johnson (1977) reported that food availability seemed to be the main factor limiting the density of colonies in forests. Similarly, a study in Southeast Asia detected that food (mainly pollen) was the factor with the highest impact on nest densities. However, the availability of nesting sites also had a significant effect, albeit to a lesser extent (Eltz et al. 2002). In contrast, in a different forest of Costa Rica, Brosi (2009) found a more marked link between suitable nesting sites and density of colonies. In Mexico, Macías-Macías and Quezada-Euán (2015) in a cloud tropical forest of western Jalisco, found that the availability of large cavities in trees (183 cm or greater in diameter) was positively related to the density of *M. colimana* colonies. Thus, nest and food availability are probably linked. In a given area, if food is scarce even with sufficient nesting sites, colony density would decline, and *vice versa*. Nonetheless, habitat conservation should increase food availability and the opportunities for nesting. Indeed, the highest diversity of colonies of different species has been found in preserved forests (Hubbell and Johnson 1977; Batista et al. 2001; Samejima et al. 2004). In addition to the negative effect of deforestation, human practices, like honey hunting, could represent an additional threat to the survival of stingless bee colonies (Batista et al. 2001; Kerr 2002; Slaa 2006).

It seems that the best strategy for the conservation of stingless bees is the preservation of species diversity by maintaining sufficient numbers of colonies. Diversity could be greatly enhanced by preserving patches of forest, even in highly deforested regions, that serve as corridors connecting distant areas, allowing the transit of reproductives and the survival of colonies (Brosi et al. 2007; Fischer et al. 2008; Landaverde-González et al. 2017).

6.3 Sexual Dimorphism

The concept of sexual size dimorphism (SSD) refers to the consistent difference in average body size of males and females within a species (Fairbairn 1997). SSD can be estimated using variables related to body mass (weight) and also linear measurements (dimension). In animals, SSD could be male or female biased, depending on which sex is larger. In species in which both sexes are of similar size SSD is therefore, close to zero (Fairbairn 1997).

It is assumed that body size differences can be subject to selection (natural or sexual), that differentially affect one sex over the other (Fairbairn 1997). If the pressure of selection upon fertility is strong, then females become larger than males. However, if sexual selection acting upon males is relatively more important, then SSD should be male biased (Teder 2005). In mammals and birds where males control a group of females, SSD is male biased. In contrast, in most insect species SSD is female biased, which reflects more intense selection for fertility on females, than sexual selection on males (Blanckenhorn et al. 2007).

Not much is known about the patterns of SSD in bees, and this is most evident in the highly social species. In a comparative study, Shreeves and Field (2008) evaluated the effect of parental care on SSD in solitary bees and wasps. The studied species were classified into two groups, those that build their own nest, and parasite species that do not build nests nor forage. The results showed that in both groups SSD was female biased, but that it was more pronounced in species that build nests, than in parasites (Shreeves and Field 2008). It seems that the selection pressure acting on female fertility is stronger in females that build nests. Parasite females released from the pressure of building a nest have experienced less selection for large body size. Interestingly, the study also revealed that female-biased SSD is less intense in bees compared with other Hymenoptera, possibly because of the comparatively minor load represented by pollen in relation with prey (Shreeves and Field 2008).

Recently, the pattern of SSD was evaluated in eusocial corbiculate bees (*A. mellifera, M. beecheii*, and *E. viridissima*) to test if, similar to solitary species, SSD is female biased (Medina et al. 2016). The results, using fresh weight and dimensions of virgin queens and males at emergence, showed that the highly eusocial species had a moderate male-biased SSD (Medina et al. 2016). In the case of the primitively eusocial *E. viridissima*, males and females were of similar dimensions, but males were slightly heavier. It is interesting that in both highly eusocial species (*A. mellifera, M. beecheii*), males were larger than females. One possible explanation for these results is that, similar to parasite solitary females, the queens of highly eusocial species have lost the pressure of nest building and provision, tasks that are now performed by the workers (Medina et al. 2016). Nonetheless, these results are preliminary, and more species should be studied to determine if this pattern is similar across stingless bees, and other highly eusocial species.

One proximate cause of SSD in corbiculate bees is the relative rate of development of males and females. In species where females develop faster (protogyny), females are usually smaller than males, resulting in male-biased SSD (Teder 2014). Both, the queens of *A. mellifera* and *M. beecheii* develop faster and emerge before males. In contrast, in *Bombus*, another member of the corbiculate bees, queens develop slower and are larger than males (Duchateau and Velthuis 1988; Cueva del Castillo and Fairbairn 2012). Interestingly, stingless bee gynes other than *Melipona* are larger than males and, it is suggested that they also take longer to develop (Wille 1983; Gonzalez-Acereto 2008).

Although preliminary, the evidence suggests two patterns of SSD in stingless bees (Fig. 6.14). On one side, *Melipona* gynes develop faster and are of smaller size and weight than males. On the other side, gynes of species different to *Melipona* are larger than the males (and workers), and seem to take longer to develop (Figs. 6.14 and 6.15). These differences result in male-biased SSD in *Melipona,* but female-

Fig. 6.14 Two patterns of sexual dimorphism in stingless bees: in *M. beecheii* the male on the left is larger than the gyne on the right. The reverse is found in *Scp. pectoralis*

Fig. 6.15 Differences in body size among sexes and castes of various species of Meliponini from Yucatan. *V* virgin queen, *F* physogastric queen

biased SSD in other stingless bees (Fig. 6.15). Finding explanations to these contrasting patterns should be important to better understand the evolution of reproductive strategies in stingless bees.

6.4 Commercial Reproduction of Colonies

Stingless bee colonies normally reproduce slowly and they need a few months to be self-sufficient (Slaa 2006). However commercial use of colonies requires methods to facilitate a sustainable mass production over time (Quezada-Euán et al. 2001; Slaa 2006).

One way to rapidly increase colony numbers is through the division of more than one daughter from a single mother colony. For commercial purposes, an effective way would be using few workers and combs in what are caled mini-divisions (Imperatriz-Fonseca 1977). In Yucatan, studies with *M. beecheii,* have shown five main aspects to consider in the production of mini-colonies in this species (Quijano et al. 2008):

- The mother colonies used in the production of mini-divisions should have a young queen and large population and reserves.
- The mother colony should not have been harvested in the previous season, and should be fed with sugar syrup (and possibly some pollen supply), for at least 3 months to maintain large adult population and good rates of brood production.
- A worker number for a successful mini-division is ~200 young workers and a comb with ~100 pupae. It is recommended that young naive workers are used, because forager bees will return to the mother colony.
- Thermoregulation is crucial, so excess space in mini-hives should be avoided or reduced. Small boxes (10 × 15 × 7 cm) made of thick wood (of at least 3 cm) are recommended.
- Divisions should be well protected against the invasion of phorid flies. Mosquito nets may be used to keep them closed for 2 or 3 days, and use vinegar traps if necessary (see Chap. 8). Artificial feeding with pasteurized syrup is also recommended after 1–2 days (sugar in water 1:1 or 2:1).
- If possible provide cerumen for the construction of structures.
- Supplying food pots should be avoided for a few days until the division is organized.
- On a commercial scale, the use of temperature chambers could increase the success of the mini-divisions (Pers. Obs.). It would be important to further evaluate this alternative method of mass production of stingless bee colonies.

In the case of *Melipona* species, queen availability for the mini-divisions is not a limiting factor; normally, there are surplus queens. However, for species other than *Melipona,* the presence of a queen or a queen cell is essential for the success of mini-divisions. The workers of most non-*Melipona* bees cannot produce emergency

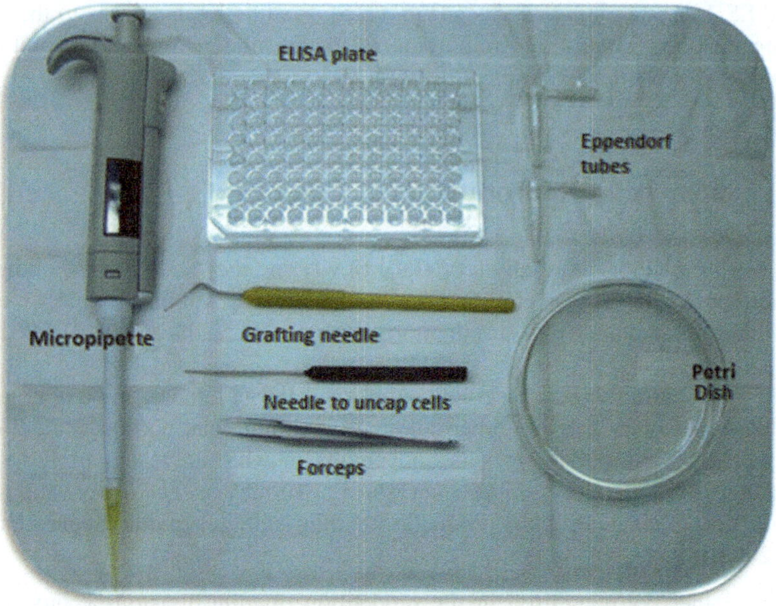

Fig. 6.16 Materials used for artificial queen rearing in non-*Melipona* stingless bees. The materials should be sterilized prior to use

queens (Sakagami 1982). Queen availability is therefore an important factor limiting the mass production of colonies in such species.

Recently, methods to artificially rear queens of non-*Melipona* stingless bees have been developed. Prato (2010) successfully raised queens after transferring young *Ttr. angustula* larvae to ELISA acrylic plate wells containing 55 μL of larval food (Fig. 6.16). Baptistella et al. (2012) evaluated the survival and size of queens in *Fr. varia* produced using 61–62 μL of larval food (2.3 times the amount needed to rear a worker). They found that grafting larvae older than 1 day produced small low quality queens (Menezes et al. 2013). In *Scp. depilis* artificial queen rearing was successful using 134 μL of larval food (Menezes et al. 2013).

It seems that one crucial factor for artificial queen rearing in stingless bees is humidity. The recommended humidity is 75% and temperature is 28° ± 0.5 °C (although the latter requires adjustment considering the ideal temperatures for each species). Another consideration is the care of larvae when transferring them from the cell to the acrylic plate. Larvae have soft bodies that can be easily damaged. In addition, they need to be placed on the food in the same position that they were when removed from their cells.

It is important remembering that, when larvae hatch, they lie on the larval food on one of their sides. The spiracles in contact with the food are obstructed and the larvae use the spiracles on the opposite side to breath. If larvae during transfer are

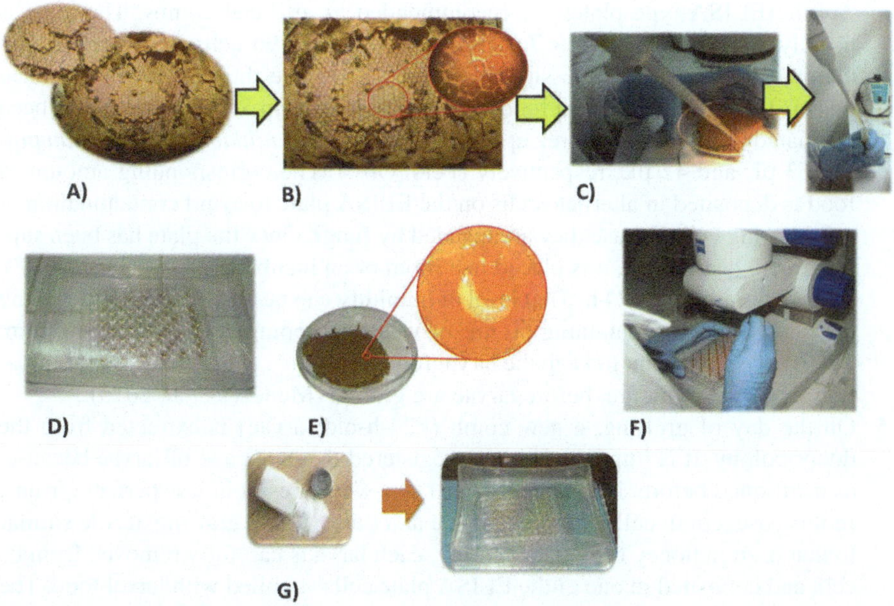

Fig. 6.17 Artificial production of queens of *Scaptotrigona pectoralis*. The letters refer to the different steps described in the text.

placed on the side that has not been in contact with the food, they will not be able to breath and will die (Baptistella et al. 2012; Menezes et al. 2013).

In Yucatan, we have used the basic methods of Baptistella et al. (2012) and Menezes et al. (2013), with some modifications (Figs. 6.17 and 6.18), to rear queens of *Scp. pectoralis*, *N. perilampoides*, and *L. niitkib*. The general process is described in the following sections (see Fig. 6.17); the amount of food and the temperature are the main factors that vary between species:

1. Two strong colonies must be used; one will serve as the donor of larval food and the other of young larvae. It is not recommended to obtain larvae and food from the same colony because it can reduce its strength.
2. All materials should be sterilized using hot water or UV light for 15 min. For the collection of larval food, one or two new combs (<2 days sealed) should be removed from the donor colony, and transferred to the lab or a room under clean conditions. The caps of the cells are removed using forceps, opening four to six cells at a time to avoid desiccation of the liquid food. The eggs on top of the food should be discharged.
3. The larval food is collected using a micropipette. The food is deposited in Eppendorf tubes (1.5 mL), and homogenized using a vortex apparatus (at ca. 3200 rpm). If the larval food is not used immediately it can be stored for a few days at −20 °C (Prato 2010).

4. Acrylic ELISA-type plates are recommended as artificial combs. The plate is usually 1.5 cm thick (8.4 × 7.6 cm) with a total of 96 cells, each with 4 mm diameter and 8 mm depth, similar to the cells of the three species mentioned above. The amount of larval food used to rear queens in *Scp. pectoralis* has been estimated at 86 μL (Gutiérrez et al. 2016), and for *L. niitkib* and *N. perilampoides* 63 μL and 42 μL, respectively (Pers. Obs.). The corresponding amount of food is deposited in alternate cells on the ELISA plate to avoid contamination of neighboring cells in case they are invaded by fungi. Once the plate has been supplied with larval food, it is placed in a room or an incubator at 28 °C with 100% relative humidity for 24 h. That level of humidity can be best achieved by placing the plate in a tray containing 50 mL of water, wrapping both with plastic film. This step is important to keep the larval food moistened and also for its stratification as occurs in nature, before larvae are grafted (Menezes et al. 2013).
5. On the day of grafting, a new comb (<24-h-old larvae) is extracted from the donor colony. It is important to have registered the exact age of larvae because, as mentioned before, larvae older than 1 day would result in less perfect queens.
6. In this case, comb cells are opened, one at a time. With a grafting needle similar to that used in honey bee queen rearing, each larva is carefully removed from its cell, and deposited in one of the ELISA plate cells supplied with larval food. The grafting needle should have a smooth surface to avoid damaging the larva. After all the larvae had been grafted, the plate is kept for additional 4 days at 28 + 2 °C and 100% relative humidity, which is achieved as explained in step D. The larval food should be moistened to allow adequate consumption by the larvae. If the food desiccates, larvae will not be capable of consuming the total amount and workers or imperfect queens are produced (pers. Obs.).
7. If larval food has been totally consumed after 4 days, the humidity is decreased to ca. 75%, by adding NaCl (18 g of common salt to 50 mL of water in the tray). If the food has not been consumed, one more day could be given to the larvae at 100% humidity. However, monitoring the humidity is important. Under excess humidity, fungi could proliferate, but if humidity is insufficient, the larval tegument may dry (loss of the creamy, shiny aspect), in which case adding more water to the tray is recommended. Larvae should be monitored every day.

When individuals enter the pupal stage, the plates can be checked every 3 days (Fig. 6.18). After approximately 44 days, gynes of *Scp. pectoralis* complete their development. Artificially produced gynes can be placed in small plastic boxes, with a group of workers and syrup in Eppendorf tubes, where they can live up to 2 weeks (pers. Obs.). Significant advances have been made to understand the reproductive strategies of stingless bees. Manipulating reproduction opens the possibility of preserving species and populations, and also for commercial use. However, significant gaps are still left to fully comprehend reproduction in these insects that may support sustainable management.

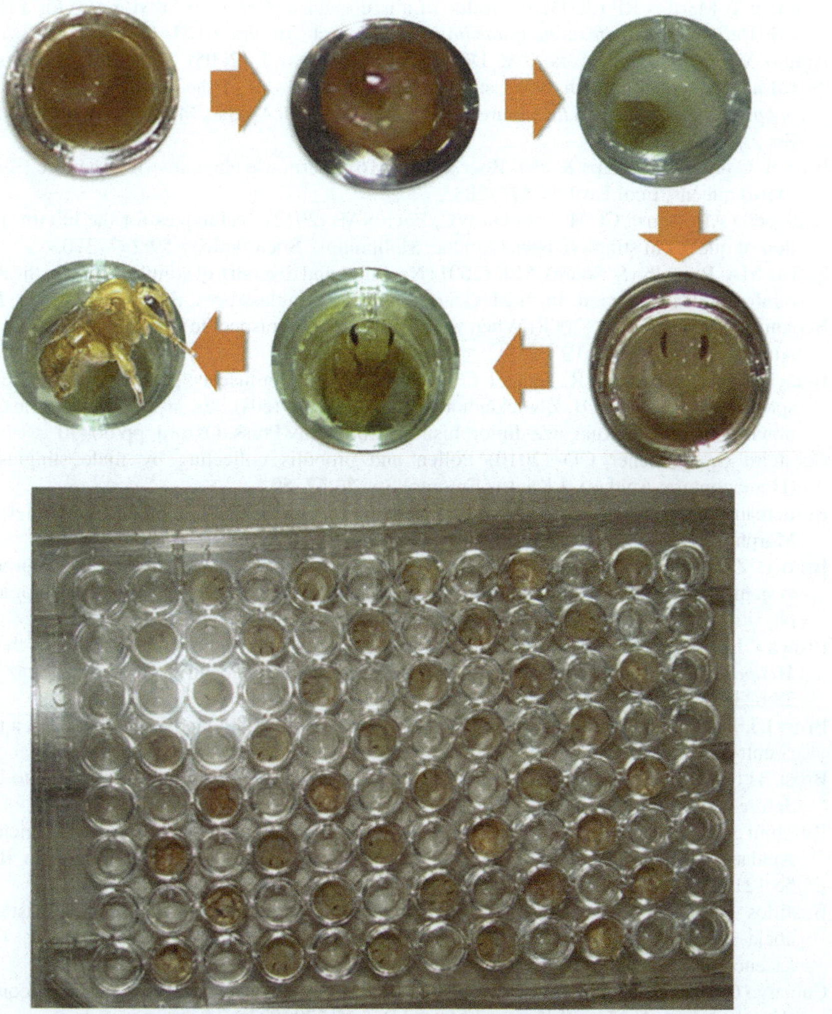

Fig. 6.18 Above: Different phases of artificial *S. pectoralis* gyne production. Below: ELISA plate with *S. pectoralis* gynes in their final pupal stage

References

Abreu SAB, Penna MNP (1992) Biologia reprodutiva de meliponíneos. Naturalia, Ed. Especial Encontro Brasileiro sobre Biologia de Abelhas e Outros Insetos Sociais

Alcock J (1997) Competition from large males and the alternative mating tactics of small males of Dawson's burrowing bee (*Amegilla dawsoni*) (Apidae, Apinae, Anthophorini). J Insect Behav 10:99–113

Alves DA, Imperatriz-Fonseca VL, Francoy TM, Santos-Filho PS, Nogueira-Neto P, Billen J, Wenseleers T (2009) The queen is dead—long live the workers: intraspecific parasitism by workers in the stingless bee *Melipona scutellaris*. Mol Ecol 18:4102–4111

Antonini Y, Martins RP (2003) The value of a tree species (*Caryocar brasiliense*) for a stingless bee *Melipona quadrifasciata quadrifasciata*. J Insect Conserv 7:167–174
Araújo VA, Zama U, Neves CA, Dolder H, Lino-Neto J (2005) Ultrastructural, histological and histochemical characteristics of the epithelial wall of the seminal vesicle of mature *Scaptotrigona xanthotricha* Moure males (Hymenoptera, Apidae, Meliponini). Braz J Morphol Sci 22:193–201
Baer B, Collins J, Maalaps K, den Boer SPA (2016) Sperm use economy of honeybee (*Apis mellifera*) queens. Ecol Evol 6:2877–2885
Baptistella AR, Souza CCM, Santana WC, Egea SAE (2012) Techniques for the in vitro production of queens in stingless bees (Apidae, Meliponini). Sociobiology 59:297–310
Batista MA, Ramalho S, Soares AEE (2001) Nest sites and diversity of stingless bees in the Atlantic rainforest, Bahia, Brazil. In: II Mexican seminar on stingless bees. F.M.V.Z. U.A.D.Y, Mérida
Beekman M, Oldroyd BP (2008) When workers disunite: intraspecific parasitism in eusocial bees. Annu Rev Entomol 53:19–37
Blanckenhorn WU, Meier R, Teder T (2007) Rensch's rule in insects: patterns among and within species. In: Fairbairn DJ, Blanckenhorn WU, Szekely T (eds) Sex, size & gender roles, evolutionary studies of sexual size dimorphism. Oxford Univ Press, Oxford, pp 60–70
Boongird S, Michener CD (2010) Pollen and propolis collecting by male stingless bees (Hymenoptera: Apidae). J Kansas Entomol Soc 83:47–50
Bordereau C (1982) Ultrastructure of the fat body of the reproductive pair in higher termites. J Morphol 172:313–322
Brito P, Zama U, Dolder H, Lino-Neto J (2010) New characteristics of the male reproductive system in the Meliponini bee, *Friesella schrottkyi* (Hymenoptera: Apidae): histological and physiological development during sexual maturation. Apidologie 41:203–215
Brown CJ, Albrecht C (2001) The effect of tropical deforestation on stingless bees of the genus *Melipona* (Insecta: Hymenoptera: Apidae: Meliponini) in central Rondonia. Braz J Biol 28:623–634
Brosi BJ, Daily GC, Ehrlich PR (2007) Bee community shifts with landscape context in a tropical countryside. Ecol Appl 17:418–430
Brosi BJ (2009) The complex responses of social stingless bees (Apidae: Meliponini) to tropical deforestation. For Ecol Manag 258:1830–1837
Buschini MLT, Campos LO (1995) Caste determination in *Trigona spinipes* (Hymenoptera, Apidae); influence of available food and juvenile hormone. Revista Brasileira de Biología 55:121–129
Bustillos A (2016) Diferencias en tamaño y asimetría fluctuante en individuos y castas de la abeja sin aguijón *Melipona beecheii* Benett (Hymenoptera: Apidae: Meliponini). Tesis de Licenciatura. Universidad Autónoma de Yucatán, Mérida
Camargo CA (1972) Mating of the social bee *Melipona quadrifasciata* under controlled conditions (Hymenoptera, Apidae). J Kansas Entomol Soc 45:520–523
Camargo CA (1984) Spermatozoa numbers and migration to the seminal vesicles in haploid and diploid males of *Melipona quadrifasciata* Lep. J Apic Res 23:15–17
Cameron EC, Franck P, Oldroyd BP (2004) Genetic structure of nest aggregations and drone congregations of the southeast Asian stingless bee *Trigona collina*. Mol Ecol 13:2357–2364
Camposeco F (2005) Conducta y parámetros reproductivos en machos de Melipona beecheii (Hymenoptera: Meliponini) en Yucatán, México. Tesis de Maestría, F.M.V.Z. Universidad Autónoma de Yucatán, Mérida
Carvalho-Zilse GA, Kerr WE (2004) Substituição natural de rainhas fisogástricas e distância de vôo dos machos em Tiuba (*Melipona compressipes fasciculata* Smith, 1854) e Uruçu (*Melipona scutellaris*) Latreille, 1811 (Apidae, Meliponini). Acta Amazon 34:649–652
Chapman RF (1998) The insects: structure and function. Cambridge University Press, Cambridge
Chinh TX, Grob GBJ, Meeuwsen FJAJ, Sommeijer MJ (2003) Patterns of male production in the stingless bee *Melipona favosa* (Apidae, Meliponini). Apidologie 34:161–170

Colonello NA, Hertfelder K (2005) She's my girl - male accessory gland products and their function in the reproductive biology of social bees. Apidologie 36:231–244

Cortopassi-Laurino M (1979) Observações sobre atividades de machos de *Plebeia droryana* Friese (Apidae, Meliponinae). Revista Brasileira de Entomología 23:177–191

Couvillon MJ, Hughes WOH, Perez-Sato JA, Martin SJ, Roy GGF, Ratnieks FLW (2010) Sexual selection in honey bees: colony variation and the importance of size in male mating success. Behav Ecol 21:520–525

Crozier RH, Pamilo P (2003) Evolution of social insect colonies: sex allocation and kin selection. Oxford University Press, Oxford

Cruz-Landim C (2000) Ovarian development in Meliponine bees (Hymenoptera: Apidae): the effect of queen presence and food on worker ovary development and egg production. Genet Mol Biol 23:83–88

Cruz-Landim C (2009) Abelhas: Morfologia e função de sistemas. Editora UNESP, São Paulo

Cruz-Landim C, Reginato RD, Imperatriz-Fonseca VL (1998) Variation on ovariole number in Meliponinae (Hymenoptera, Apidae) queen's ovaries, with comments on ovary development and caste differentiation. Papéis Avulsos de Zoologia 40:289–296

Cruz-Landim C, Yabuki AT, Iamonte M (2003) Ultrastructure of the spermatheca of *Melipona bicolor bicolor* LEP (Hymenoptera, Apinae, Meliponini). Biosci J Uber 19:57–64

Cueva del Castillo R, Fairbairn DJ (2012) Macroevolutionary patterns on bumblebee body size: detecting the interplay between natural and sexual selection. Ecol Evol 2:46–57

Dallacqua RP, Cruz-Landim C (2003) Ultrastructure of the ducts of the reproductive tract of males of *Melipona bicolor* Lepeletier (Hymenoptera, Apidae, Meliponini). Anat Histol Embryol 32:276–281

de Bruijn MML, Sommeijer MJ (2006) How the production patterns of sexuals may contribute to genetic variation in *Melipona favosa*. Proc Nether Entomol Soc Meet 17:79–85

den Boer SAP, Boomsma JJ, Baer B (2009) Honey bee males and queens use glandular secretions to enhance sperm viability before and after storage. J Insect Physiol 55:538–543

dos Santos CF, Menezes C, Vollet-Neto A, Imperatriz-Fonseca VL (2014) Congregation sites and sleeping roost of male stingless bees (Hymenoptera: Apidae: Meliponini). Sociobiology 61:115–118

dos Santos CF, Ferreira-Caliman MJ, Nascimento FS (2015) An alien in the group: eusocial male bees sharing nonspecific reproductive aggregations. J Insect Sci 15:157

dos Santos CF, Imperatriz-Fonseca VL, Arias MC (2016) Relatedness and dispersal distance of eusocial bee males on mating swarms. Entomol Sci 19:245–254

Duchateau JM, Velthuis HHW (1988) Development and reproductive strategies in *Bombus terrestris* colonies. Behaviour 107:186–207

Eltz T, Bruhl AC, Van Der KS, Linsenmal EK (2002) Determinants of stingless bee density in lowland dipterocarp forest of Sabah, Malaysia. Oecologia 131:27–34

Engels W (1987) Pheromones and reproduction in Brazilian stingless bees. Memorias Instituto Oswaldo Cruz, Rio de Janeiro 82(Suppl III):35–45

Engels W, Imperatriz-Fonseca VL (1990) Caste development, reproductive strategies, and control of fertility in honey bees and stingless bees. In: Engels W (ed) Social insects: an evolutionary approach to castes and reproduction. Springer, Berlin, pp 167–230

Fairbairn DJ (1997) Allometry for sexual size dimorphism: patterns and process in the coevolution of body size in females and males. Annu Rev Ecol Syst 28:659–687

Ferreira A, Abdalla FC, Kerr WE, Cruz-Landim C (2004) Comparative anatomy of the male reproductive internal organs of 51 species of bees. Neotrop Entomol 33:569–576

Fischer J, Brosi BJ, Daily GC, Ehrlich PR, Goldman R, Goldstein J, Lindenmayer DB, Manning AD, Mooney HA, Pejchar L, Ranganathan J, Tallis H (2008) Should agricultural policies encourage land sparing or wildlife-friendly farming? Front Ecol Environ 6:380–385

Flores-Grande N, Quezada-Euán JJG., Medina-Medina L (2003) Vuelos de fecundación e inicio de postura en reinas de *Melipona beecheii* en el estado de Yucatán, México. Memorias X Congreso Internacional de Actualización Apícola mayo 2003 Tlaxcala, Tlaxcala

Galindo-López JC, Kraus BF (2009) Cherchez la femme? Site choice of drone congregations in the stingless bee *Scaptotrigona mexicana*. Anim Behav 77:1247–1252

Gonzalez-Acereto JA (2008) Cría y manejo de abejas nativas sin aguijón en México. Universidad Autónoma de Yucatán. Secretaria de Fomento Agropecuario y Pesquero. Fundación Produce Yucatán, Yucatán

Gutiérrez E, Ruiz D, Solís T, May-Itzá WJ, Moo-Valle H, JJG QE (2016) Does larval food affect cuticular profiles and recognition in eusocial bees? A test on *Scaptotrigona gynes* (Hymenoptera: Meliponini). Behav Ecol Sociobiol 70:781–789

Hölldobler B, Wilson EO (1990) The ants. Springer, Berlin

Hubbell SP, Johnson LK (1977) Competition and nest spacing in a tropical stingless bee community. Ecology 58:949–963

Imperatriz-Fonseca VL, Zucchi R (1995) Virgin queens in stingless bees (Apidae, Meliponinae) colonies: a review. Apidologie 26:231–244

Imperatriz-Fonseca VL (1977) Studies on *Paratrigona subnuda* (Moure) (Hymenoptera, Apidae, Meliponinae)—II: behaviour of the virgin queen. Bol Zool Univ São Paulo 2:169–182

Jarau S, van Veen JW, Aguilar I, Ayasse M (2009) Virgin queen execution in the stingless bee *Melipona beecheii*: the sign stimulus for worker attacks. Apidologie 40:496–507

Kärcher MH, Menezes C, Alves DA, Beveridge OS, Imperatriz-Fonseca VL, Ratnieks FLW (2013) Factors influencing survival duration and choice of virgin queens in the stingless bee *Melipona quadrifasciata*. Naturwissenschaften 100:571–580

Kerr WE (1949) Algumas comparações entre a abelha européia (*Apis mellifera*) e as abelhas Brasileiras (Meliponini). O Solo 1:40–47

Kerr WE (1987) *Biologia, manejo e genética de* Melipona compressipes fasciculata *Smith (Hymenoptera, Apidae)*. Tese de Professor Titular. Universidade Federal de Maranhão, São Luíz

Kerr WE (1996) Biologia e manejo da tiuba: a abelha do Maranhão. EDUFMA, São Luíz

Kerr WE (2002) Extincão de especies: a grande crise biologica do momento e como afeta os meliponinios. Anais do V encontro sobre abelhas, Riberão Preto, pp 4–9

Kerr WE, Zucchi R, Nakadaira JT, Butolo JE (1962) Reproduction in the social bees (Hymenoptera, Apidae). J NY Entomol Soc 70:265–276

Kerr WE, Carvalho GA, Nascimento VA (1996) Abelha uruçu: biologia, manejo e conservação. Fundação Acangaú, Paracatu

Koedam D, Aguilar MI, Sommeijer MJ (1995) Social interactions of gynes and their longevity in queenright colonies of *Melipona favosa* (Apidae, Meliponinae). Neth J Zool 45:480–494

Koedam D, Contrera FAL, Imperatriz-Fonseca VL (1999) Clustered male production by workers in the stingless bee *Melipona subnitida* Ducke (Apidae, Meliponinae). Insect Soc 46:387–391

Koeniger G (1986) Reproduction and mating behavior. In: Rinderer TE (ed) Bee genetics and bee breeding. Academic Press, New York, pp 255–282

Koffler S, Meneses HM, Peixoto Kleinert AdM, Jaffé R (2016) Competitive males have higher quality sperm in a monogamous social bee. BMC Evol Biol 16:195

Kraus FB, Weinhold S, Moritz RFA (2008) Genetic structure of drone congregations of the stingless bee *Scaptotrigona mexicana*. Insect Soc 55:22–27

Landaverde-González P, Enríquez E, Ariza MA, Murray T, Paxton RJ, Husemann M (2017) Fragmentation in the clouds? The population genetics of the native bee *Partamona bilineata* (Hymenoptera: Apidae: Meliponini) in the cloud forests of Guatemala. Conserv Genet 18(3):631–643

Lima MAP, Lino-Neto J, Campos LAO (2006) Sexual maturation in *Melipona mondury* males (Apidae: Meliponini). Braz J Morphol Sci 23:369–375

Lisboa LCO, Serrão JC, Cruz-Landim C, Campos LAO (2005) Effect of larval food amount on ovariole development in queens of *Trigona spinipes* (Hymenoptera, Apinae). *Anat Histol Embryol* 34:179–184

Mac Arthur RH, Wilson EO (1967) The theory of island biogeography. Princeton University Press, New Jersey

Macías-Macías JO, Quezada-Euán JJG (2015) Quantification of brood emergence and distribution of individuals in *Melipona colimana* (Hymenoptera: Meliponini) in temperate climate. Revista Mexicana de Ciencias Pecuarias 6:233–241

Macías-Macías JO, Quezada-Euán JJG, Tapia-Gonzalez JM, Contreras-Escareño F (2014) Nesting sites, nest density and spatial distribution of *Melipona colimana* Ayala (Hymenoptera: Apidae: Meliponini) in two highland zones of Western, Mexico. Sociobiology 61:423–427

Martins GF, Serrâo JE (2004) A comparative study of the ovaries in some Brazilian bees (Hymenoptera; Apoidea). Papéis Avulsos de Zoologia 44:45–53

Medina RG, Fairbairn DJ, Bustillos A, Moo-Valle H, Medina S, Quezada-Euán JJG (2016) Variable patterns of intraspecific sexual size dimorphism and allometry in three species of eusocial corbiculate bees. Insect Soc 63:493–500

Melo G, Buschini MLT, Campos L (2001) Ovarian activation in *Melipona quadrifasciata* queens triggered by mating plug stimulation (Hymenoptera, Apidae). Apidologie 32:355–361

Meneses HM, Koffler S, Freitas BM, imperatriz-Fonseca VL, Jaffé R (2014) Assessing sperm quality in stingless bees (Hymenoptera: Apidae). Sociobiology 61:517–522

Menezes C, Vollet-Neto A, Imperatriz-Fonseca VL (2013) An advance in the in vitro rearing of stingless bee queens. Apidologie 44:491–500

Michener CD (1974) The social behavior of the bees: a comparative study. Belknap Press, Harvard University, Cambridge

Moo-Valle H, Quezada-Euán JJG, Wenseleers T (2001) The effect of food reserves on the production of sexual offspring in the stingless bee *Melipona beecheii* (Apidae: Meliponini). Insect Soc 48:398–403

Moo-Valle H, Quezada-Euán JJG, Canto-Marin J, Gonzalez-Acereto JA (2004) Caste ontogeny and the distribution of reproductive cells on the combs of *Melipona beecheii* (Apidae: Meliponini). Apidologie 35:587–594

Mueller MY, Moritz RFA, Kraus FB (2012) Outbreeding and lack of temporal genetic structure in a drone congregation of the neotropical stingless bee *Scaptotrigona mexicana*. Ecol Evol 2:1304–1311

Nogueira-Neto P (1997) Vida e criação de abelhas indígenas sem ferrão. Editora Nogueirapis, São Paulo

Paes de Oliveira VT, Cruz-Landim C (2003) Size of fat body trophocytes and the ovarian development in workers and queens of *Melipona quadrifasciata anthidioides*. Sociobiology 41:701–709

Paes de Oliveira VT, Roat TC, Berger B, Cruz-Landim Cd (2012) Cytochemistry of fat body trophocytes and ovaries of workers and queens of *Melipona quadrifasciata anthidioides* (Hymenoptera: Apidae: Meliponini) during vitellogenesis. Microsc Res Tech 75:1623–1631

Palmer KA, Oldroyd BP, Quezada-Euán JJG, Paxton RJ, May-Itzá WdJ (2002) Paternity frequency and maternity of males in some stingless bee species. *Mol Ecol* 11:2107–2113

Peters JM, Queller DC, Imperatriz-Fonseca VL, Roubik DW, Strassmann JE (1999) Mate number, kin selection and social conflicts in stingless bees and honeybees. Proc R Soc Lond Ser B 266:379–384

Paxton RJ (2005) Male mating behaviour and mating systems of bees: an overview. Apidologie 36:145–156

Paxton RJ, Weißchuh N, Engles W, Hartfelder K, Quezada-Euan JJG (1999) Not only single mating in stingless bees. Naturwissenschaften 86:143–146

Paxton RJ, Ruhnke H, Shah M, Bego LR, Quezada-Euán JJG, Ratnieks FLW (2001) Social evolution in the stingless bees: are the workers or is the queen in control of male production? In: II Seminario Mexicano sobre abejas sin aguijón. Universidad Autonóma de Yucatán Mérida, Mérida, pp 104–107

Pech-May FG, Medina-Medina L, May-Itza´W d J, Paxton RJ, Quezada-Euán JJG (2012) Colony pollen reserves affect body size, sperm production and sexual development in males of the stingless bee *Melipona beechei*. Insect Soc 59:417–424

Prato M (2010) Ocorrência natural de sexuados, produção in vitro de rainhas e multiplicação de colônias em *Tetragonisca angustula* (Hymenoptera, Apidae, Meliponini). Tesis Universidad de Sao Paulo, Ribeirão Preto

Quezada-Euán JJG, May-Itzá W d J, González-Acereto JA (2001) Meliponiculture in México: problems and perspective for development. Bee World 82:160–167

Quezada-Euán JJG, May-Itzá WdJ, Rincón M, De la Rúa P, Paxton RJ (2012) Genetic and phenotypic differentiation in endemic *Scaptotrigona hellwegeri* (Apidae: Meliponini): implications for the conservation of stingless bee populations in contrasting environments. Insect Conserv Divers 5:433–443

Quezada-Euán JJG, May-Itzá Wd J, Montejo E, Moo-Valle H (2015) Isometric worker size variation in relation to individual foraging preference and seasonal colony growth in stingless bees. Insect Soc 62:73–78

Quijano E, González-Acereto JA, Quezada-Euán JJG (2008) Desarrollo de divisiones de colonias de *Melipona beecheii* (Hymenoptera, Meliponini) a partir de tres tamaños de población. Bioagrociencias 1:4–11

Rinderer TE, Hellmich RL II (1991) The process of Africanization. In: Spivak M, Fletcher DJC, Breed MD (eds) The "African" honey bee. Westview Press, Boulder, pp 95–118

Roubik DW (1982) Seasonality in colony food storage, brood production and adult survivorship: studies of *Melipona* in tropical forest (Hymenoptera: Apidae). J Kansas Entomol Soc 55:789–800

Roubik DW (1983) Nest and colony characteristics of stingless bees from Panamá (Hymenoptera: Apidae). J Kansas Entomol Soc 56:327–355

Roubik DW (1989) Ecology and natural history of tropical bees. Cambridge University Press, Cambridge

Roubik DW (1990) Mate location and mate competition in males of stingless bees (Hymenoptera: Apidae: Meliponinae). Entomolia Generalis 15:115–120

Roubik DW (2006) Stingless bee nesting biology. Apidologie 37:124–143

Ruttner F (1985) Reproductive behavior in honeybees. In: Hölldobler B, Lindauer M (eds) Experimental behavioral ecology and sociobiology, Fortschritte der Zoologie, no. 31. Sinauer Associates, Sunderland, pp 225–236

Sakagami SF (1982) Stingless bees. In: Hermann HR (ed) Social insects, vol III. Academic Press, London, pp 361–423

Samejima H, Marzuki M, Nagamitsu T, Nakashizuka T (2004) The effects of human disturbance on a stingless bee community in a tropical rainforest. J Biol Conserv 120:577–587

Santiago LR, Francisco FO, Jaffé R, Arias MC (2016) Genetic variability in captive populations of the stingless bee *Tetragonisca angustula*. Genetica 144:397–405

Santos LAC (2006) Distribución espacial de los sitios de anidación de abejas eusociales (Hymenoptera: Apidae: Meliponini y Apini) en Sudzal, Yucatán, México. Tesis de Maestría, Universidad Autónoma de Yucatán, Mérida

Schlüns H, Schlüns EA, van Praagh J, Moritz RFA (2003) Sperm numbers in drone honeybees (*Apis mellifera*) depend on body size. Apidologie 34:577–584

Schorkopf DLP (2016) Male meliponine bees (*Scaptotrigon*a aff. depilis) produce alarm pheromones to which workers respond with fight and males with flight. J Comp Physiol A 202:667–678

Seeley TD (1985) Honeybee ecology, a study of adaptation in social life. Princeton University Press, Princeton

Shreeves G, Field J (2008) Parental care and sexual size dimorphism in wasps and bees. Behav Ecol Sociobiol 62:843–852

Slaa EJ (2006) Population dynamics of a stingless bee community in the seasonal dry lowlands of Costa Rica. Insect Soc 53:70–79

Sommeijer MJ, de Bruijn LLM (1995) Drone congregations apart from the nest in *Melipona favos*a. Insect Soc 42:123–127

Sommeijer MJ, de Bruijn LLM (2003) Why do workers of *Melipona favosa* chase their sistergynes out of the nest? Proc Sect Exp Appl Entomol Amst 14:45–48

Sommeijer MJ, de Bruijn LLM, Meeuwsen FJAJ (2003) Reproductive behaviour of stingless bees: solitary gynes of *Melipona favosa* (Hymenoptera: Apidae, Meliponini) can penetrate existing nests. Entomologische Berichten 63:31–35

Sommeijer MJ, de Bruijn LLM, Meeuwsen FJAJ (2004) Behaviour of males, gynes and workers at drone congregation sites of the stingless bee *Melipona favosa* (Apidae: Meliponini). Entomologische Berichten 64:10–15

Teder T (2005) Male-biased size dimorphism in ichneumonine wasps (Hymenoptera: Ichneumonidae)—the role of sexual selection for large male size. Ecol Entomol 30:342–349

Teder T (2014) Sexual size dimorphism requires a corresponding sex difference in development time: a meta-analysis in insects. Funct Ecol 28:479–486

Tóth E, Strassmann JE, Nogueira-Neto P, Imperatriz-Fonseca VL, Queller DC (2002) Male production in stingless bees: variable outcomes of queen-worker conflict. Mol Ecol 11:2661–2667

Van Benthem FDJ, Imperatriz-Fonseca VL, Velthuis HHW (1995) Biology of the stingless bee *Plebeia remota* (Holmberg), observations on evolutionary implications. Insect Soc 42:71–87

van Oystaeyen A, Alves DA, Oliveira RC, do Nascimento DL, do Nascimento SF, Billen J, Wenseleers T (2013) Sneaky queens in *Melipona* bees selectively detect and infiltrate queenless colonies. Anim Behav 86:603–609

van Veen JW (1999) Colony reproduction in stingless bees. PhD Thesis, University of Utrecht, Utrecht

van Veen JW, Sommeijer MJ (2000) Observations on gynes and drones around nuptial flights in the stingless bees *Tetragonisca angustula* and *Melipona beecheii* (Hymenoptera, Apidae, Meliponinae). Apidologie 31:47–54

van Veen JW, Sommeijer MJ, Meeuwsen F (1997) Behaviour of drones in *Melipona* (Apidae, Meliponinae). Insect Soc 44:435–447

van Veen JW, Sommeijer MJ, Aguilar Monge I (1999) Behavioural development and abdomen inflation of gynes and newly mated queens of *Melipona beecheii* (Apidae, Meliponinae). Insect Soc 46:361–365

Velthuis HHW, Koedam D, Imperatriz-Fonseca VL (2005) The males of *Melipona* and other stingless bees, and their mothers. Apidologie 36:169–185

Villanueva-Gutiérrez R, Roubik DW, Colli-Ucán W, Güemez-Ricalde FJ, Buchmann SL (2013) A critical view of colony losses in managed Mayan honey-making bees (Apidae: Meliponini) in the heart of Zona Maya. J Kansas Entomol Soc 86:352–362

Wenseleers T, Hart AG, Ratnieks FLW, Quezada-Euán JJG (2004) Queen execution and caste conflict in the stingless bee *Melipona beecheii*. Ethology 110:725–736

Wenseleers T, Alves DA, Francoy TM, Billen J, Imperatriz-Fonseca VL (2011) Intraspecific queen parasitism in a highly eusocial bee. Biol Lett 7:173–176

Wille A (1983) Biology of the stingless bees. Annu Rev Entomol 28:41–64

Wilson EO (1971) The insect societies. Harvard University Press, Cambridge

Winston ML (1987) The biology of the honey bee. Harvard University Press, Cambridge

Chapter 7
Services Provided by Stingless Bees

Melipona beecheii collecting pollen on achiote (*Bixa orellana*) flower

Social insects have their higher diversity in the tropics (Wilson 1971). The highly eusocial bees (most *Apis* species and meliponines) are almost exclusively tropical, being the most abundant flower visitors in these areas. Both groups of highly eusocial bees are characterized by their large, perennial colonies, and their capacity to store surplus amounts of food in wax made receptacles (Roubik 1989). Colonies of highly eusocial bees can contain thousands of individuals and, because they are perennial, require large supplies of food. To obtain food, stingless bees and honey bees visit a large diversity of flowers, making them the most important generalist pollinators in the tropics (Roubik and Moreno-Patiño 2013) (Table 7.1). However, compared with honey bees, stingless bee colonies have smaller flight ranges, probably a few km^2 (Roubik and Aluja 1983; van Nieuwstadt and Ruano-Iraheta 1996; Kuhn-Neto et al. 2009). Notably, palynological analyses of the diet reveal differences in the use of the flora among stingless bees, indicating preferences for some plant species, even within bees considered floral generalists (Roubik and Moreno-Patiño 2013).

J. J. G. Quezada-Euán, *Stingless Bees of Mexico*,
https://doi.org/10.1007/978-3-319-77785-6_7

Table 7.1 Diversity of plant species visited by *M. beecheii* in Yucatan identified by palynology of honey samples. Stingless bees are generalists, although preferences for some kinds of plants can be evident (Parra-Canto unpublished data)

Family	Species	Life form
Bixaceae	*Bixa orellana*	Bush
Burseraceae	*Bursera simaruba*	Tree
Compositae	*Bidens pilosa*	Herb
Compositae	*Viguiera dentata*	Herb
Euphorbiaceae	*Croton humilis*	Bush
Fabaceae	*Cassia fistula*	Tree
Fabaceae	*Pithecellobium albicans*	Tree
Fabaceae	*Senna racemosa*	Bush
Fabaceae	*Acacia gaumeri*	Bush
Fabaceae	*Delonix regia*	Tree
Fabaceae	*Mimosa bahamensis*	Bush
Fabaceae	*Cassia racemosa*	Bush
Fabaceae	*Senna atomaria*	Tree
Fabaceae	*Piscidia piscipula*	Tree
Malpighiaceae	*Byrsonima crassifolia*	Tree
Myrtaceae	*Psidium guajava*	Tree
Nyctaginaceae	*Boerhavia erecta*	Herb
Polygonaceae	*Gymnopodium floribundum*	Bush
Polygonaceae	*Antigonon leptopus*	Creeper
Rhamnaceae	*Gouania lupuloides*	Creeper
Sapindaceae	*Thouinia paucidentata*	Tree
Solanaceae	*Solanum nigrum*	Bush
Solanaceae	*Solanum rudepanum*	Bush
Verbenaceae	*Aloysia virgata*	Bush

In their constant search for pollen and nectar, the highly eusocial bees provide one of the most important ecosystem services, that is, the pollination of plants (Roubik 1989). Pollination is the transfer of pollen from anthers to stigma between flowers of the same species, being essential to the reproduction of most angiosperms (plants with flowers) (Free 1993; Delaplane et al. 2013). In the tropics, the role of eusocial bees as pollinators is even more relevant because of the higher diversity of flowering plants, and the wider spatial distances between individuals, compared with temperate areas (Bawa 1990). Indeed, the proportion of animal-pollinated plants rises from a mean of 78% in temperate-zone communities, to 94% in tropical communities (Ollerton et al. 2011). In Mexico, an estimated 85% of plants whose fruit and seed are used by humans, depend to some degree on animal pollination (Ashworth et al. 2009). Of the many animal pollinators in tropical areas, bees are the most abundant and diverse, contributing to the reproduction of 50–70% of neotropical plants (Bawa 1990; Biesmeijer 1997).

When visiting flowers to collect resources (pollen, nectar, resins), stingless bees interact with many other species (and conspecifics), which has probably resulted in a diversification of strategies for resource collection (Hrncir and Maia-Silva 2013). Curiously, some species also rob the resources produced by plants, with no apparent benefit to these. For instance, *Tr. ferricauda*, *Tr. fulviventris*, and *Tr. fuscipennis*, use their mandibles to open small holes at the base of the corollas, from where they can obtain nectar, but without making contact with the reproductive parts of the flower. These species can affect resource availability and, can indirectly change the frequency and behavior of legitimate pollinators. Other bees use the holes produced by robber bees to extract nectar too. Although potentially negative, nectar robbing may sometimes be neutral, or even beneficial to plants (Roubik 1982; Murphy and Breed 2008).

In addition to their service as pollinators, stingless bee colonies also produce large amounts of waste (feces and dead individuals). Most species of stingless bee disperse debris over the forest, contributing to the recycle of nutrients. An estimation, using cleptobiotic *Lestrimelitta*, indicates that a colony can produce almost 3 kg (representing 8–10 g per day) of waste annually (Roubik 1989). Thus, a population of stingless bees in a given area, can produce several hundred kg of residues per km^2 annually.

Stingless bees, like other pollinators, have a significant impact on ecosystems. They are considered keystone species for the reproduction of plants and the life of many animals, which depend on them for food in the form of fruit and seed. They are support of the food chain (Biesmeijer 1997). The disappearance of bees could have major consequences on ecosystems, but also on the provision of food, and other commodities on which humankind depends (IPBES 2016).

7.1 Pollination of Crops

Besides their essential role as pollinators of wild plant species, stingless bees are frequent visitors of many crops, which are used for food, medicine, fiber, and fuel (Ashworth et al. 2009). In a first assessment, it was estimated that stingless bees could contribute to the production of ca. 250 plants used by humans (Heard 1999). Noteworthy, to date estimates of the pollination efficiency of stingless bees to different crops are limited. Data on at least 18 cultivated plants have shown a positive effect of stingless bee visitation, but no crops exclusively pollinated by stingless bees have been found (Heard 1999; Slaa et al. 2006). However, in general, information on the pollination requirements for many tropical crops is not available (Roubik 1995), and the effect that stingless bees may have as pollinators of cultivated plants in the tropics, could well be underestimated.

Some attributes make stingless bees effective pollinators, and suitable for commercial use (Roubik 1995; Heard 1999; Slaa et al. 2006); among these:

Floral constancy and specialization: Although stingless bees are generalists, species and colonies tend to visit some plants with higher frequencies than others (Slaa et al. 2006). Individually, foragers also behave as specialists, showing preference for a few flower types. This behavior is possibly related to the capacity of stingless bees to learn, and may increase their efficiency in terms of energy expenditure during the collection of food (Chittka et al. 1999). Stingless bee workers are markedly constant in their daily pollen collection, and up to 97% pollen collected in individual trips come from one plant type. Importantly, this constancy is also evident across different days, and foragers recurrently visit (>70% collecting trips) the same plant species (White et al. 2001). In practical terms, floral constancy enhances pollination efficacy by increasing the chances of pollen being transferred from anthers to stigmas of the same plant. This implies a clear benefit to the productivity of a given crop, and avoids the risk of non-conspecific pollination (Dafni et al. 2005).

One type of behavioral specialization for food collection is the mechanism of pollen acquisition in species of the genus *Melipona.* Workers of *Melipona,* similar to bumble bees, use sonication for the collection of pollen. During flower visitation, the muscles of the thorax vibrate and, through a powerful pulse (sonication), the flower's anthers are shaken, releasing concealed pollen. In various plants (approximately 8–9% of seed-bearing species), pollen is maintained in specialized conical or tubular anthers that do not experience dehiscence when the pollen matures (Buchmann 1983; Free 1993). These anthers bare a pore or orifice on their extreme (poricidal anthers), through which the pollen can be expelled. Plants with this pollination system are called *buzz-pollinated* because sonication or "buzzing" is necessary for effective pollination (De Luca and Vallejo-Marín 2013). Buzz pollination is an example of plant adaptation to restrict access of floral visitors as pollen is a valuable resource (Solís-Montero et al. 2015). The mechanism of buzz pollination is present in at least 21 plant families (De Luca and Vallejo-Marín 2013). In Mexico, economically important buzz-pollinated crops are tomatoes (*Lycopersicon*) and peppers (*Capsicum*), plus crops of regional importance like annatto (*Bixa*) (Caro et al. 2017).

Adaptation to artificial lodgings: Many species of stingless bee can be adapted to nest in wooden hives, which allows their relocation to crops when needed. Many species also adapt well to working in confined areas like greenhouses (Cauich et al. 2004). However, it is important to note that many species do not adapt to wooden lodgings, possibly due to thermoregulatory problems, or behave erratically in confined environments (pers. Obs). In general, species that in natural conditions live in termite mounds or underground experience greater difficulties in adapting to boxes (pers. Obs.). Another important consideration when maintaining nests in artificial lodgings is the availability of queens in some species; this has been discussed in Chap. 6.

Perennial colonies and large populations per colony: Stingless bee colonies are perennial. The queen may be naturally superseded, but the colony persists for a long time (Slaa et al. 2006). However, it is recommended that species considered for pollination should have at least a few hundred adult workers, as small forager populations may limit the capacity to visit significant flower numbers (Roubik 1995). Perhaps small body size could also be an impediment to move across large distances on field crops, but not so in greenhouses (Cauich et al. 2004).

Pollination efficiency: This is one of the most important features of pollinators, however, it is perhaps the least studied. The concept of pollination efficiency refers to the capacity of floral visitors to positively contribute to the reproduction of the plant (Ne'eman et al. 2010). In terms of crop productivity, pollination efficiency is, thus, directly related to the amount of fruit and seed produced and their quality.

Under field conditions, it is common that flowers of any given crop are visited by different insect species. Therefore, assessing the relative pollination efficiency of a particular visitor is no easy task. However, the evaluation of efficiency is important in crop management because not all flower visitors may contribute equally to plant reproduction (Delaplane et al. 2013).

Most angiosperms need pollination to set seed, but benefit differently from animal vectors, depending on their reproductive system (Delaplane et al. 2013). One system of reproduction, called xenogamy (outcrossing), requires the transfer of pollen between flowers of different plants of the same species (cross-pollination). Another system called autogamy (selfing) occurs when fertilization takes place on the same flower (self-pollination), or between different flowers on the same plant (geitonogamy). Mixed mating systems, in which plants use outcrossing and autogamy, are common too (Meléndez-Ramírez et al. 2004; Rizzardo et al. 2012). Evidently, strict xenogamic plants need pollen vectors that move between plants, while geitonogamic or autogamic plants might obtain more benefit from vectors visiting flowers on the same plant. Thus, a flower visitor must meet the specific reproductive needs of a plant to qualify as a legitimate pollinator (Delaplane et al. 2013).

Evaluation of the pollination efficiency of individual species requires carefully conducted experiments to test their capacity to produce fruit and/or seed (Roubik 1995). In nature, fruit set usually happens after repeated flower visits, but when evaluating different candidate pollinators an equitable way to compare them is using single-flower visits (Delaplane et al. 2013). It is also important to compare the results of single floral visits with flowers that were not visited and flowers visited without restriction. A standard method to conduct such comparisons involves the use of bags to control the access of different visitors to "virgin flowers," i.e., not previously visited (Free 1993; Delaplane et al. 2013).

Bagging experiments start when flowers are in the bud stage. On the day of flower anthesis, bags are removed and flowers are constantly observed to detect when they are first visited. This entails some flowers being visited by insects different to the one(s) being assessed, so sufficient virgin flowers should be available. Visits are genuine only if the vector touches the stigma. After a single visit by the insect under evaluation, the flower is bagged again to stop other visitors from diluting the effect (Dafni et al. 2005; Ne'eman et al. 2010). Variables used as indicators of efficiency are fruit set (an initial number of formed fruit), number and weight of ripe fruit, as well as number and weight of seeds plus germination rate (Free 1993; Delaplane et al. 2013). For unbiased comparisons, it is important to include replicates of each treatment (restricted, unrestricted and vector visited flowers) on each of several plants. To calculate pollination efficiency, the results of individual floral visitors are compared with unrestricted and non-visited flowers (Fig. 7.1).

Fig. 7.1 Evaluation of pollination efficiency. A crop with sufficient number of flowering plants (**a**) is identified. Select at least ten plants to bag flower buds. When flowers open, three treatments can be used: unrestricted visits, permanently bagged (zero visits), and single visits by the pollinator under study; a minimum of 3–5 flowers of each treatment should be used on each experimental plant (**b**). Flowers are unbagged on the day of anthesis, and after being visited, they are bagged again (**c**). Different treatments should be unmistakenly identified (**d**). The quantity and quality of fruit and seed produced (**e**), can be compared among treatments

These parameters have been combined in an index that is useful to compare the effect of different floral visitors (Spears 1983). The formula of Spear's index is

$$(PE)=(Pi)(Z)/(U)(Z)$$

Where

Pi is the average number of seed in flowers that received a single visit by potential pollinator *i*.
Z is the average number of seed produced in flowers that received no visits (bagged control).
U is the average number of seed in flowers with unrestricted visits.

One could also introduce a treatment "hand cross-pollination," which is the theoretical maximum of a plant. Replacing *U* in the Spear's index with this new treatment gives a measure of the "ecosystem service of pollination" (Landaverde et al. 2017).

For a comprehensive comparison of efficiency, it is also possible to include data on visitation rates per unit time, and the speed at which each visitor works on flowers (Essenberg 2012). Such additional parameters could indicate the numerical importance of visitors. For instance, a visitor with a modest individual pollination index could significantly contribute to crop pollination due to sheer abundance, or by visiting more plants per unit time, compared with more efficient, but less frequent visitors.

As pollinator efficiency is not easily assessed, the frequency and behavior of different visitors may serve as indicators of pollination effectiveness (Ne'eman et al. 2010). Useful indicators of pollination service are, the time of contact with the stigma, number of pollen grains transported on the body, and the quantity of pollen deposited on stigmas after single visits, among others (Dafni et al. 2005; Ne'eman et al. 2010; Delaplane et al. 2013).

7.2 Stingless Bees as Pollinators of Crops in Mexico

The geographic distribution of many stingless bee species in Mexico overlap with important agricultural areas (Ayala 1999). However, accurate estimation of their presence on crops across different geographic areas is still lacking (Fig. 7.2). Nevertheless, the number of studies assessing plant pollination needs and the effect of native bees is rapidly increasing, as well as the possibility of using stingless bees in commercial pollination (Quezada-Euán 2005; Cauich et al. 2004; Caro et al. 2017). In the following paragraphs, information is presented on various crops on which the effect stingless bees has been evaluated in Mexico.

*7.2.1 Avocado (*Persea americana*)*

Mexico is the main world producer of this native fruit with ca. 1467 thousand tons per year, which represented almost 53% of the world production in 2013. In Mexico, avocado cultivated area is close to 170,000 ha and is growing by ca. 1% annually (SHCP 2014).

Avocado flowers are hermaphrodite and dichogamous (flowers have stigma and anthers maturing at different times). Avocado flowers have asynchronic protogyny, meaning that first open as female and, after a period of time they close, opening later as males. Because both sexes are not active at the same time, self-fertilization (and geitonogamy) is unlikely, and cross-pollination is required (Ish-Am and Eisikowitch 1992). In addition to such flowering peculiarity, some trees will open first in the

Fig. 7.2 Crops in the Yucatan on which the efficiency and effectiveness of stingless bees have been assessed. (**a**) Avocado: *F. nigra* transports large amounts of pollen. (**b**) Achiote: *M. beecheii* sonicates the anthers of flowers making it an efficient pollinator. (**c**) Sikil-té (*J. curcas)*: *F. nigra* is a frequent and efficient visitor of this crop used as biofuel. (**d**) Habanero pepper: *N. perilampoides* is an efficient pollinator in greenhouses. (**e**) Seeds of habanero peppers with and without pollination by *N. perilampoides*. (**f**) Other cultivars of local importance are visited by stingless bees, like Yuca (*Pachyrhizus erosus*), but they probably act as commensals of more efficient visitors

morning as functionally female, close, and reopen in the afternoon as functionally male (type A flowering tree). Other trees open first in the afternoon as functionally female, close at night, and reopen the next morning as functionally male (type B flowering sequence). Because of this particular system of reproduction, it is necessary to have plants of both types, A and B, in the same plantation to ensure adequate levels of cross-pollination (Lahav and Gazit 1991).

The main region for avocado production in Mexico is the state of Michoacán on the Pacific coast with almost 90% of the national production. The variety known as Hass is the most important, in terms of planted area and economic value (Fig. 7.3, SHCP 2014). Hass avocado is of temperate origin, but there are many other avocado varieties in Mexico; some of them tropical and well adapted to poor soils, like those of the Yucatan Península (Barrientos et al. 1991). There are some accounts of the insect fauna visiting avocado in Mexico. In avocado cultivars on the Pacific coast (Michoacán and Chiapas), and along the Gulf coast (Puebla and Veracruz), the diversity of flower visitors was studied. A high frequency of stingless bees was detected in both areas, as well as different wasps and flies (Castañeda-Vildózola et al. 1999; Ish-Am et al. 1999). Nine species of stingless bee were seen on avocado flowers, however, only *Pt. bilineata* was frequent at all locations, while *Geotrigona acapulconis* and *N. perilampoides* were present on both coasts. Visitors only seen in

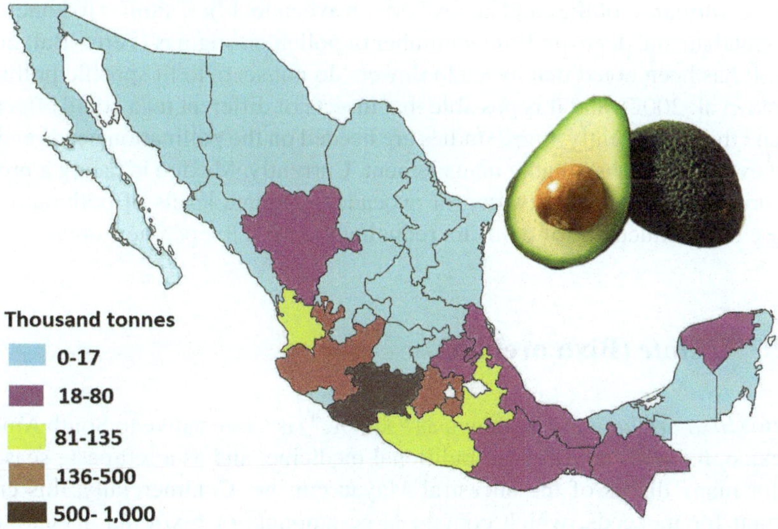

Fig. 7.3 Annual production in thousands of tons of the states that cultivate avocado in Mexico (SIAP 2014)

Veracruz include *Pb. frontalis*, *Scp. mexicana*, *Scp. pectoralis*, *Tr. fulviventris*, *Tr. nigerrima*, and *Fr. nigra*.

Due to their possible contribution to avocado pollination, several attempts have been made to import colonies of *Scp. mexicana* to Israel, one of the world's top avocado exporters. There, the bees frequently visited the flowers, but the intense heat (>43 °C) caused a collapse of their colonies (Gazit and Ish-Am 2007).

In Yucatan, Can-Alonzo et al. (2005) surveyed the bee fauna on "criollo" avocados, a West Indian variety. They found that *Fr. nigra* and *N. perilampoides* were the two most frequent species in those avocado cultivars (Fig. 7.2). The number of pollen grains was counted on the bodies of both bee species, resulting in significantly larger quantities on *Fr. nigra* compared with *N. perilampoides.* In addition, the intense activity even under conditions of high humidity, suggests that *Fr. nigra* could be a good avocado pollinator. Interestingly, in a further study in the same region, no stingless bees were observed visiting avocado flowers, although two colonies of *Fr. nigra* were found nesting in the area (Pérez-Balam et al. 2012). It would be important to analyze the floral preferences of these bee species in the presence of potentially competitive blossoms. For instance, honey bees abandon avocado flowers when more attractive flowers are present (Ish-Am and Eisikowitch 1998). This seems related to the high concentration of potassium salts in avocado nectar that repels honey bees. Interestingly, some native pollinators (including *Fr. nigra*) better tolerate salt concentration of avocado nectar (Afik et al. 2014). There is also a possibility that salts and other compounds in the nectar could act differently on visitors. Flies seem particularly attracted to avocado flowers and can be frequently found at high numbers (Castañeda-Vildózola et al. 1999; Ish-Am et al. 1999; Pérez-Balam et al. 2012). A study in the Yucatan Peninsula, found evidence of similar pollinator performance of flies and honey bees on avocado. Flies, similar to honey bees, were abundant and deposited equal number of pollen on stigmas (Pérez-Balam et al. 2012). It has been noted that avocado flowers do not seem to fit specific pollinators (Wysoki et al. 2002), and it is possible that insects of different taxa could effectively pollinate them. Evidently, more studies are needed on the pollination needs and visitors of avocado for sustainable management. Currently, Mexico is facing a problem of deforestation for the cultivation of avocado, adequate levels of pollination may increase the productivity of cultivars reducing the need to open new areas.

*7.2.2 Achiote (*Bixa orellana*)*

Annatto (*Bixa orellana*) (also known as "achiote") is a tree native to South America. In Mexico, its seeds are used in traditional medicine, and as a red paste seasoning base for many dishes of the ancestral Mayan cuisine. Commercially, this crop is important for its seeds, which contain large amounts of bixin, an apocarotenoid widely used as a natural colorant in the food industry (Caro et al. 2017). In Yucatan, as in most of Mexico, annatto production is predominantly small scale, crop yields are mainly for local consumption, and no sufficient production is obtained to cover the demand (Rivera-Madrid et al. 2006).

This plant has perfect (hermaphrodite) flowers and is visited by many species of bee, although it does not produce nectar (Roubik, 1995). Annatto's anthers (ca. 300) have a unique curved horseshoe-like theca with slit-like openings on top that require sonication for pollen release (Venkatesh 1956). Other flower characteristics, such as heterostylia (style longer and above anthers), make autogamy difficult (Caro et al. 2017). Controlled pollination experiments have shown that this plant is predominantly xenogamic, requiring cross-pollination for reproduction (Rivera-Madrid et al. 2006), but some level of auto-compatibility could also exist (Joseph et al. 2012; Caro et al. 2017).

It has been reported that, in various regions of the neotropics, achiote is not attractive to honey bees (Roubik 1995); however, in southern Mexico, honey bees are numerically dominant on flowers. Nevertheless, honey bees do not have the capacity to sonicate, and could be less efficient on achiote (Caro et al. 2017). In Yucatan, a study was conducted to compare the pollination efficiency of honey bees and native buzz-pollinating *M. beecheii* on achiote. The results, comparing the amount of pollen grains adhering to stigmas and that develop tubes, the number and weight of fruit, as well as the number and weight of seed, showed that *M. beecheii,* on individual visits, is a more efficient pollinator of this crop (Caro et al. 2017) (Figs. 7.2 and 7.4). However, given the sheer abundance on the crops, honey bees could act as pollination facilitators. Notably, honey bees were 3–4 times more abundant than stingless bees on the crop (Caro et al. 2017). A synergy between *M. beecheii* releasing pollen from achiote flowers, and honey bees gleaning and moving it across flowers, could be possible.

In general, the relationship between genuine visitors like *M. beecheii,* and facilitators like the honey bee, has been little studied on any crop. On one side, it is possible

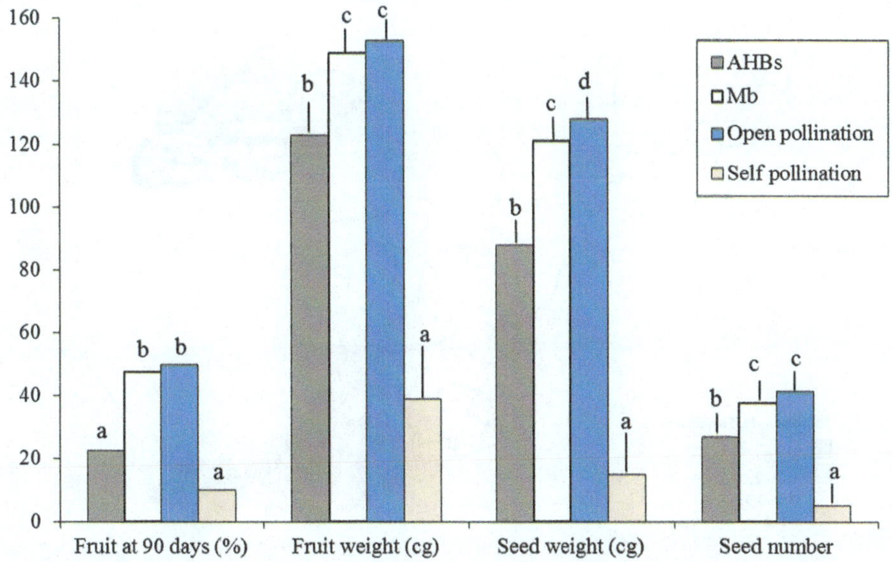

Fig. 7.4 Pollination efficiency of *M. beecheii* (Mb) compared with Africanized honey bees (AHBs), open pollination, and bagged flowers in achiote (*B. orellana*) in Yucatan (Caro et al. 2017). Different letters indicate significant differences among treatments at $p < 0.01$

that in spite of being less efficient on individual visits, honey bees may contribute to pollination as a result of sheer number on flowers. However, such large numbers could also disturb the behavior of more efficient pollinators and negatively affect productivity (Roubik 2000; Pacheco-Filho et al. 2015; Solís-Montero et al. 2015).

7.2.3 *Coffee (*Coffea arabica *and* C. canephora*)*

Mexico is the fifth largest world producer of coffee and the third largest exporter of organic coffee, almost 1.5 million people depend economically on coffee cultivation (Vergara and Badano 2009). In Mexico, the state of Chiapas, on the Isthmus of Tehuantepec, is the largest producer of coffee (Fig. 7.5).

Coffee has hermaphrodite flowers and is self-compatible. It has been generally assumed that coffee is predominantly wind pollinated; however, cross-pollination mediated by bees has a significantly positive effect on the productivity of some varieties, especially *C. canephora* (Roubik 2002; Klein et al. 2003). In Mexico, the main floral visitor of coffee grown under shade, is *A. mellifera*, reaching up to 84% of total visits. However, the stingless bee *Tr. corvina* has also been reported as an important visitor, with 6% of total visits. Other species of stingless bee also visit coffee flowers, but at lower frequency, among them, *Scp. mexicana, Tr. niguerrima*, and *Pb. frontalis* (Vergara and Badano 2009; Badano and Vergara 2011). In general, the diversity of bees found in different studies on coffee in Mexico, is low (between

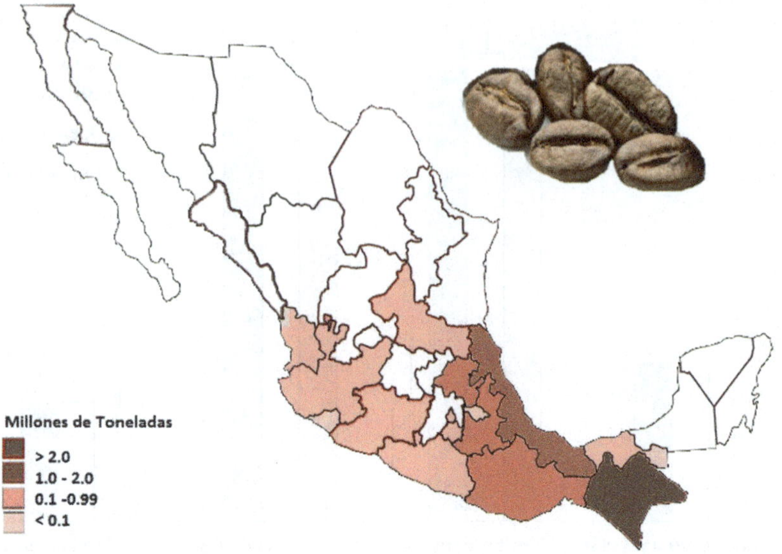

Fig. 7.5 Annual production in millions of tons (Millones de toneladas) of the main states cultivating coffee in Mexico (SIAP 2014)

five and seven species), compared with over a few dozen registered in Central and South America (Roubik 2002; Klein et al. 2003; Ricketts 2004; Veddeler et al. 2008).

One possible explanation of the low frequency of native bees in Mexican orchards could be the sheer abundance of honey bees, but also the fragmented landscape around them (Badano and Vergara 2011; Saturni et al. 2016). The effect of other factors, such as the use of pesticides, has not been documented. This aspect is of relevance because a low frequency of native bees has been correlated with low coffee productivity (Vergara and Badano 2009; Badano and Vergara 2011). The introduction of colonies of stingless bees seems a good alternative to increase the number of flower visitors. Indeed, Arzaluz et al. (2002) divided colonies of *Scp. mexicana* in coffee plantations, and colonies grew rapidly, suggesting a good contribution of resources from this plant. Notably, in South America it is the high frequency of honey bees which has been associated with large production of this crop (Roubik 2002). More studies are needed in Mexico to understand the relative effect of native bees and honey bees, plus different environmental factors on coffee production.

7.2.4 Chillies and Peppers (Capsicum)

The different species and varieties of pepper are among the most important crops in Mexico. The country is the center of origin of *C. annuum*, of which there are at least 120 different varieties, including cultivated (*C. annuum* var. *annuum*) and wild

Fig. 7.6 Annual production in thousands of tons (Miles de toneladas) of the main states cultivating *Capsicum annuum* in Mexico (SIAP 2014). *NS* Not significant

types (*C. annuum* var. *glabriusculum*). Introduced species also have great economic and cultural importance, like habanero pepper (*C. chinense*; Conabio 2016). Hot peppers are an essential part of Mexican gastronomy and one distinctively recognized trait worldwide. The most important pepper-producing regions in Mexico are divided in accordance with the species and varieties cultivated. The central, coast and northern states produce mainly *C. annuum* with its different types (serrano, poblano, chipotle, guajillo, etc.) and, the Yucatan Peninsula is the center of production of habanero pepper (Fig. 7.6).

The Mexican production of hot peppers is only sufficient to cover half of the country's demand, the rest is imported from China and India. In spite of the clear importance of chillies and peppers for the national economy, there is only little information on the pollination needs and flower visitors (Macías-Macías et al. 2009). In other countries, like Brazil, a diverse bee fauna has been recorded on *C. annuum* (Raw 2000).

In the Yucatan Peninsula, two native varieties of *C. annuum* are most consumed, one domesticated (*xkat-ik*) and the other wild (*max-ik*). However, the most representative and economically important species in this region is the South American native habanero pepper (*C. chinense*). In 2008, the Yucatan Peninsula obtained a certificate of origin for the cultivation of habanero pepper, which triggered field and greenhouse cultivation, the latter mainly for export.

Pepper flowers are hermaphrodite and potentially self-compatible. However, flowers have poricidal anthers that require specialized visitors for pollen release (Delaplane and Mayer 2000). Bees are frequent visitors of hot pepper flowers (Raw 2000; Macía-Macías et al. 2009). In the case of *C. chinense*, studies have been conducted to evaluate pollination needs under greenhouse and open-field conditions

(Cauich et al. 2006; Macías-Macías et al. 2009). In greenhouses, the stingless bee *N. perilampoides* was an efficient pollinator of *C. chinense*, comparatively similar to bumble bees and artificial vibration (Palma et al. 2008). The mechanism that *N. perilampoides* bees use to collect the pollen from Solanaceae (including tomato as explained in a following section) is not clear because, apparently, they are not capable of sonication (Slaa et al. 2006). It seems that bees may "milk" anthers with their mandibles to extract pollen (pers. Obs.) (Fig. 7.2). This is similar to the behavior of honey bees to obtain pollen from tomato (Banda and Paxton 1991). Although excessive biting of the anthers can cause flower abortion in Solanaceae (Palma et al. 2008), the action by small *Nannotrigona* bees may not inflict great damage.

Under open-field conditions, stingless bees have been seen visiting habanero pepper flowers. One study in Dzununcán, south of Mérida, reported the presence of *Pb. frontalis, Fr. nigra, N. perilampoides, Pt. bilineata*, and *Tr. fulviventris* (Ancona-Xiu 2007). However, stingless bees were less frequent compared with specimens of *Ceratina, Exomalopsis*, and halictids. Nonetheless, a more extensive survey encompassing 37 localities, revealed that the frequency of stingless bees on habanero flowers is variable depending on the degree of fragmentation of the habitat around crops (Landaverde-González et al. 2017). Presumably, habitat fragmentation reduces the opportunities for nesting of stingless bees, compared to solitary species. In addition to the stingless bee species reported by Ancona-Xiu (2007), Landaverde-Gonzáles et al. (2017) also found *Scp. pectoralis, Tr. fuscipennis*, and *Trigonisca* on habanero pepper flowers.

The efficiency of different pollinators on habanero pepper was studied in field conditions by Macías-Macías et al. (2009); the most efficient bee visitors were various halictids and the anthophorid *Exomalopsis*, all capable of sonication (Macías-Macías et al. 2009; Landaverde-González et al. 2017). However, these species are ground nesting, and therefore, difficult to rear for commercial pollination, especially under greenhouse conditions. In the Yucatan Peninsula. For the pollination of greenhouse *C. chinense*, a good option is the stingless bee *N. perilampoides* (Cauich et al. 2006; Palma et al. 2008), or possibly *M. beecheii*, although this species seems more difficult to adapt to closed spaces (pers. Obs.). The introduction of bumble bees from other regions or from abroad is not a good alternative for the Yucatan Peninsula (where there is only one native species of bumble bee), because of the risk of introducing exotic species that could scape and outcompete the native ones, or bring pathogens in (Winter et al. 2006; Torres-Ruiz et al. 2013).

7.2.5 Cucurbits

The cucurbits encompass a large group of climbing plants whose fruits, seeds, and flowers are greatly appreciated in Mexico. Of particular importance are the domesticated species and varieties of squashes of the genus *Cucurbita* native to Mesoamerica: calabaza (*C. moschata*), calabacín (*C. pepo*), calabaza pipiana or rayada (*C. argyrosperma*), and chilacayote (*C. ficifolia*). In addition to species of

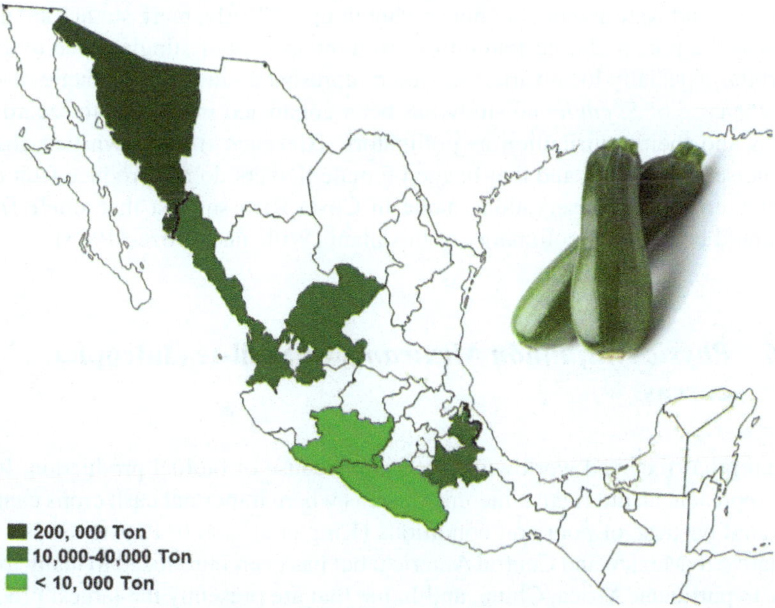

Fig. 7.7 Annual production (in thousands of tons) of the main states cultivating calabacín (*Cucurbita pepo*) in Mexico (SIAP 2014)

the genus *Cucurbita*, species of the genus *Sechium*, especially *S. edule*, locally known as "chayotes," are grown in different parts of the country. Mexico is the center of origin and leading world exporter of chayotes (SIAP 2014). Of great importance are also introduced cucurbit species like watermelon (*Citrullus lanatus*), cucumber (*Cucumis sativus*), and cantaloupe melon (*Cucumis melo*). The main states for the production of cucurbits are Sonora and Sinaloa on the Northern Pacific coast (Fig. 7.7).

In the case of the genus *Cucurbita*, all species are monoecious (separate female and male flowers on the same plant) and, thus, necessarily require pollen vectors for fertilization. Bees are effective pollinators of cucurbits. In particular the genera *Peponapis* and *Xenoglossa* evolved with *Cucurbita* and are adapted specialized pollinators (Canto-Aguilar and Parra-Tabla 2000; López-Uribe et al. 2016). Commercially, honey bees are regarded as efficient pollinators of these plants (Free 1993; Delaplane and Mayer 2000).

Interestingly, although specialized bees are highly efficient on cucurbits, many nonspecialized visitors are frequent on flowers and, could serve as alternative pollinators when specialized visitors are scarce (Canto-Aguilar and Parra-Tabla 2000; Winfree et al. 2007).

In Yucatan, a study of the bee fauna visiting different *Cucurbita,* showed that stingless bees (*Pt. bilineata* and *Tr. fulviventris*) are frequent visitors. However, the dominant bees across diverse cucurbit crops were species of *Ceratina* and halictids (Meléndez-Ramírez et al. 2002). Nonetheless, *Pt. bilineata* was present on squash, cucumber, and watermelon, but not on cantaloupe. Clearly, more studies are needed to assess the potential contribution of different species of stingless bee to various cucurbits, especially local varieties, which represent an important genetic resource.

In the case of *S. edule,* no study has been conducted in Mexico to record floral visitors and their contribution as pollinators. However, it is known that chayotes have unisexual flowers and that bagged female flowers do not produce fruit (Wille and Orozco 1983). Observations made in Costa Rica suggest that small *Trigona* bees could be efficient pollinators of this plant (Wille and Orozco 1983).

7.2.6 Physic Nut, Pinón Mexicano, or **Sikil-té** *(***Jatropha curcas***)*

This crop has received worldwide attention because of biofuel production. Its oily seeds represent an alternative income in areas where important cash crops cannot be cultivated because of poor soil conditions (King et al. 2009). Physic nut is a small tree native to Mexico and Central America, but has been introduced to many tropical areas, in particular Africa, China, and India, that are presently the largest producers (Ye et al. 2009). In Mexico, extensive cultivation of *J. curcas* started recently, and the states of Sinaloa and Tamaulipas, on both coasts, have the largest cultivated area (Fig. 7.8; SIAP 2014).

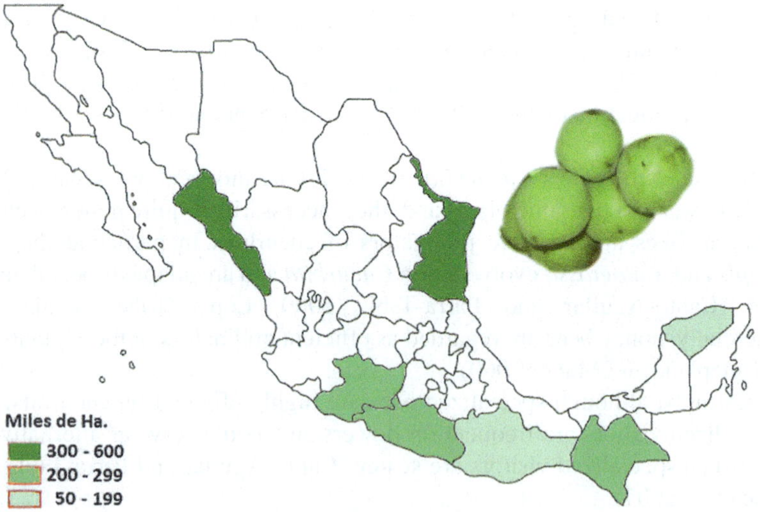

Fig. 7.8 Cultivated area in thousands of ha (Miles de Ha.) of the main states cultivating physic nut in Mexico (SIAP 2014)

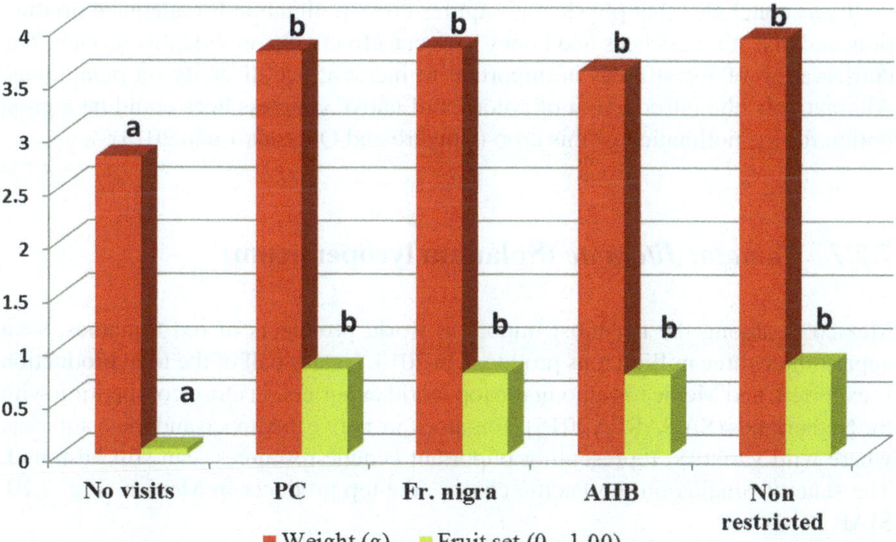

Fig. 7.9 Pollination efficiency of *Fr. nigra* compared with Africanized bees (AHB), unrestricted pollination, no pollination, and hand cross-pollination (PC) of *J. curcas*. The red bars indicate weight (g) and the yellow bars fruit set (%). Different letters indicate significant differences at $p < 0.01$ (data from Romero and Quezada-Euán 2013)

Sikil-té is a monoecious plant (unisexual flowers present on the same plant). The plant produces panicles with flowers of both sexes but also exhibits protandry; male flowers opening before female ones (Achten et al. 2010). It seems that insect-mediated pollination is needed for effective fertilization. Apomixis (seed production without fertilization) may be common in some varieties of physic nut, too (Abdelgadir et al. 2012).

In the Old World, different species of honey bee are the main visitors of physic nut. In a first study in Mexico, Romero and Quezada-Euán (2013) surveyed the insect fauna visiting flowers of *sikil-té* in Yucatan. Honey bees and stingless bees were the two main insect groups present on the flowers. Among the stingless bees, the most abundant species was *Cph. zexmeniae*, followed by *Fr. nigra*, *Tr. fulviventris*, and *N. perilampoides*. However, *Cph. zexmeniae* was not present at all surveyed localities, and higher bee diversity was found in plantations surrounded by secondary forest. Comparison of the quality of fruit and seed produced after single visits by Africanized bees and the stingless bee *Fr. nigra* (Fig. 7.9), suggested that both bee species were similarly efficient. Moreover, their effect was similar to that found in flowers visited without restriction (Fig. 7.8). Nonetheless, *Fr. nigra* was significantly more frequent on flowers, being potentially more important in providing pollination services (Romero and Quezada-Euán 2013).

It was concluded that physic nut requires cross-pollination for adequate production, and that stingless bees and honey bees can effectively provide this service. The conservation of forest areas is important to increase bee diversity on plantations. Alternatively, the introduction of colonies of native stingless bees could be a good option for the pollination of this crop (Romero and Quezada-Euán 2013).

7.2.7 *Tomato: Jitomate (***Solanum lycopersicum***)*

Mexico is among the ten most important world producers of red tomatoes, with approximate three million tons produced in 2014. Nearly half of the total production is exported, and Mexico is among the top world exporters, in close competition with the Netherlands (SAGARPA 2015). Tomatoes are native to Mexico and Mesoamerica, where wild varieties, representing important genetic resources, can still be found. The state of Sinaloa on the Pacific coast is the top producer in Mexico (Fig. 7.10, SIAP 2014).

Tomato flowers are self-compatible, but the anthers are poricidal and require vibration for pollen release and effective pollination. Positive contribution of insect visitors has been demonstrated for tomato production (Delaplane and Mayer 2000). In Mexico, tomato cultivation in greenhouses has experienced a dramatic increase and, presently it represents almost 75% of protected crops. As a result of this rapid growth, the demand for pollinators of tomato in greenhouses has increased in parallel. Commercially, honey bees have been used for tomato pollination in green-

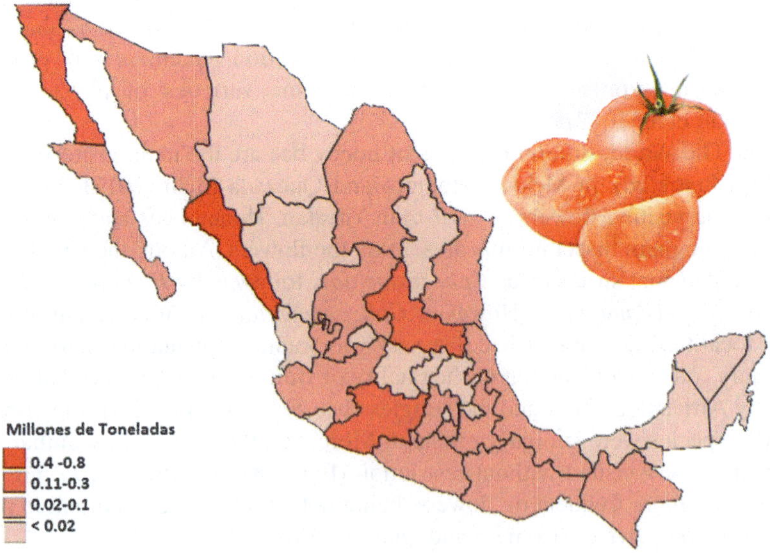

Fig. 7.10 Annual production in millions of tons (Millones de toneladas) of the main states cultivating tomato in Mexico (SIAP 2014)

houses. Interestingly honey bees are effective tomato pollinators, in spite of their inability to sonicate flowers. Nonetheless, buzz-pollinating bees, like bumble bees, are comparatively more efficient and preferred for tomato production (Banda and Paxton 1991). Initially, importing bumble bees from Europe and the United States seemed like a good and rapid alternative to supply the demand for greenhouse tomato pollinators. Nonetheless, there are negative aspects of importing exotic bumble bees, like pathogen spread, naturalization, and competition with native species (Arbetman et al. 2013; Schmid-Hempel et al. 2014). In Mexico, there is also a potential risk of hybridization, as commercially used species from the United States are related to Mexican species (Duennes et al. 2017). As a possible alternative, attempts to rear native species of bumble bee have been made in Mexico, with varying success (Torres-Ruiz et al. 2013). However, bumble bees are not native to all areas of the country, or native species are not suitable for commercial management. In order to find alternative pollinators for greenhouse tomatoes in the Yucatan Peninsula, Cauich et al. (2004) assessed the pollination efficiency of the stingless bee, *N. perilampoides*. These bees easily adapted to confinement and started foraging the following day after being introduced to the greenhouse. The results in terms of quality and quantity of fruit and seed were similar to those obtained by artificial (hand) vibration. It is not clear if *N. perilampoides* can sonicate tomato (Slaa et al. 2006). However, it seems that they can "milk" the anthers, similar to the behavior of honey bees for obtaining pollen in this crop (Banda and Paxton 1991).

Another potential species to use in buzz-pollinated crops in the Yucatan Peninsula is *M. beecheii*, provided that its adaptation to greenhouses is improved. To date few attempts have been made to introduce *M. beecheii* in greenhouses, particularly because colonies are more difficult to obtain (pers. Obs.). In Brazil, *M. quadrifasciata* is an efficient pollinator of tomato in greenhouses, although open-field crops were seldom visited by these bees. It is argued that the lack of nectar in tomato flowers could make them less attractive compared with other blossoms nearby (dos Santos et al. 2009). Perhaps this feature of tomato flowers can also explain the low frequency of *N. perilampoides, M. beecheii*, and other stingless bees in open fields in the Yucatan Peninsula.

In open-field tomato, halictid and anthophorid bees are common on the flowers (Macías-Macías et al. 2009). Some of these bees are capable of flower sonication. In a study in the Yucatan, bees of the genera *Exomalopsis* and *Augochloropsis* were significantly more efficient in the production of fruit and seed in open-field tomatoes compared with Africanized honey bees (Macías-Macías et al. 2009).

7.2.8 Other Crops

Stingless bees are frequent visitors of crops of regional economic importance. Many of these crops have yet unexploited potential.

Stingless bees (and honey bees) are seen frequently visiting flowers of pitahaya (*Hylocereus undatus*). The flowers of this plant are large (~15 cm long) and nocturnal, adapted for bat pollination (Ferral-Piña et al. 2012). Nonetheless, it is possible that bees could complement the pollination services on this and other bat-pollinated plants, like agave.

In Mexico one native tuber consumed locally is jicama (pronounced *heecama*), or yam bean (*Pachyrhizus erosus*). This climbing plant belongs to the family Papilionidae of the Fabaceae, and produces flowers in panicles (Sørensen 1996). The flowers of jicama are hermaphrodite, but the sex organs are concealed. Specialized flower visitors open the flower and expose the stigma and anthers (Fig. 7.2). A valvular system of pollen exposition requires flower visitors to exert pressure on the flower and push down the bottom petals to expose the nectaries. During nectar collection the anthers and stigma of the flower rub against the ventral parts of the insect. After visitation, the flower returns to its original form, preventing nonspecialized visitors from gaining access (Cuanalo 2017). Although the predominant reproductive system in jicama seems to be autogamy (Sørensen 1996), its flowers produce large quantities of nectar that attract various species of bee (Cuanalo 2017). Controlled pollination experiments in Yucatan showed that autogamy in jicama seems facultative, and that bee visitation increases the production of seed (Cuanalo 2017). Stingless bees of the species *Cph. zexmeniae, Fr. nigra*, and *N. perilampoides* are frequent flower visitors of jicama. However, stingless bees are not capable of exposing the sex organs of the flower, and are therefore non-legitimate visitors, compared with anthophorid and megachilid bees. However, the relative contribution of non-legitimate flower visitors to pollination of this crop has not been investigated (Cuanalo 2017).

Beans, native (*Phaseolus lunatus*) and introduced (*Vigna unguiculata*), are visited by bees. The stingless bees *Tr. fulviventris, Fr. nigra, Cph. zexmeniae*, and *N. perilampoides* have been observed on flowers of *V. unguiculata* in milpa crops of southern Yucatan (Farfán-Morales et al. 2015), but no study has been conducted to assess their relative frequency and importance as pollinators.

Another crop on which stingless bees have been observed is nance (*Byrsonima crassifolia*). In Yucatan, foragers of *N. perilampoides* are attracted in great numbers to flowers of this plant where they avidly collect pollen (pers. Obs.). Although oil-collecting bees of the tribe Centridini (*Centris*, *Epicharis*) are considered efficient specialized pollinators of this tree, no comparison with non-specialized floral visitors has been conducted to assess their relative contribution as pollinators (Sigrist and Sazima 2004; Ribeiro et al. 2008).

An introduced crop of increasing importance in Mexico is rambutan (*Nephelium lappaceum*). This tree originates from Southeast Asia. Rambutan produces inflorescences and most commercial cultivars are hermaphrodite and self-fertile. However, the lack of sufficient pollen results in fruit malformation, making insect-mediated pollination important (Tindall 1994). In its native region cross-pollination is achieved by honey bees, which are predominant visitors, but some species of stingless bee are also reported on the inflorescences, such as *Ttgn. iridipennis* (Shivaramu et al. 2012). Rambutan was first introduced to Mexico in the 1950s as an alternative crop in severely deteriorated habitats in Chiapas. Today this state has become the most important rambutan producer in the country (Pérez-Romero and Pohlan 2004). The flowers of rambutan do not seem to require specialized pollinators, and the introduction of colonies of honey bees during blossom has resulted in adequate levels of fruit production in Asia (Shivaramu et al. 2012). Interestingly, in Mexico it has been reported that honey bees are not attracted to rambutan flowers. As an alternative, colonies of the stingless bee *Scp. mexicana* have been introduced to the crop, which resulted in twofold increase in fruit production per ha. In Chiapas, the presence of other species of stingless bees, like *Ttr. angustula* and *N. perilampoides*, on rambutan flowers is also reported, but at lower frequencies (Rincón-Rabanales et al. 2015).

In the Yucatan Peninsula, coconut palm (*Cocos nucifera*) is grown along the coast, and flowers are visited in great numbers by honey bees. Although stingless bees are not commonly seen on coconut flowers, it is thought that they can contribute to pollination (Meléndez-Ramirez et al. 2004).

Many other crops are visited by stingless bees in tropical and subtropical areas of Mexico for which reports are lacking. Conducting more studies to determine the bee fauna visiting crops is essential, as well as evaluation of their contribution as pollinators. That information is basic to plan strategies for sustainable management and conservation (Quezada-Euán et al. 2001).

7.3 Facilitating Pollination Services

Many species of bee are not adaptable to artificial nests and the best option to ensure adequate populations for pollination services is the conservation of habitats (Romero and Quezada-Euán 2013; Landaverde et al. 2017). In addition to sufficient nesting sites, food availability is essential. Crop flowers are normally available during short periods; thus, bees may require additional food sources when a crop is not in blossom. Additional plants in and around crops are important sources of additional food for bees and other pollinators to maintain their populations. Flowers of these additional plants may also attract pollinators when a crop is in bloom, potentially increasing the diversity of visitors and pollination rates of a crop (Kremen and M'Gonigle 2015). Likewise, cultivation systems containing a mixture of interplanted primary and minority crops seem to increase bee richness (Norfolk et al. 2016). In addition, it is becoming evident that an important aspect to consider for sustainable agriculture is the management of ruderal plants. Apart from restricting the use of pesticides and destroying potential nesting sites that will reduce flower visitors, it is important to preserve additional flowering plants (some of them considered weeds today), because they could facilitate the pollination of crops (Kremen and M'Gonigle 2015; Norfolk et al. 2016).

References

Abdelgadir HA, Johnson SD, Van Staden J (2012) Pollen viability, pollen germination and pollen tuve growth in the biofuel seed crop *Jatropha curcas* (Euphorbiaceae). S Afr J Bot 79:132–139

Achten WMJ, Nielsen LR, Aerts R, Lengkeek AG, Kjae RED, Trabucco A, Hansen JK, Maes WH, Graudal L, Akinnifesi FK, Muys B (2010) Towards domestication of *Jatropha curcas*. Biofuels 1:91–107

Afik O, Delaplane KS, Shafir S, Moo-Valle H, Quezada-Euán J, Javier G (2014) Nectar minerals as regulators of flower visitation in stingless bees and nectar hoarding wasps. J Chem Ecol 40:476–483

Ancona-Xiu P (2007) Evaluación de la toxicidad aguda del malatión y diazinon para *Nannotrigona perilampoides* y efecto de su uso en la polinización de Capsicum chinense en Dzununcán, Yucatán. Tesis de Maestría, Universidad Autónoma de Yucatán, Mérida

Arbetman M, Meeus I, Morales C, Aizen M, Smagghe G (2013) Alien parasite hitchhikes to Patagonia on invasive bumblebee. Biol Invasions 15:489–594

Arzaluz A, Obregón F, Jones R (2002) Optimum brood size for artificial propagation of the stingless bee *Scaptotrigona mexicana*. J Apic Res 41:62–63

Ashworth L, Quesada M, Casas A, Aguilar R, Oyama K (2009) Pollinator-dependent food production in Mexico. Biol Conserv 142:1050–1057

Ayala R (1999) Revisión de las abejas sin aguijón de México (Hymenoptera: Apidae: Meliponini). Folia Entomológica Mexicana 106:1–123

Badano EI, Vergara CH (2011) Potential negative effects of exotic honey bees on the diversity of native pollinators and yield of highland coffee plantations. Agric For Entomol 13:365–372

Banda HJ, Paxton RJ (1991) Pollination of greenhouse tomatoes by bees. Acta Hortic 288:194–198

Barrientos-Priego AF, Boris MW, Ben-Ya'acov AD, De la Cruz Torres E, López-López L (1991) El germoplasma de aguacate en Yucatán; México. Memoria del Centro de investigaciones científicas y tecnológicas del aguacate en el estado de México. Coatepec, México, pp 97–104

Bawa KS (1990) Plant-pollinator interactions in tropical rain forests. Annu Rev Ecol Syst 21:399–422

Biesmeijer JC (1997) Abejas sin aguijón, su biología y la organización de la colmena. University of Utrecht, Utrecht

Buchmann SL (1983) Buzz pollination in angiosperms. In: Jones CE, Little JR (eds) Handbook of experimental pollination biology. Van Nostrand Reinhold, New York, pp 73–113

Can-Alonzo C, Quezada-Euán JJG, Ancona-Xiu P, Moo-Valle H, Valdovinos-Nunez GR, Medina-Peralta S (2005) Pollination of 'criollo'avocados (*Persea americana*) and the behaviour of associated bees in subtropical Mexico. J Apic Res 44:3–8

Canto-Aguilar A, Parra-Tabla V (2000) Importance of conserving alternative pollinators: assessing pollination efficiency of the squash bee *Peponapis limitaris* in *Cucurbita moschata* (Cucurbitaceae). J Insect Conserv 4:201–208

Caro A, Moo-Valle H, Alfaro R, Quezada-Euán JJG (2017) Pollination services of Africanized honey bees and native *Melipona beecheii* to buzz-pollinated annatto (*Bixa orellana* L.) in the neotropics. Agric For Entomol 19:274

Castañeda-Vildózola A, Equihua MA, Valdés CJ, Barrientos Priego AF, Ish-Am G, Gazit S (1999) Insectos polinizadores del aguacatero en los estados de México y Michoacán, México. Revista Chapingo Ser Hortic 5:129–136

Cauich O, Quezada-Euán JJG, Macias-Macias JO, Reyes-Oregel V, Medina-Peralta S, Parra-Tabla V (2004) Behavior and pollination efficiency of *Nannotrigona perilampoides* (Hymenoptera: Meliponini) on greenhouse tomatoes (*Lycopersicon esculentum*) in subtropical Mexico. J Econ Entomol 97:475–481

Cauich O, Quezada-Euán JJG, Meléndez-Ramírez V, Valdovinos-Nuñez GR, Moo-Valle H (2006) Pollination of habanero pepper (*Capsicum chinense*) and production in enclosures using the stingless bee *Nannotrigona perilampoides*. J Apic Res 45:125–130

Chittka L, Thomson JD, Waser NM (1999) Flower constancy, insect psychology, and plant evolution. Naturwissenschaften 86:361–377

Conabio (2016) Chiles. http://www.biodiversidad.gob.mx/usos/alimentacion/chile.html

Cuanalo C (2017) Biología reproductiva y apifauna visitante de la jícama *Pachyrhizus erosus* (L) Urban: (Fabaceae-Papilionoideae) en Yucatán. Tesis Maestría, Universidad Autónoma de Yucatán, Mérida

Dafni A, Kevan PG, Husband BC (2005) Practical pollination biology. Enviroquest, Ontario

De Luca PA, Vallejo-Marín M (2013) What's the 'buzz' about? The ecology and evolutionary significance of buzz-pollination. Curr Opin Plant Biol 16:429–435

Delaplane KS, Mayer DF (2000) Crop pollination by bees. CABI, Oxon

Delaplane KS, Dag A, Danka RG, Freitas BM, Garibaldi LA, Goodwin RM, Hormaza JI (2013) Standard methods for pollination research with *Apis mellifera*. J Apic Res 52(4):1–28

dos Santos SAB, Roselino AC, Hrncir M, Bego LR (2009) Pollination of tomatoes by the stingless bee *Melipona quadrifasciata* and the honey bee *Apis mellifera* (Hymenoptera, Apidae). Genet Mol Res 8:751–757

Duennes MA, Petranek C, Diez de Bonilla EP, Mérida-Rivas J, Martinez-López O, Sagot P, Vandame R, Cameron SA (2017) Population genetics and geometric morphometrics of the *Bombus ephippiatus* species complex with implications for its use as a commercial pollinator. Conserv Genet 18:553–572

Essenberg CJ (2012) Explaining variation in the effect of floral density on pollinator visitation. Am Nat 180:153–166

Farfán-Morales E, Mena-Avila M, Jímenez-Gutiérrez G (2015) Apifauna asociada a la flor del cultivo de espelón (*Vigna unguiculata*) en Oxcutzcab, Yucatán. Universidad Autónoma de Yucatán. Trabajo de Investigación Formativa, Licenciatura en Agroecología. Facultad de Medicina Veterinaria y Zootecnia

Ferral-Piña J, Trejo-Rivero JAS, Mez BG, Mez G (2012) Insectos polinizadores en el cultivo de Pitahaya (*Hylocereus undatus*). EAE Editorial Academia Española

Free JB (1993) Insect pollination of crops. Academic Press, New York

Gazit A, Ish-Am G (2007) Stingless bees can serve as efficient avocado pollinators. In: Gardner CAC et al (eds) 9th International pollination symposium on plant-pollination relationships-diversity in action. Iowa State University, Ames

Heard TA (1999) The role of stingless bees in crop pollination. Annu Rev Entomol 44:183–206

Hrncir M, Maia-Silva C (2013) On the diversity of foraging-related traits in stingless bees. In: Vit P, Pedro SRM, Roubik D (eds) Pot-honey: a legacy of stingless bees. Springer, New York, pp 201–215

IPBES (2016) In: Potts SG et al (eds) Summary for policymakers of the assessment report of the intergovernmental science-policy platform on biodiversity and ecosystem services on pollinators, pollination and food production. Secretariat of the Intergovernmental Science-Policy Platform on Biodiversity and Ecosystem Services, Bonn, p 36

Ish-Am G, Eisikowitch D (1992) New insight into avocado flowering in relation to its pollination. Calif Avo Soc Yearb 75:125–137

Ish-Am G, Eisikowitch D (1998) Low attractiveness of avocado (*Persea americana* Mill.) flowers to honeybees (*Apis mellifera* L.) limits fruit set in Israel. J Hortic Sci Biotechnol 73:195–204

Ish-Am G, Barrientos-Priego AF, Castañeda-Vildózola A, Gazit S (1999) Avocado (*Persea americana*) pollinators in its region of origin. Revista Chapingo Serie Horticultura 5:137–143

Joseph N, Siril EA, Nair GM (2012) Reproductive characterization and preliminary studies on controlled breeding of annatto (*Bixa orellana* L.). Plant Syst Evol 298:239–250

King AJ, He W, Cuevas JA, Freudenberger M, Ramiaramanana D, Graham IA (2009) Potential of *Jatropha curcas* as a source of renewable oil and animal feed. J Exp Bot 60:2897–2905

Klein AM, Steffan-Dewenter I, Tscharntke T (2003) Fruit set of highland coffee increases with the diversity of pollinating bees. Proc R Soc Lond Ser B 270:955–961

Kremen C, M'Gonigle LK (2015) Small-scale restoration in intensive agricultural landscapes supports more specialized and less mobile pollinator species. J Appl Ecol 52:602–610

Kuhn-Neto B, Contrera FAL, Castro MS, Nieh JC (2009) Long distance foraging and recruitment by a stingless bee, *Melipona mandacaia*. Apidologie 40:472–480

Lahav E, Gazit S (1991) World listing of avocado cultivars according to flowering type. Fruits 49:299–313

Landaverde-González P, Quezada-Euán JJG, Theodorou P, Murray TE, Husemann M, Ayala R, Moo-Valle JH, Vandame R, Paxton RJ (2017) Sweat bees on hot chillies: provision of pollination services by native bees in traditional slash-and-burn agriculture in the Yucatan peninsula of tropical Mexico. J Appl Ecol 54:1814

López-Uribe MM, Cane JH, Minckley RL, Danforth BN (2016) Crop domestication facilitated rapid geographical expansion of a specialist pollinator, the squash bee *Peponapis pruinosa*. Proc R Soc Ser B 283:20160443

Macias-Macias O, Chuc P, Ancona-Xiu P, Cauich O, Quezada-Euán JJG (2009) Contribution of native bees and Africanized honeybees (Hymenoptera: Apoidea) to Solanaceae crop pollination in tropical México. J Appl Entomol 133:456–465

Meléndez-Ramirez V, Magaña-Rueda S, Parra-Tabla V, Ayala R, Navarro J (2002) Diversity of native bee visitors of cucurbit crops (Cucurbitaceae) in Yucatán, México. J Insect Conserv 6:135–147

Meléndez-Ramírez V, Parra-Tabla V, Kevan P, Ramírez-Morillo I, Harries H, Zizumbo D, Harries H, Fernandez-Barrera M, Zizumbo-Villareal D (2004) Mixed mating strategies and pollination by insects and wind in coconut palm (*Cocos nucifera* L. (Arecaceae)). Agric For Entomol 6:155–163

Murphy CM, Breed MD (2008) Nectar and resin robbing in stingless bees. Am Entomol 54:36–44

Ne'eman G, Jürgens A, Newstrom-Lloyd L, Potts SG, Dafni A (2010) A framework for comparing pollinator performance: effectiveness and efficiency. Biol Rev 85:435–451

Norfolk O, Eichhorn MP, Gilbert F (2016) Flowering ground vegetation benefits wild pollinators and fruit set of almond within arid smallholder orchards. Insect Conserv Divers 9:236–243

Ollerton J, Winfree R, Tarrant S (2011) How many flowering plants are pollinated by animals? Oikos 120:321–332

Pacheco-Filho AJDS, Verola CF, Lima Verde LW, Freitas BM (2015) Bee-flower association in the Neotropics: implications to bee conservation and plant pollination. Apidologie 46:530–541

Palma G, Quezada Euán JJG, Meléndez-Ramírez V, Irigoyen J, Valdovinos-Nuñez GR, Rejón M (2008) Comparative efficiency of *Nannotrigona perilampoides*, *Bombus impatiens* (Hymenoptera: Apoidea), and mechanical vibration on fruit production of enclosed habanero pepper. J Econ Entomol 101:132–138

Pérez-Balam J, Quezada-Euán JJG, Alfaro-Bates R, Medina S, McKendrick L, Soro A, Paxton RJ (2012) The contribution of honey bees, wasps and flies to avocado (*Persea americana*) pollination in southern Mexico. J Polli Ecol 8:42–47

Pérez-Romero A, Pohlan HAJ (2004) Prácticas de cosecha y poscosecha del Rambután en el Soconusco, Chiapas, México. *Leisa Revista de Agroecología* volumen desconocido:24–26

Quezada-Euán JJG, May-Itzá W d J, González-Acereto JA (2001) Meliponiculture in México: problems and perspective for development. Bee World 82:160–167

Quezada-Euán JJG (2005) Biología y uso de las abejas nativas sin aguijón de la Península de Yucatán, México (Hymenoptera: Meliponini). Universidad Autónoma de Yucatán, Mérida

Raw A (2000) Foraging behaviour of wild bees at hot pepper flowers (*Capsicum annuum*) and its possible influence on cross pollination. Ann Bot 85:487–492

Ribeiro EKMD, Rêgo MMC, Machado ICS (2008) Cargas polínicas de abelhas polinizadoras de *Byrsonima chrysophylla* Kunth. (Malpighiaceae): fidelidade e fontes alternativas de recursos florais. Acta botánica Brasileira 22:165–171

Ricketts TH (2004) Tropical forest fragments enhance pollinator activity in nearby coffee crops. Conserv Biol 18:1262–1271

Rincón-Rabanales M, Roubik DW, Guzmán MA, Salvador-Figueroa M, Adriano-Anaya L, Ovando I (2015) High yields and bee pollination of hermaphroditic rambutan (*Nephelium lappaceum* L.) in Chiapas, Mexico. Fruits 70:23–27

Rivera-Madrid R, Escobedo GMRM, Balam-Galera E, Vera-Kú M, Harries H (2006) Preliminary studies toward genetic improvement of annatto (*Bixa orellana* L). Sci Hortic 109:165–172
Rizzardo RAG, Milfont MO, Silva EMS, Freitas BM (2012) *Apis mellifera* Pollination improves agronomic productivity of anemophilous castor bean (*Ricinus communis*). Ann Braz Acad Sci 84:605–608
Romero MJ, Quezada-Euán JJG (2013) Pollinators in biofuel agricultural systems: the diversity and performance of bees (Hymenoptera: Apoidea) on *Jatropha curcas* in Mexico. Apidologie 44:419–429
Roubik DW (1982) The ecological impact of nectar-robbing bees and pollinating hummingbirds in a tropical shrub. Ecology 63:354–360
Roubik DW (1989) Ecology and natural history of tropical bees. Cambridge University Press, New York
Roubik DW (1995) Pollination of cultivated plants in the tropics. Agricultural Service Bulletin 118, Food and Agriculture Organization of the United Nations, Italy
Roubik DW (2000) Pollination system stability in tropical America. Conserv Biol 14:1235–1236
Roubik DW (2002) Tropical agriculture: the value of bees to the coffee harvest. Nature 417:708
Roubik DW, Aluja M (1983) Flight ranges of *Melipona* and *Trigona* in tropical forest. J Kansas Entomol Soc 56:217–222
Roubik DW, Moreno-Patiño JE (2013) How to be a botanist using pollen spectra. In: Vit P, Pedro SRM, Roubik DW (eds) Pot-honey: a legacy of stingless bees. Springer, New York, pp 295–314
SAGARPA (2015) http://www.sagarpa.gob.mx/saladeprensa/2012/Paginas/2015B466.aspx
Saturni FT, Jaffé R, Metzger JP (2016) Landscape structure influences bee community and coffee pollination at different spatial scales. Agric Ecosyst Environ 235:1–12
Schmid-Hempel R, Eckhardt M, Goulson D, Heinzmann D, Lange C, Plischuk S, Escudero LR, Salathé R, Scriven JJ, Schmid-Hempel P (2014) The invasion of southern South America by imported bumblebees and associated parasites. J Anim Ecol 83:823–837
Shivaramu K, Sakthivel T, Rami Reddy PV (2012) Diversity and foraging dynamics of insect pollinators on rambutan (*Nephelium lappacum* L.). Pest Manag Hortic Ecosys 18:158–160
Sigrist MR, Sazima M (2004) Pollination biology of twelve species of Neotropical Malpighiaceae: stigma morphology and its implications for the breeding system. Ann Bot 94:33–41
SHCP (2014) Panorama del aguacate. Secretaría de Hacienda y Crédito Público- Financiera Nacional de Desarrollo Agropecuario, Rural, Forestal y Pesquero
SIAP (2014) Anuarios estadísticos de la Producción Agrícola. http://infosiap.siap.gob.mx/aagricola_siap/icultivo/index.jsp
Slaa EJ, Sánchez-Chaves LA, Malagodi-Braga KS, Hofstede FE (2006) Stingless bees in applied pollination: practice and perspectives. Apidologie 37:293–315
Solís-Montero L, Vergara CH, Vallejo-Marín M (2015) High incidence of pollen theft in natural populations of a buzz-pollinated plant. Arthropod Plant Interact 9:599–611
Sørensen M (1996) Yam bean (*Pachyrhizus* DC.). In: Heller et al (eds) Promoting the conservation and use of underutilized and neglected crop, Institute of Plant Genetics and Crop Plant Research, vol 2. International Plant Genetic Resources Institute, Rome
Spears EE (1983) A direct measure of pollinator effectiveness. Oecologia 57:196–199
Tindall HD (1994) Rambutan cultivation. Food and Agriculture Organization of the United Nations, Rome
Torres-Ruiz A, Wallace RW, Ayala-Barajas R (2013) Present and potential use of bees as managed pollinators in Mexico. Southwestern Entomol 38:133–148
van Nieuwstadt MGL, Ruano Iraheta CE (1996) Relation between size and foraging range in stingless bees (Apidae, Meliponinae). Apidologie 27:219–228
Veddeler D, Olschewski R, Tscharntke T, Klein AM (2008) The contribution of non-managed social bees to coffee production: new economic insights based on farm-scale yield data. Agrofor Syst 73:109–114
Venkatesh CS (1956) The curious anther of *Bixa*—its structure and dehiscence. Am Midl Nat 55:473–476

Vergara CH, Badano EI (2009) Pollinator diversity increases fruit production in Mexican coffee plantations: the importance of rustic management systems. Agric Ecosyst Environ 129:117–123
White D, Cribb BW, Heard TA (2001) Flower constancy of the stingless bee *Trigona carbonaria* smith (Hymenoptera: Apidae: Meliponini). Aust Entomol 40:61–64
Wille A, Orozco E (1983) Polinización del chayote *Sechium edule* (Jacq) Swartz en Costa Rica. Rev Biol Trop 31:145–154
Wilson EO (1971) The insect societies. Belknap Press of Harvard University Press, Cambridge
Winfree R, Williams NM, Dushoff J, Kremen C (2007) Native bees provide insurance against ongoing honey bee losses. Ecol Lett 10:1105–1113
Winter K, Adams L, Thorp R, Inouye D, Day L, Ascher J, Buchmann S (2006) Importation of non-native bumble bees into North America: potential consequences of using *Bombus terrestris* and other non-native bumble bees for greenhouse crop pollination in Canada, Mexico, and the United States. North American Pollination Protection Campaign, San Francisco
Wysoki M., van den Berg M.A., Ish-Am G. Gazit S., Peña J.E., Waite G. K. (2002) Pests and pollinators of avocado. Peña J.E., Sharp J.L., Wysoki M. (ed.) Tropical fruit pests and pollinators: biology, economic importance, natural enemies and control, pp. 223–293. CABI, New York
Ye M, Li C, Francis G, Makkar HPS (2009) Current situation and prospects of *Jatropha curcas* as a multipurpose tree in China. Agrofor Syst 76:487–497

Chapter 8
Managing and Preserving Stingless Bees

In collaboration with Humberto Moo-Valle

A colony division of *Melipona beecheii* after 3 weeks in its new hive

8.1 The Api-Botanical Cycle

Stingless beekeeping (or meliponiculture) depends closely on the different floral resources to support the growth of colonies and the production of surplus reserves of honey, pollen, and resins for a profitable harvest. The adequate management of colonies, through periods of abundance of resources and draught, is a key aspect to keep them in good condition and to obtain maximum production.

Management of the colonies for commercial production needs to be planned and conducted in accordance to the annual cycle of blossoms and the periods of resource scarcity. The integration of management practices in relation with the availability of floral resources is known as the api-botanical cycle (De Araujo Freitas et al. 2010).

In the Yucatan Peninsula, there are three clear flowering periods that define three phases of the api-botanical cycle: pre-harvest, harvest, and postharvest.

J. J. G. Quezada-Euán, *Stingless Bees of Mexico*,
https://doi.org/10.1007/978-3-319-77785-6_8

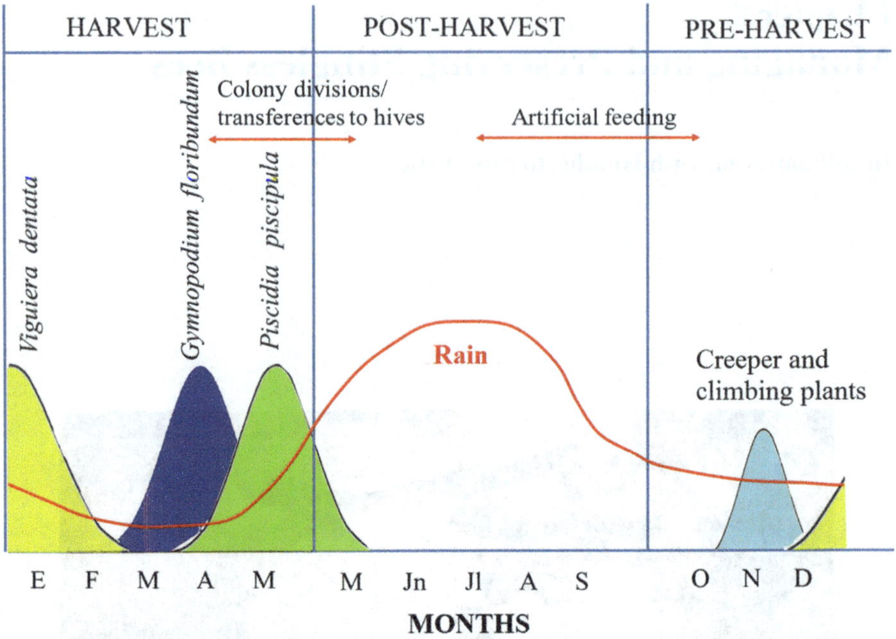

Fig. 8.1 The api-botanical cycle in the Yucatan Peninsula

The pre-harvest period roughly encompass the period between November and January (Fig. 8.1). During this season some cold and rainy weather (called "norte") frequently takes place and the lowest temperatures for this region are recorded (10° C in January–February). In this lapse, various species of climbing plants bloom (Table 8.1). The flow of nectar during the preharvest is not abundant, but stimulates population growth after a period of draught that occurs in the postharvest. It is important to consider that colonies with large populations can better exploit the nectar abundance during the harvest season, and it is in the interest of the beekeeper to promote growth right before the main blossoms. This may be a good time to transfer colonies from their natural lodgings to man-made ones, as colonies do not have abundant food reserves that may attract parasites, and there is also plenty of food to recover rapidly.

The harvest season in the Yucatan Peninsula takes place between January and May (Fig. 8.1). In this season, the conditions become dry and some low temperatures can still occur. However, temperature by the end of the harvest season could reach up to 43° C, especially during April and May. It is in this period when the main honey plants bloom. These plants are found in high densities, mainly in altered secondary forests across the region, and produce vast amounts of nectar, with a high sugar content. The three most important nectar blossoms are known by their Mayan names as: *Tahonal* (*Viguiera dentata*), *Ts'i ts'il che'* (*Gymnopodium floribundum*), and *Habin* (*Piscidia piscipula*). In optimal years, the three blossoms occur in succession making it possible to obtain several honey harvests in this period. The weather during the

Table 8.1 Flowering months of the main nectar- and pollen-producing plants in the Yucatan Peninsula (Alfaro-Bates et al. 2010; Flores-Guido and Vermont-Ricalde 2011)

Species	J	F	M	A	M	J	J	A	S	O	N	D
Gymnopodium floribundum		x	x	x								
Viguiera dentata	x	x										
Piscidia piscipula			x	x	x							
Waltheria americana			x	x								
Cordia gerascanthus			x	x								
Vitex gaumeri				x	x							
Neomillspaughia emarginata				x	x							
Coccoloba uvifera				x	x							
Caesalpinia yucatanensis			x	x								
Caesalpinia gaumeri			x	x								
Mimosa bahamensis				x								
Haematoxylum campechianum				x								
Spondias mombin						x	x					
Melanthera nivea	x	x	x	x	x	x	x	x	x	x	x	x
Wissadula amplissima								x	x	x	x	x
Boerhavia erecta				x	x	x	x	x	x	x		
Solanum rudepannum								x	x	x	x	x
Cassia fistula					x	x	x	x				
Lysiloma latisiliquum				x	x	x						
Byrsonima crassifolia				x	x	x						
Bidens pilosa						x	x	x				
Acacia gaumeri						x	x					
Dalbergia glabra						x	x					
Sabal yapa						x	x	x	x			
Leucaena leucocephala						x	x	x	x			
Hampea trilobata						x	x	x	x			
Havardia albicans				x	x	x	x	x	x	x		
Bursera simaruba				x	x	x	x	x				
Conocarpus erectus				x								
Croton fragilis			x	x	x							
Croton humilis					x	x	x	x	x	x	x	x
Merremia aegyptia											x	x
Ipomoea nil											x	x
Jacquemontia pentantha											x	x
Ipomoea crinicalyx											x	x
Ipomoea triloba											x	x
Lonchocarpus longistylus										x	x	
Tournefortia gnaphalodes											x	x
Tridax procumbes											x	x
Gayoides crispum											x	x
Tithonia rotundifolia										x	x	
Turbina corymbosa											x	x
Tohuinia paucidentata											x	x
Senna racemosa	x	x	x							x	x	x

The names in white indicate species that flower during the period of harvest, in green the ones that flower during the postharvest, and in yellow the plants that flower in the pre-harvest

harvest season greatly influences the availability of nectar. Abundant nectar flow would depend on sufficient rain before the main blossoms and high temperatures when the flowers open. Sporadic showers during bloom can increase the period of flower and nectar production by these plants (De Araujo Freitas et al. 2010).

During January–February, it is important to verify the occurrence of brood mortality or excessive build of layers of involucrum, which may indicate a detrimental effect of low temperatures on the colony. In such case, it may be necessary to reinforce thermal isolation of colonies using polystyrene or cardboard to cover the hives. This measure enhances control of the temperature and reduces the risk of brood mortality. At the end of the harvest season, colonies should be populous and with abundant reserves allowing the production of new ones. Divisions are better carried out at the end of the harvest, when only scattered rains occur, as high humidity is linked with the abundance of phorid flies (González-Acereto 2008).

The last part of the api-botanical cycle is the postharvest season between the months of June and October (Fig. 8.1). It is important to maintain sufficient honey and pollen in the colonies to help them survive the postharvest because the availability of resources, mainly nectar, drastically decreases in this period.

The postharvest in the Yucatan concur with the rainy and hurricane seasons. Many large trees flower during this season, but the total nectar production per area is not sufficient to grant a harvest, at least not in the northern part of the Peninsula. Normally, becasue of reduced food income, the reserves of colonies decrease at this time of the year. One essential aspect of management in this season is watching over the amount of brood and reserves avoiding a decrease at levels that may compromise colony survival. If population and reserves drastically reduce, measures should be taken to avoid colony collapse such as artificial feeding. In particular, small colonies and colony divisions may require feeding with sugar syrup during this period.

The api-botanical cycle may vary between regions and geographic areas, depending on the flowering cycles. However, the general activities for colony management during abundance and scarcity can apply to different regions. The main objective is to keep colonies in good conditions across the different periods. In other areas of México, for instance, there are not such clear pre-harvest, harvest, and postharvest phases as in the Yucatan. In that case, the beekeeper can consider phases of abundance and scarcity to plan activities (Fig. 8.1).

One basic aspect of the api-botanical cycle is information on the different blossoms and the period of the year in which they occur. It is also important to classify the different blossoms regarding their potential for honey production or as support for the colonies. Honey production depends on plant species that not only produce nectar of good quality, but are also found in large numbers in an area (Table 8.1).

8.2 Establishing a Stingless Bee Yard

Essential to stingless beekeeping is the selection of the area where the colonies will be maintained. It is important that the area includes diversity and abundance of nectar- and pollen-producing species that supply food for the colonies across the whole api-botanical cycle (Table 8.1). It is also important to select an area where potential risks, such as the frequent use of insecticides, are absent or can be kept to a minimum.

The beekeeper should detect when nectar and pollen could be in low supply in order to provide them to colonies if necessary. One cheap and ecological form of feeding colonies during dearth is through management of flowering species around the bee yard. Some good nectar- and pollen-producing species can be planted around the meliponario, and with irrigation, they can provide food all year round. In the Yucatan Peninsula, species like San Diego (*Antigonon leptopus*) or Lippia (*Aloysia virgata*) can be especially important during the rainy season or when few local plants bloom (see Annex 2). This kind of plants are excellent food providers and represent a natural method to maintain colonies without additional investment in sugar or pollen supplements.

Stingless bee colonies should be kept under shade as direct exposure to sunlight and heat could cause overheating of the brood. To avoid sun exposure a hut or meliponario could be build, to shelter the colonies. In the meliponario, the colonies can be kept on individual bases, or in shelves that allow better use of the space (Fig. 8.2). In some areas where high trees provide shade, colonies can be kept on individual bases, but using a roof on top of each hive for sun protection, may be necessary.

Most stingless beekeepers start with a few colonies that are frequently obtained from their natural habitats. However, it is important to avoid overexploiting feral populations to supply colonies for meliponarios (see Chap. 6). In particular, rare or endangered species can be severely affected by this type of practice. A way to reduce the exploitation of feral populations is to increase the number of colonies in meliponarios using methods for artificial division. One method to obtain colonies from the wild without affecting feral populations is to withdraw only part of the colony leaving the rest in its original lodging in the forest. For species living in trees, an opening can be made with the use of an electric saw to expose part of the nest. From this, some combs and workers can be withdrawn to start a new colony in a hive box. After this process, the removed bark section is placed back and sealed with mud or tar (Colleto-Silva 2005). Using this method, the mother and daughter colonies can recover with little damage, increasing managed nests and preserving feral populations at the same time.

It is also important to consider that several species, especially those nesting in termite mounds and below ground, are difficult to adapt to hives, so it is not recommended to extract these species from their natural lodging. On the other hand, recovering colonies from areas where the forest has been logged down, could be an important way to save them from destruction.

A basic aspect of management is to keep records of all colonies and the conditions around the meliponario. Constantly registering information concerning the state of each colony is essential to make decisions and plan the different activities. Each colony should be assigned an individual number or identification code. Individual record sheets should be designed to register all relevant information, as well as performed or required management. The format can be designed in accordance to the beekeeper's needs, but it is important that all relevant information (like colony number, food and brood condition, presence of queen, etc.), is included (see an example in Annex 3). The information should be easy to read, quick to find, and

Fig. 8.2 A stingless bee hut or meliponario with a roof made with palm leaves, metal or cardboard sheets as protection against direct sunlight. The colonies could be stacked on shelves or like in the Maya traditional system using inclined wooden supports for logs (hobones)

unambiguous to avoid confusion. As the number of colonies increases, managing individual records may require the use of digital formats. For this, several software are available and are relatively easy to obtain and use. Colony records should contain information on the past and present situation of colonies, and are therefore basic sources of information, not only to make immediate decisions, but also to plan future actions. One valuable application of records is the identification of productive colonies, as well as those recurrently experiencing problems of different type. The information on honey production, weight, brood production, and other productive traits, could be useful to develop selection programs including colonies from different meliponarios (Silva-Barros 2006). It is important remembering that many productive traits could have a genetic basis. The information registered across seasons and years can be helpful to identify colonies which might be disease resistant, and good honey producers, and use them in selection programs. When dividing colonies, it is also important to keep track of their ancestry, in order to avoid overrepresenting a few genotypes in the meliponario (see Chap. 4).

Ideally, beekeepers should be registered in some organization, and their records could be useful to develop networks to increase the opportunities of economic support and fair trade. Moreover, schemes for conservation and maintenance of stingless beekeeping could be better designed if based on standardized data.

8.3 Transferring Colonies to Hive Boxes

Transferring colonies from their natural habitat to hive boxes is one important aspect of stingless beekeeping, albeit one which may represent great difficulty. There are several different models of hive boxes to keep stingless bees. Collectively they are usually called "rational" hives (Nogueira-Neto 1997; Sommeijer 1999; González-Acereto 2008; Venturieri 2008; Greco et al. 2010; Villas-Bôas 2012; Kiatoko et al. 2016).

The different rational hive models basically differ in the position of the food pots in regard to the brood. There are two main patterns of hive boxes: horizontal and vertical. In vertical models, the brood is placed in the bottom part of a tower-like hive, in a section called the brood chamber. As the colony grows, most food pots (and also part of the brood) develop on the top section of the brood chamber. In horizontal models, the food pots are mainly located next to the brood area. As the colony expands, food pots grow on both sides of the brood and eventually on top of it. In both types of model, supers can be added on top to accomodate the expanding brood and food reserves. The history and design of the different rational hive models including the Mayan "jobon" can be found in different texts and won't be covered in this book (see Nogueira-Neto 1997; González-Acereto 2008).

In nature, both patterns of nest construction (nest expanding horizontally or vertically) can be found (Fig. 8.3). This means that colonies can adapt to both model types similarly. What is most important is providing the adequate space to avoid deficient control of temperature. This is perhaps the most important aspect to consider when transferring a colony to a modern hive. We have seen cases in which colonies are transferred to boxes that more than double their size, with the fallacy that as the colony grows, it will eventually fulfil such excess space. However, if the colony does not have adequate temperature conditions from the start, it may suffer from brood deaths and reduction of the adult population, which could cause its collapse (González-Acereto 2008).

Because stingless bee colonies mostly depend on passive mechanisms for thermoregulation, the thickness of the wood used for the hives should be of at least 2.5–3 cm. The width of hive walls is important to ensure isolation from extreme temperatures. It is also important to use wood that has not been treated against woodworm or termites, because these insecticides are harmful for the bees and can kill the colony.

One problem with the incorporation of hive box technology in peasant farmer communities from the Yucatán Peninsula is that many designs can be complex and expensive. For the Yucatan, Jorge González-Acereto designed a simple and cheap horizontal hive model with hinges on the lateral walls (TIBGA), to allow easy inspec-

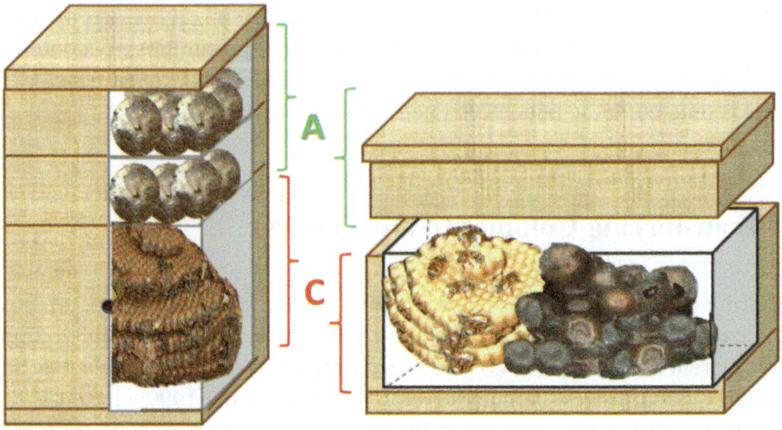

Fig. 8.3 Representation of sectioned "rational" hive boxes for stingless bees showing the general disposition of the brood and food pots, in vertical hives (on the left), and horizontal hives (on the right): A) supers and C) brood chamber

tion of the nest (Fig. 8.4). As the colony grows, supers can be added on top as in other horizontal models. The size of the brood chamber and the super varies in accordance with the species; but it is important to use a size that fits the needs of each species. One way of calculating the adequate hive space for species that have not been previously kept in hives is by taking measures of the brood and pot areas of several colonies in natural conditions. From these measurements, it is possible to obtain an average volume and, then estimate length, width and height of the hive, accordingly (see Annex 3). In vertical models, it is recommended that the interior space of the brood chamber is 2–3 cm wider than the largest comb recorded in natural nests (Villas-Bôas 2012).

Useful elements of different hive models are the movable wooden strips on top of each hive section (Fig. 8.4). The bees affix the nest structures to these strips allowing detachment of the different hive sections with only little damage to brood cells and pots.

The strips allow easy inspection of colonies and removing supers for colony harvest (Fig. 8.4).

Once a model hive has been selected, and the adequate size calculated, the beekeeper can start transferring colonies from their natural lodgings. It is important to bear in mind that colony transference to a new lodging is inevitably damaging, but the aim is keeping this damage to a minimum.

The steps for the transference of colonies to a wooden box are presented in Fig. 8.5, and described in the following sections.

If the colony is in a log, it is recommended to make two opposite longitudinal incisions with a chainsaw. Sawing can damage some structures and should be made carefully. If substantial damage was made, the colony should be sealed using mud or tar and returned to its original position, for at least 2 weeks, for recovery (Fig. 8.5a). In case the colony was not damaged, the trunk can be opened straight away and proceed to transfer the colony to the hive.

Fig. 8.4 Above on the left, a horizontal wooden hive model TIBGA showing the movable walls. The holes on the posterior wall are potential entrances; after the colony has been transferred, three holes are sealed with wax or cerumen and only one entrance is maintained. On the right, a Portugal Araujo vertical hive with its different parts. Below, a close-up of the Portugal Araujo super, with wooden strips on top, and a partial wooden floor to support the construction of food pots

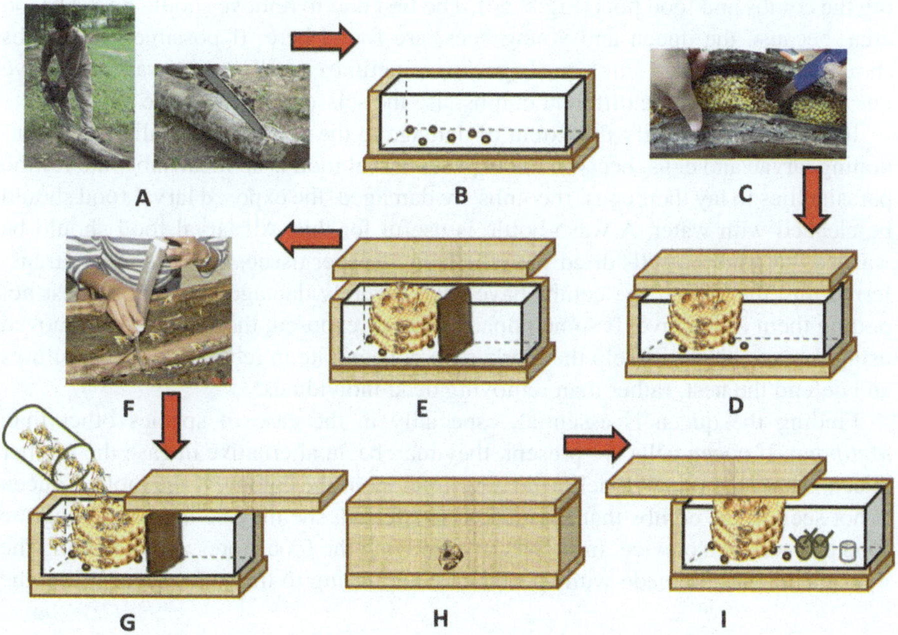

Fig. 8.5 Transferring stingless bee colonies to wooden hives. Letters correspond to the description made in Sect. 3

When transferring colonies, the first thing before opening the trunk is to prepare the hive box that will accommodate the colony. No spaces should be left between the box and the floor or the lid. Any gaps should be filled using wax or cerumen. Different kinds of plastic tape can also be used for the same purpose. Blocking gaps is necessary because after transference the bees are not yet organized, and parasite flies can enter the colony. In a following step, small spheres made with wax or cerumen (between 0.5 and 1 cm depending on the size of the adult worker bees), are placed on the hive floor. These spheres will serve as supports for the combs or cells, maintaining space (bee space) through which the bees can transit. The brood should not be directly placed on the floor because the bees won't be able to reach cells containing dead or damaged individuals, and remove them from the colony (Fig. 8.5b). The cerumen spheres also grant emergence of individuals, which could be difficult if combs are directly placed on top of each other.

Once the box is ready, the trunk with the colony can be opened. The structure of the entrance should be removed with the help of a sharp knife and keep it aside to use at the end of the process. After this, the trunk is opened with the help of chisels to fully expose the cavity. It is important to avoid excessive hammering to reduce damaging the young brood. As the cavity is exposed, it is important to detect the position of the brood and the food reserves, and place the trunk with the brood cells facing upwards, to avoid drowning of the larvae in the liquid food.

Slowly, sections of the trunk are separated to expose as much of the nest as possible. The connectives attaching the different nest structures to the walls should be carefully cut, using scissors and thin blade knives. It is important to avoid perforating the combs and food pots (Fig. 8.5c). The first part to remove should be the brood area, because the queen and young bees, are found here. If possible, the combs should be removed as a single package, but sometimes it will be necessary to remove one by one. To separate different combs, use the scissors or thin blade knives.

It is essential to verify the extent of damage to the combs, especially those containing larvae and eggs, because the exposed larval food is an ideal substrate for the parasite flies to lay their eggs. If combs are damaged, the exposed larval food should be cleaned with water. A wash bottle is useful for this. All larval food should be washed out, and the cells dried with the help of paper tissue, before they are transferred into the box. If the combs have been severely damaged, it may be best not putting them in the hive. If some pupae had been exposed, they should be removed using forceps; this will help the workers to concentrate in rebuilding the structures and defend the nest, rather than removing dead individuals.

Finding the queen is essential, especially in the case of species other than *Melipona*. If queen cells are present, they may be an alternative in case the mother queen is lost, but this will delay the development of the colony. If the mother queen is not seen in the combs that have been transferred, she may be hidden among the food reserves or a crevice. In such case, removing the food stores and collecting the workers, should be made with special care and trying to find the queen. When the

Fig. 8.6 Entomological sucking pumps to collect bees from their nest during transference to hive boxes. Left: Plastic tubes with caps on both ends; on the right car vacuum cleaner adapted to a plastic container to collect bees

queen is found, one can use a piece of wood or a sheet of paper where she can climb and take her to the hive, on the combs.

Once the combs and the queen have been transferred to the hive box, the lid should be placed on top, to stop reducing the temperature of the brood and possible entrance of flies (Fig. 8.5d).

Next is the collection of the adult population that remain in the trunk. This is done using manual or electric pumps to aspire the individuals into a tube, or some kind of container, where they are temporarily collected (Figs. 8.5f and 8.6). Attention should be put in collecting most young workers (detected by the pale color of the cuticle in non-*Melipona* bees), as they will not be able to find the new nest, and would be lost otherwise.

When several individuals have been collected in the containers, they are transferred to the hive. This should be made with a quick shake and strong hit on the container that would make the bees fall inside the hive. After this, the lid should be rapidly placed back to stop them from flying out. This procedure is performed several times until most bees have been collected from the old nest and transferred to the hive (Figs. 8.5g and 8.6).

The food reserves are last to be removed from the trunk. By then, most bees should have been transferred to the box. Otherwise, many bees can drown in honey if the pots break. If the mass of food pots can be completely removed, it should be placed in a plastic container. The bees remaining in the spaces between pots should be collected with the pumps too. When most bees have been collected, the food pots can be cut to harvest the honey and pollen, or alternatively taken to a fridge to do this later (see Sect. 2.4).

Before finishing the transference, an evaluation of the size of the colony and the space occupied in the box should be made. If an excess space is left, it would be necessary to reduce it, by means of plywood or thick cardboard. Narrowing the space will allow better control of temperature and defense. An extremely large volume can increase the loss of heat through convection, which could be potentially more harmful in vertical hives (Quezada-Euán 1988; Moo-Valle et al. 2000). Lack of adequate temperature control can lead to brood death and an eventual colony collapse (Quezada-Euán and González- Acereto 1994).

If there were any bees hidden in crevices of the log, they should be collected and transferred to the hive as explained before or using forceps. Once most adults are inside the hive, the lid is sealed with plastic tape. Finally, the hive should be placed where the trunk was located, trying to match the original position and height of the entrance. Flying bees can only find the hive if the hive entrance is in the same position and at a similar height of the original trunk entrance. To further help them in finding their nest, the structure of the entrance cut off the trunk (that was removed beforehand), should be attached to the box entrance using a bit of wax or cerumen (Fig. 8.5h). The field bees can identify the smell of the entrance and enter the hive box.

It is not advisable to place food pots in the hive immediately after transferring a colony because they can attract phorid flies. The workers are still disorganized and will have problems in rejecting intruders. Any small perforation of the pollen pots would be used for the flies to lay their eggs. After 2 or 3 days, food can be provided to the colony, using the honey collected or sugar syrup (see Sect. 5). The day after transference, an evaluation of the number of flies inside the hive should be made (Fig. 8.5i). If a large number of flies is detected, then it may be necessary to use some control methods (see Sect. 7).

The resins and cerumen in the original nest could be given to the colony right after transference, provided there have no pollen or honey residues. Otherwise, they should be washed and dried with a paper tissue before incorporating them to the hive. Resins and cerumen are important resources that the bees need to reconstruct and defend the nest.

In the first colony inspections (usually 24 h after), it is recommended to quickly open the hive, just to check if there might be problems with the reconstruction and flies. Workers seen actively foraging and little or no attack by fly maggots, are good indications that the transference has been successful. It is not advisable to provide pollen until the colony is well organized. When pollen pots are used to feed the colonies, it is important that only sealed pots are used to avoid fly invasion.

8.4 Dividing Colonies

The process of division will increase the number of colonies in the meliponarios, and avoid depending on feral populations.

Under natural conditions stingless bee colonies reproduce at low rates (see Chap. 6). However, the stingless beekeeper can accelerate this process by improving the conditions of the colonies. One way to do this is to leave in the harvest at least half the amount of honey and pollen obtained, in the colony. Additionally, the colonies can be artificially fed to stimulate the production of brood. When the colonies have reached a large size, they can be divided, preferably, at the end of the period of nectar flow.

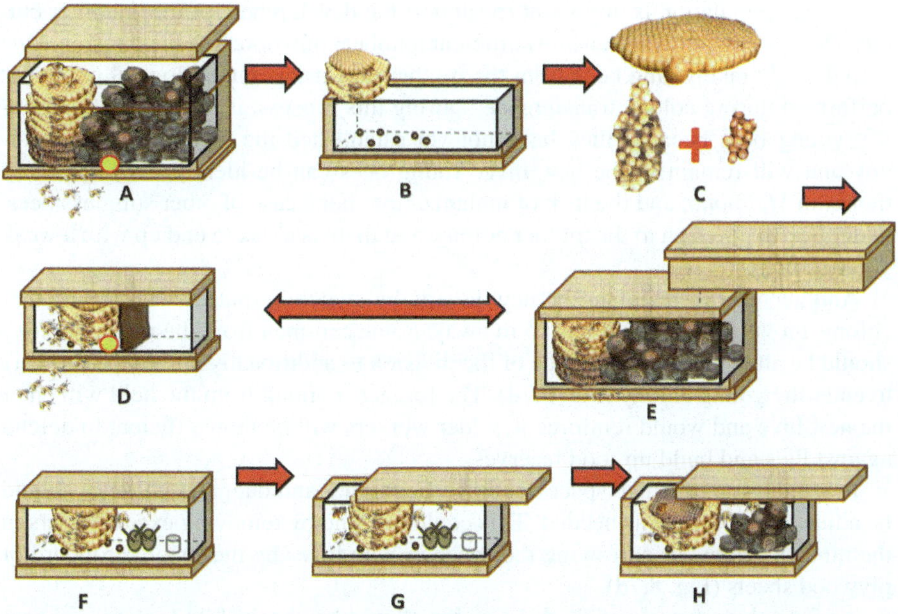

Fig. 8.7 The process of dividing stingless bee colonies. The letters correspond to the description in Sect. 4

In general, the process of dividing stingless bee colonies is similar to transferring colonies to hives (Fig. 8.7). First, the hive box should be prepared sealing the crevices and placing the sphere supports of cerumen or wax on the hive floor (Fig. 8.5).

The donor colony should be strong enough to extract four to five brood combs and keep at least a similar number in the mother nest (Fig. 8.7a). Only "mature" combs containing pupae are recommended in the production of daughter colonies. As reviewed in Chap. 2, the cells containing individuals in early stages of development (eggs and larvae) are made of cerumen. Combs made of cerumen are soft and easily damaged. Exposing the larval food from cerumen combs is one of the main causes of colony collapse by parasite flies (Fig. 8.7b). Even if these combs are intact, the eggs and larvae can sink and die in the liquid larval food. In contrast, combs containing pupae are made of cocoons, and are more resistant to manipulation. In stingless bees other than *Melipona*, it is essential to have at least one queen cell on the combs that will be transferred. These species won't produce a queen otherwise (Fig. 8.7c). The rule of using only pupae cells for colony division, does not apply completely to species that build cell clusters, such as *Fr. nigra*. In these species, as explained in Chap. 4, the production of queens results from joining two contiguous cells containing young larvae. Therefore, in addition to mature cells, a few cells containing eggs, or very young larvae, should be provided in these species (Fig. 8.7c).

Worker population is important to support the development of the daughter colony.There are two ways to ensure sufficient numbers of workers in the division. One is to directly take young bees from the mother nest using entomological pumps as performed during colony transference. During this process, it is important to identify young bees because they have not yet established the location of their colony and will remain in the new hive. Young bees can be identified by the hairy thorax in *Melipona,* and the lack of melanization in the case of other stingless bees. Older bees will return to the mother colony, and there is a risk to end up with a weak young colony.

Another way to populate the new hive is by switching places with the mother colony, moving the latter at least 5 m away. Some cerumen from the mother colony should be attached at the entrance of the division to additionally attract the foragers to enter the young colony (Fig. 8.7d). The foragers coming from the field will enter the new hive and would reinforce it. Older workers will be more efficient to defend against flies and build up food reserves.

In a final step, the hive space occupied by mother and daughter colonies should be adjusted in the hive as needed. This could be done by removing excess supers in the mother colony, or narrowing the space in both hives by means of cardboard or plywood sheets (Fig. 8.7d).

As for colony transference, the division should be checked 24 h after, and provide food accordingly. The presence of flies should be monitored, and if abundant, control methods should be applied (Fig. 8.7f). It is also essential to verify that a queen cell or virgin queen is present, and that the workers are reconstructing the nest. The frequency of artificial feeding should be adjusted in accordance with the consumption of syrup or honey; usually, it can be done twice a week (Fig. 8.7g). Eventually, new combs will be built as the new queen mates and starts laying eggs (Fig. 8.7h).

In *M. beecheii,* the new colony could take time to accept a queen. In these circumstances, the division can experience a reduction on the population of adult bees. If so, one or two combs with pupae could be extracted from the mother nest to provide new workers, one or two weeks after division. However, this must be conducted only if the survival of the mother colony is not compromised (Figs. 8.7h and 8.8).

Some hive models have been designed to facilitate the process of colony division. One such model is the Brazilian Fernando Oliveira/INPA. In this model the brood and super boxes are of equal dimensions. A special type of super is known as the module for division. The bottom section of the module of division is adapted with triangular pieces of wood attached to the corners, to where the bees will affix combs and pots (Fig. 8.9d). In large colonies, the brood grows upwards, with some combs constructed in the module of division (Fig. 8.9, Villas-Bôas 2012). When sufficient combs are being built in the division module, the colony can be divided by just separating the brood area and the module. The process produce little damage because combs and pots have been securely attached to each box. This process has the advantage that even brood combs containing young individuals can be separated without harm (Fig. 8.9a). The supers for food stores also have special supports (see Fig. 8.4), which facilitate separation (Villas-Bôas 2012). The internal

Fig. 8.8 Reducing the hive space to fit the new division is a crucial aspect of management, also recommended for weak colonies. For this, a piece of cardboard (red arrow) or plywood, could be used to confine the division

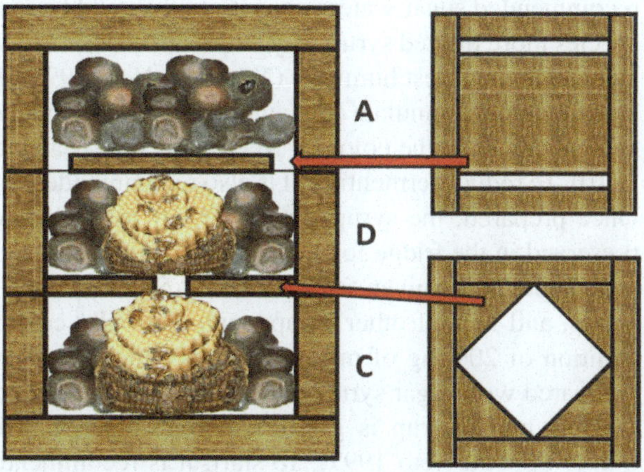

Fig. 8.9 A section of hive model Fernando Oliveira/INPA with the brood chamber (C), the module of division (D), and the super for food stores (A). On the right, views of the bottom part of the division module and the super with the position in the hive indicated with red arrows. The different sections can be moved with little damage to the colonies (in accordance to Villas-Bôas 2012)

measures of the Fernando Oliveira/INPA hive for species of *Melipona* from Brazil are 15 × 15 cm. As a rule, the height of the supers is calculated adding 2–3 cm to the average height of the food pots according to the species (Villas-Bôas 2012).

In the Yucatan Peninsula, the optimal time for making colony divisions is between February and late April, which encompass the dry season. One advantage is that uring draught, parasite flies are less frequent (Robinson 1981; Oliveira et al. 2013). The rainy season between late May and November is not recommended for dividing colonies, because the frequency of flies increases in the environment, and there is also little food in the field to support the growth of new colonies (González-

Acereto 2008). During the rainy season queen mating may take longer because the production of males decreases in this period too (Moo-Valle et al. 2001; González-Acereto 2008).

8.5 Artificial Feeding

The amount of honey reserves in the colonies can vary across seasons in accordance with the availability of nectar in the field. It is important that colonies keep a stock of honey and pollen because the reduction of reserves affects the production of individuals and strength.

At times of little nectar income, it may be necessary to artificially feed colonies (Fig. 8.1). The ideal food is honey, but it may be expensive, instead sugar syrup is a good alternative. The use of sugar syrup is also recommended after transferring or dividing colonies. Some beekeepers use *A. mellifera* honey to feed stingless bee colonies, but such practice may carry the risk of disease transmission.

The concentration of sugar in the syrup can vary. More concentrated syrups last longer without spoiling, but represent more expenditure. The percentage humidity in the honey of stingless bees is close to 25% (Catzín-Ventura et al. 2008), so a recommended sugar:water concentration should be in the range of 1:1–3:1. In some species more diluted syrup may be preferred because its evaporation may be important to preserve nest humidity (Nogueira-Neto 1997). In the honey bee, diluted syrups are used to stimulate egg laying by the queen, and more concentrated syrups are used to maintain the colonies when no nectar is available (De Araujo Freitas et al. 2010). To reduce fermenting, it is also recommended to boil the syrup for 10–15 min. Once prepared, the syrup should be left to cool down before using it. It can be preserved in the fridge for 1 week or so, but it needs to be at environment temperature before being given to the colonies. Sugar syrup is mainly a source of carbohydrates, and lack of other components normally contained in nectar. In Brazil, the addition of 200 mg of mineral salts per liter of syrup has produced good results, compared with sugar syrup alone (Costa and Venturieri 2009).

How much syrup is given to colonies will depend on the rate of consumption (Nogueira-Neto 1997). To start, it is recommended to use ca. 50–100 mL; if completely consumed, a similar amount could be provided after 2 days. The frequency of feeding is normally two or three times a week. Sometimes the bees do not consume the syrup, but store it sealing the recipient where it was supplied. In such case, no more syrup should be given because there is evidence that the bees are not using it and the potential risk that stored syrup could be mixed with nectar, when the flowering season starts.

There are various ways in which the syrup can be supplied to colonies. One is by means of plastic syringes (5–10 mL). Nogueira-Neto (1997) describes how to make feeders with syringes. Although feeders made with plastic syringes or plastic tubes work well, the cotton blocks to seal the extremes of the tubes, if not sufficiently tight, may cause some syrup spill over.

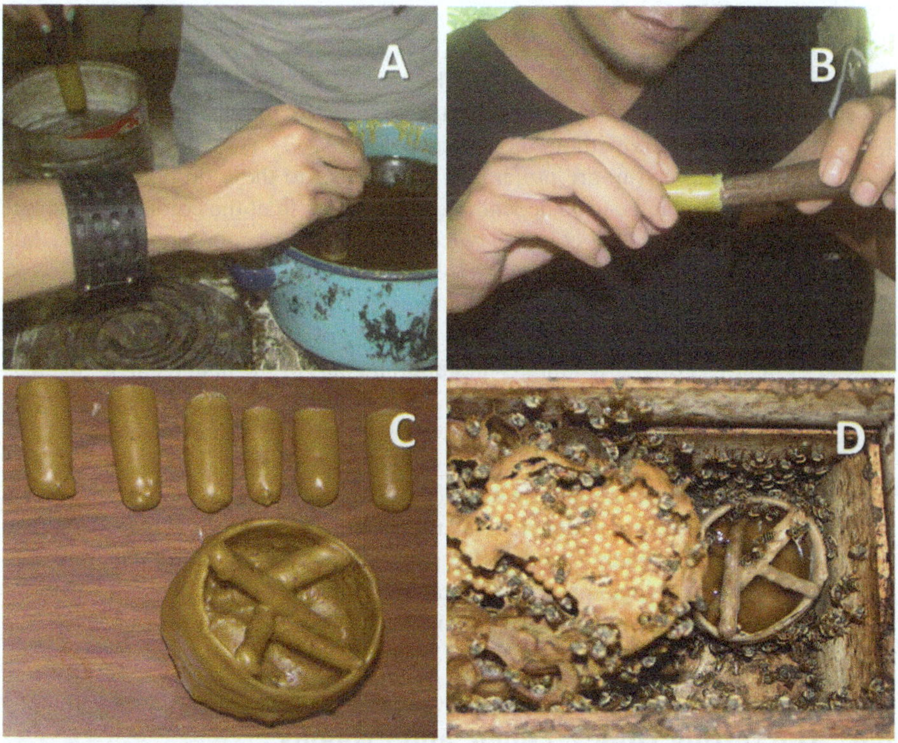

Fig. 8.10 Feeders made with cerumen. In A and B, wooden molds to make feeders with melted cerumen. In C, cerumen feeders of various sizes and a plastic feeder covered in cerumen. In D, one cerumen covered plastic container in a colony of *M. beecheii*

Other types of feeders can be made with cerumen. These feeders resemble food pots and are better accepted by the bees. To make them, wooden molds of different sizes are used, in accordance to the amount of syrup (Fig. 8.10). The wooden mold should be soaked in water for at least 48 h before using. This is important or feeder will crack and the syrup will pour out. To make these feeders, equal amounts of cerumen and water are heated at low or medium flame. It is recommended to gently stir the solution to homogenize it. If cerumen is scarce, it is also possible to add some amount of pure honey bee wax of known origin. It is important that the temperature of the mixture is only sufficient for the cerumen to melt. Above 65 °C, the mixture starts bubbling and could cause burns (Fig. 8.10).

Once the cerumen is thoroughly melted, the mixture is left to cool down for about 5 min. Next, the wooden molds are slowly immersed in the melted cerumen for about 30 s, and then retrieved and dipped in cold water to harden the cerumen adhered to it. The excess water is dried off with a tissue and the mold is immersed in the melted cerumen again. This process can be done a few times, until the feeder is about 2 mm thick. Then, the feeder can be removed from the mold and is ready to be used. Before using the molds again, they need to be soaked in water for about 5–10 min.

Another version of cerumen feeders is made using a plastic cup (30–50 mL), which is immersed in melted cerumen (Fig. 8.10). Some wooden sticks can be placed inside the cup before immersion. The wooden sticks are good support for the bees, to avoid drowning in the syrup. Plastic containers with wooden sticks can also be used as syrup feeders without the cerumen cover.

The feeders with syrup are preferably placed next to the food pots (Fig. 8.10). One or two feeders can be used depending on the strength of the colony and the rate of consumption. Feeding the colonies twice a week seems a good standard measure. As explained before, feeding should stop when the bees significantly reduce consumption, or when nectar collection becomes evident. If the bees mix the nectar with syrup, it would spoil the honey. In the case of beekeepers selling colonies instead of honey, artificial feeding is an important part of the management, and can be used all year round (Villas-Bôas 2012).

Stingless bees normally store large amounts of pollen, and it may not be necessary to provide additional pollen or pollen supplements. Nonetheless, some attempts have been made to produce protein supplements for stingless bees using soy meal (18% protein). These supplements are usually well accepted by the bees. Brazilian *M. flavolineata* bees, feeding on a pollen supplement, showed significant development of the hypopharyngeal glands (Costa and Venturieri 2009). To date, no commercial brand for pollen substitutes is available in Mexico for stingless bees.

8.6 Harvesting Honey, Pollen, and Propolis

During the flowering season, the colony reserves of honey and pollen grow considerably. Because of the abundance of resources, it is possible to harvest colonies without compromising their well-being. It is recommended that only half of the honey reserves should be harvested to allow a quick recovery of the colonies (González-Acereto 2008). In the Yucatan Peninsula, the harvest season takes place between January and April (Fig. 8.1).

The process of honey harvest has to be carefully conducted to avoid extensive damage to colonies, and to obtain a product of high quality for human consumption. In this regard, it is important that the process meets top hygienic standards, similar to those handling any other food. The harvest should be preferably made in a closed room. However, it is also possible to harvest in the field, provided that hygienic conditions are ideal. For this, a large tent with a canvas roof and mosquito net walls is good (González-Acereto 2008).

All materials in contact with the honey should be previously cleaned using hot water and neutral detergents or soap (Villas-Bôas 2012). Honey is sugar concentrated and can easily ferment if its humidity increases, so all containers and materials used for the extraction should be dry.

It is recommended using ceramic, glass, and stainless steel to avoid contamination of the honey. Utensils for the harvest include knives of different sizes, strainers, and colanders of different mesh size, cotton cloth, sterile plastic syringes (10–

20 mL), jars, and jugs. The people involved in and in contact with the honey should wear face masks, gloves and aprons as additional hygienic measures (Fig. 8.11).

A simple way to harvest the honey is to first identify the honey pots. Usually honey pots are mixed with those containing pollen, but have a wet appearance and

Fig. 8.11 Honey harvest in stingless bees: above, using a 20 ml syringe with a 14 G needle; middle, using a syringe with a plastic extension; and below, a honey extractor made with an adapted electric vacuum pump

are soft. Then, with the help of a needle the honey pots are punctured one by one. As soon as the pots are perforated, the honey will start to pour. Because honey is a viscous material, it is difficult to extract using small syringes. Syringes of volumes of at least 20 mL are more appropriate for the extraction of honey. Alternatively, needles with a gauge of 18G or below could be adapted to syringes of small volume (Fig. 8.11). Another option is to adapt a narrow plastic hose of sufficient diameter to a syringe. These simple methods of honey extraction could be lengthy when harvesting many colonies (Fig. 8.11).

A faster method to extract the honey is using an electric suction pump. A homemade device could be made using a portable vacuum cleaner adapted to a plastic container (2–3 L), where the honey will be collected (Fig. 8.11). This method is fast, but needs access to electricity, which may be difficult in remote areas. In addition, it may increase the amount of bubbles in the honey, and possibly creates more contact with air microorganisms, so it needs to be done slowly (Villas-Bôas 2012). An alternative in case of no access to electricity, could be the use of a bike floor pump (Villas-Bôas 2012).

Harvesting the honey during the process of colony transference to hive boxes involves some other considerations. Groups of pots (packages) obtained from the colonies (see Sect. 3) should be placed in clean plastic containers, while the transference of the colony is carried. Afterwards, extracting the pot contents should be done in two steps.

First, to obtain "pure" honey, the beekeeper should use a syringe or a sucking pump as explained. Pollen pots should not be opened at this time, to avoid mixing with the honey (see extraction process further).

After extraction of honey, pollen pots can be opened to scoop out their contents. However, at the end of the harvest, some pollen and honey pots remain that are difficult to extract, because they are intermingled. With the help of a knife, the pot package is sliced, and the contents (honey and pollen mixed) can be collected. This mixture should not be combined with the honey first extracted.

It may be necessary to squeeze the pot package with the hands. This mixture of pollen and honey can be commercialized too. It is becoming popular in the Yucatan Peninsula with the name of "Mayan jelly".

After the harvest, any residues of wax or debri are filtered from the honey using a clean cotton cloth. The filtered honey can be stored in polycarbonate-free plastic or glass containers, at room temperature, or in the fridge. The containers to keep the honey should be previously sterilized using hot water, but must be completely dry to avoid increasing humidity in the honey. It is also important to maintain the containers sealed to avoid exposure to environmental humidity.

Stingless bee honey can be consumed straight away after the harvest. If stored for long periods, one potential problem is fermentation because of its natural high humidity. In this case, the life of the product can be extended, and its properties maintained using various methods (Venturieri et al. 2007; Villas-Bôas 2012). Avoiding exposure to heat (direct or environmental) is essential, as this can alter sugar and nutrient contents. Some enzymes in the honey are heat sensitive and can degrade in a hot environment. The content of hydroxymethyl furfural (HMF) increases in honeys maintained in warm environments for long periods. This compound is a product of the degradation of the fructose when exposed to heat.

Honey can be preserved in the fridge at 2–4 °C. At such temperature, the honey can be kept for 1 year without altering its quality. It is important to bear in mind that some honeys crystallize or granulate when exposed to cold temperatures (Venturieri et al. 2007). This is due to the balance of the different type of sugars in the honey. In the Yucatan Peninsula, honey from *V. dentata* can easily crystallize. It is not recommended to maintain this type of honeys in refrigeration.

Another method to preserve stingless bee honey is dehydration (Venturieri et al. 2007; Villas-Bôas 2012). For this process the honey is poured on innocuous plastic or stainless steel trays to expose it to cold temperatures in an air-conditioned room for 12–48 h. The humidity is gradually removed from the honey with the help of a dehumidifier. The dry air produced by the dehumidifier absorbs moist from the honey, which is removed from the room by air conditioning. The humidity in the honey should be checked with a refractometer until it reaches 20%. After this process, the treated honey can be stored in sterile jars. Dehydrated honey can be preserved for 2 years (Villas-Bôas 2012).

Pasteurization can be used to preserve stingless bee honey too. As explained before, honey exposed to high temperature can be damaged. Because of this potential problem, when pasteurizing honey, a water bath is used to avoid direct contact with the heat source. A second condition is a maximum temperature of 65 °C in the product. Above this limit, the sugars start burning and the amino acids and vitamins deteriorate. The maximum temperature should be slowly reached. For pasteurization, only sterile glass jars should be used. The jars are filled with honey to about 85% of their capacity and placed without the lid on a metal or ceramic pot which is then deposited in a larger container with water. The container is submitted to a heat source. When temperature in the honey reaches 65 °C, the jars are left for 5 min at this temperature and then removed from the heat. The lid should be immediately sealed and the jars can then be submerged in tap water or left to cool down at room temperature. However, to better maintain taste, smell and nutritional properties pasteurized honey should be cooled down quickly (Venturieri et al. 2007). Pasteurization can extend the life of stingless bee honey between 6 months and 1 year (Villas-Bôas 2012).

Another process to preserve stingless bee honey involves partial fermenting and is called maturation. Indeed, for centuries some indigenous cultures have consumed fermented stingless bee honey (Quezada-Euán et al. 2001; Villas-Bôas 2012). For maturation, the honey is stored in glass containers at room temperature, close to but not exceeding 30 °C. At such conditions, fermentation should start. Two weeks after, the lids on the bottles are partially opened so the carbonic gas produced by fermentation can escape. After some minutes, the container is sealed. This process is repeated weekly for a period of 3–6 months. Then fermentation should have stabilized. An indicator of stabilization is when the foam on the honey sticks to the neck of the jar, and does not move when leaned (Villas-Bôas 2012).

An extensive account on stingless bee honey, its physicochemical, microbiological, and organoleptic characteristics, is presented in the book *Pot-Honey* edited by

Table 8.2 Comparison of the composition of Africanized *A. mellifera* and stingless bee honeys from the Yucatan Peninsula (Catzín-Ventura et al. 2008)

	Species					
	A. mellifera (n = 10)		*M. beecheii (n = 10)*		*Scp. pectoralis (n = 10)*	
Parameters	$\bar{x}$	St. dv.	$\bar{x}$	St. dv.	$\bar{x}$	St. dv.
Humidity (%)	17.3[a]	1.6	24.1[b]	1.2	24.1[b]	1.3
pH	4.5[a]	0.4	4.3[a]	1.1	4.1[a]	0.5
Acidity (mg/kg^{-1})	38.1[a]	4.0	40.3[a]	10.5	44.7[a]	15.5
Glucose (%)	36.7[a]	8.3	10.0[b]	1.8	13.4[c]	2.5
Sucrose (%)	0.4[a]	0.1	0.1[b]	0.1	0.01[c]	0.0
HMF (mg/kg^{-1})	8.6[a]	2.9	5.7[a]	8.3	1.2[b]	0.9

Different letters in the same row indicate significant differences at $p < 0.05$

Vit et al. (2013), and will not be analyzed in detail in this book. In general, the so-called pot honey (from stingless bees) shows contrasting differences with the honey of *A. mellifera.* In the Yucatan, Catzín-Ventura et al. (2008) compared *M. beecheii, Scp. pectoralis* and *A. mellifera* honeys (Table 8.2). Interestingly, the honey of both stingless bees was only slightly more acid, but not significantly so, compared with the honey bee. However, significant differences were found in the percentage of humidity, the content of different sugars, and HMF (Table 8.2). Stingless bee honeys had significantly lower amounts of glucose, sucose, and HMF.

In contrast with honey, stingless bee pollen has not been widely consumed and exploited. There is a potential for pollen harvest because many species store large quantities, and it is nutritionally rich (Menezes et al. 2012). For example, pollen analyses of two Brazilian species of *Melipona* showed protein contents between 24 and 38% (Rebelo et al. 2016). In the Yucatan Península, pollen is traditionally discarded by traditional beekeepers, but in other parts of Mexico, pollen is mixed with juice, milk, or water. More recently, pollen is gaining attention as an ingredient of exotic and traditional cuisine (pers. Obs.; VII seminario Mesoamericano abejas nativas). In the Yucatan honey mixed with pollen is becoming popular to make Mayan jelly.

Stingless bee pollen is normally obtained directly from the pots during the harvest of honey. No process to exclusively harvest pollen has been developed for stingless bees. Attempts to use conventional pollen traps, as in the honey bee, have failed in stingless bees because of their small entrances, and the tendency to deposit propolis on objects in front of the nest (Menezes et al. 2012). Scooping the pollen out from pots in the colony carries the problem of infiltration of phorid flies. However, during colony transference the use of a spoon is a rapid method to collect pollen. A technique to obtain fresh pollen from stingless bees has been proposed in Brazil (Menezes et al. 2012). At the peak of the foraging activity, a strong colony is selected and relocated 5–10 m away, placing an empty hive in its site. This empty hive is previously supplied with cerumen, if possible from the mother colony. The entrance tube from the original hive, or a part of it, should also be placed at the entrance of the empty hive. The foragers returning from the field should start building new pots and store honey and pollen in the new, empty hive. After a week, the pollen and honey can be harvested from the new pots without the risk of brood dam-

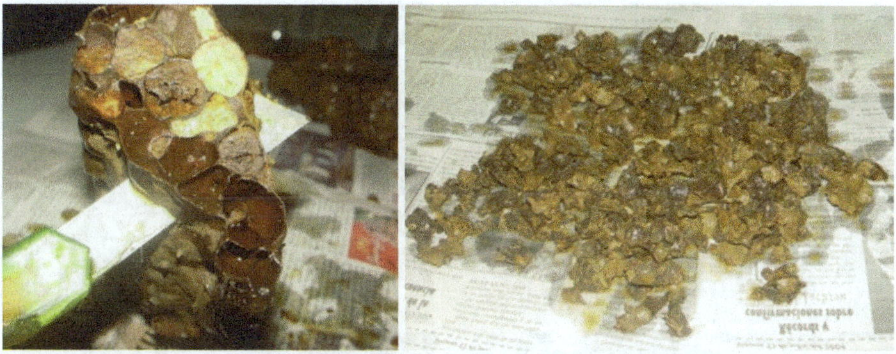

Fig. 8.12 On the left slicing pot packages to collect the pollen after the harvest of honey. On the right cerumen washed and drying on paper

age. In a week, an average of 54 g of pollen was obtained per colony in *Scp. depilis* (Menezes et al. 2012). Although the procedure could be effective to obtain fresh pollen, it involves sacrificing some part of the population.

Pollen should be stored in amber glass containers to reduce exposure to light and preserved in the fridge at 2–4 °C (Fig. 8.12).

After the harvest of honey and pollen, the cerumen should be thoroughly washed to remove residues. Then layers of cerumen can be extended on paper and left to dry for 1 or 2 days under shade. After this, the cerumen can be used again in the colonies or melted to make artificial feeders. If not used immediately, the cerumen can be stored in the fridge or at room temperature. In this case, it should be important to store it in plastic bags tightly sealed to avoid destruction by insects.

A product with potential use is resin. However, presently the harvest of stingless bee resin is limited, and there is little information on its medicinal properties (Fig. 8.13).

Some species like *Fr. nigra* build large resin stores (Fig. 8.13). Part of these reserves can be harvested. For this, a sharp knife is used to remove the propolis. The process should be made carefully to avoid contaminating the resin with other materials from the colony (Fig. 8.13). Only about half of the resin stores should be removed because it is an important material for the bees, basic for construction and defense.

8.7 Control of Parasites and Predators

The large number of colonies concentrated in a meliponario could potentially attract predators and increase the risks of disease transmission. Keeping colonies with strong populations is probably the best way to deal with the attacks of predators and to resist infection. An important basic recommendation is to also avoid large disruptions when managing the colonies.

Fig. 8.13 Collection of resins in *Fr. nigra*: (**a**, **b**) deposits of resin; (**c**) a sharp knife is used to remove them; (**d**) the resins are deposited in a clean container

Interestingly, stingless bees are resistant to some major parasites of the honey bee. Stingless beekeepers in Mexico frequently worry about the effects of the Asian Varroa mite (*Varroa destructor*) or the African small hive beetle (*Aethina tumida*). In the case of Varroa, the method of brood production in stingless bees involving immediate cell capping after oviposition, prevents these mites from entering the cells to reproduce. In the honey bee, mites are attracted by larval pheromones, and can enter the brood cells that remain open throughout the development of the larvae (Rickli et al. 1992). In addition, there seems to be a high degree of host specificity in *Varroa* mediated by semiochemicals (Pernal et al. 2005), meaning that it may not be attracted to stingless bee colonies. Indeed, the Varroa mite has coexisted with stingless bees in tropical Asia, with no evidence of the mite reproducing in their colonies. To our knowledge, no report of Varroa mites in any species of stingless bee in the Americas has been made. Interestingly, some mites can pose a risk to stingless bee colonies. In México and other parts of Latin America, the cosmopolitan parasitoid mite *Pyemotes tritici* has been reported destroying colonies of *M. subnitida* (Nogueira-Neto 1997), *M. colimana* (Macías-Macías and Otero-Colina 2004), *Fr. varia* and *Ttr. angustula* (Menezes et al. 2009). Infested ants or beetles could have been a source of infestation for *M. colimana* (Macías-Macías and Otero-Colina 2004). Nonetheless, it is noted that infestation occurred after transferring a colony to a new hive, and did not spread to other colonies (Macías-Macías and Otero-Colina 2004). This indicates that a large disruption may have created conditions for mite invasion or reproduction in the colony.

In the case of the small hive beetle *Aethina tumida* the potential of colony invasion exists, but seemingly only for weak colonies. Small hive beetles originally come from Africa, but are now spread across Mexico. They are opportunists, and are known to invade weak honey bee colonies, where their larvae can cause severe damage. The same seems to apply to stingless bees, but invasion rates seem low. In Cuba, the rate of *M. beecheii* colonies parasitized by the beetle is below 3%, and is associated with previous damage of the colonies during management (Lóriga-Peña et al. 2014). Thus, the best method to prevent invasion by these beetles is to maintain the colonies with good populations and avoid practices that extensively damage them, during the harvest or transference to hive boxes.

Other insects can be particularly destructive to stingless bee colonies. That is the case of the fly *Pseudohypocera kerteszi* (family: *Phoridae*), locally known as "*Nenem*" (in Maya), or humpback fly. This fly is not a specific plague of stingless bees; like the small hive beetle it is an opportunist (Robinson 1981; Wolff and Nava 2007). In strong colonies, the flies are rapidly chased away by guards, but weak or disturbed colonies can be attacked and destroyed. Maggots are the real cause of the destruction. They use their mandibles to perforate pots and combs feeding on their contents including immature stages (Fig. 8.14).

Adult flies are small, between 2 and 3 mm long, the females being larger than the males (Fig. 8.15; Robinson 1981). Only gravid females are usually seen inside the colonies, suggesting that mating presumably takes place outside the colony

Fig. 8.14 Above left a gravid female of *Pseudohypocera kerteszi*, the humpback fly whose voracious maggots can destroy stingless bee colonies (above right). Below, vinegar traps inside a colony (left) and with several drown flies (right)

Fig. 8.15 An attack of army ants on a colony of *M. beecheii.* A few guards have blocked the entrance to stop the ants from entering the colony. It can be seen that the ants have decapitated some guards and bee foragers have been left outside the colony. The block is removed from inside the colony after the danger has passed (photograph by Humberto Moo)

(Robroek et al. 2003). The females are attracted by the acid smell of exposed pollen and larval food where they lay their eggs (Robinson 1981; Ramos et al. 2003). Eggs eggs hatch rapidly, only 12–18 h after laid (Robroek et al. 2003). Female flies are prolific, producing up to 34 eggs during their life span of 35–45 days. The total time for the development from egg to adult is approximately 10 days, varying in accordance to environmental temperatures (Robroek et al. 2003).

In the Yucatan, the stingless bees *Scp. pectoralis* and *Fr. nigra* are particularly resistant to humpback fly invasions. These two species can successfully repel the invasions, apparently due to their high levels of hygienic behavior. Eliminating maggots in such species can be quick and the bees pile them in waste dumps before removing from the colony. Other species are more susceptible to the invasion of flies, among them *M. beecheii, Cph. zexmeniae, Tr. fulviventris*, and *Pb. frontalis* (González-Acereto 2008).

Traditional methods are applied by local beekeepers to repel humpback flies. The leaves of some plants have been used, like "*chaca'j*" (*Bursera simaruba*), "*nabanché*" (*Elaphrium pubescens*), and the toxic herb "*ik aban*" (*Croton humilis*). Before a colony is transferred or divided, the interior of the hive is smeared with leaves of those plants. Various degrees of success have been reported, but the effectiveness of the different plants to repel flies is still not clear (González-Acereto 2008).

An efficient method to control the number of flies in stingless bee colonies is the use of traps with acetic acid (Ramos et al. 2003; Oliveira et al. 2013). In the Yucatan, Ramos et al. (2003) found that commercial apple vinegar (5% of acetic acid) was highly attractive to female humpback flies (Fig. 8.14). They designed a trap using a plastic cup adding 5–10 mL of apple vinegar and sealed with a plastic cap. Small holes (about 2 mm) are made on the cap to allow passage of the adult female flies.

The flies, attracted by the smell, enter the cup and drown in the acid (González-Acereto et al. 2006). These traps catch many adult female flies inside colonies reducing the potential number of maggots invading pots and cells (González-Acereto 2008). Depending on the severity of fly invasion, one or two traps can be used per hive. Because of the volatility of the acetic acid, it is recommended replacing it every two days.

If traps donot reduce the number of adult flies and maggots, it may be necessary to clean the colony. The beekeeper would have to remove all the combs and pots out of the hive. The hive should be thoroughly cleaned with a damp cloth (only water should be used). The pots and combs should be inspected to remove maggots with the help of forceps. Any residues of pollen or larval food should also be cleaned with a wash pump or a damp cloth. The space occupied by the colony in the hive should be adjusted accordingly, using a block of plywood or cardboard. The affected colony could be swapped with a strong one to increase the number of guard and forager bees. It is important to evaluate the severity of the damage; if colonies are too weak, they may not recover despite these measures (González-Acereto 2008).

Other members of the phorid flies develop as internal parasites (parasitoids) of various Apids, including stingless bees. Approximately 170 species are now recognized in the neotropical genus of parasitoid flies *Melaloncha* (Brown 2016). The female of these bee-killing flies lays eggs inserting her ovipositor between the soft membranes of the exoskeleton. The hatching larva feeds on the organs and tissues of the host (Brown 2004, 2005, 2016). There are two major groups of *Melaloncha*: the *M. cingulata* group seems to prefer *Partamona* and *Scp. mexicana*, while flies of the *M. furcata* group frequently attack species of *Plebeia* (Brown 2004, 2005). The extent of the damage caused by bee-killing flies to the colonies of stingless bees is yet unclear.

Stingless bees are also at risk by one of the most important predators of neotropical forests, the army ants (Formicidae: Ecitoninae). Army ants form nomadic colonies frequently preying on other social insects (Hölldobler and Wilson 1990) and, thus, are important agents in shaping the structure of tropical communities (Kaspari and O'Donell 2003). In the Yucatan Peninsula *Eciton burchelli parvispinum* is the most common species of army ant, but at least 12 other species in the genera *Neivamyrmex, Nomamyrmex, Cheliomyrmex*, and *Labidus,* have been reported (Watkins 1982). A common strategy used by stingless bees against the attacks of army ants is blocking the entrance with resin and cerumen, and in some species like *M. beecheii* with the body of guards (Fig. 8.15). Although statistics are lacking, we have not seen a dramatic effect of army ants on our meliponarios. Annually, one or two colonies (out of a 100) can be lost to ant raids. Perhaps the fact that colonies are kept on bases full of water, and above ground, ameliorates the impact of ants. Nonetheless, it is possible that colonies in the wild are more frequently raided and destroyed (Roubik 1989). In contrast with their effect on stingless bee hives, army ants could be a severe problem for apiaries in the Yucatan (pers. Obs.).

Interestingly, frequent attackers of stingless bees are other stingless bees. A recent review suggests that at least 60 species of stingless bees may be involved in heterospecific conflicts. However, only a few species seem particularly prone to robbing during such attacks (Grüter et al. 2016). In the Yucatan, we have seen that

Fig. 8.16 An attack of *L. niitkib* to a colony of *Fr. nigra*. On the left robber bees opening brood cells to steal their contents and the cerumen. On the right a detail of a worker of *L. niitkib*

Scp. pectoralis and *M. beecheii* tend to more frequently rob colonies of the same and other species. In meliponaries, the proximity of colonies can trigger robbing, particularly in periods of scarcity. It is therefore recommended that artificial feeding and other management during dearth are carefully conducted, to avoid honey or syrup spilling over and trigger pillage. Feeding colonies late in the afternoon can reduce the risk of robbing (pers. Obs.).

An obligate robber stingless bee is *Lestrimelitta*. The colonies of this neotropical species survive exclusively by plundering other colonies for food and building materials. Even for reproduction, *Lestrimelitta* swarms invade the nest of other species (Quezada-Euán and González-Acereto 2002). *Lestrimelitta* shows preference for certain hosts (Sakagami et al. 1993; Quezada-Euán et al. 2013) and, within a species, some colonies (apparently weak or with food of better quality), are preferred over others (Sakagami et al. 1993; Grüter et al. 2016).

Once *Lestrimelitta* has invaded a host, it is difficult to stop the attack (Fig. 8.16). In the Yucatan one method that can prevent attacks by local *L. niitkib* is the introduction of three or four colonies of *Scp. pectoralis,* distributed across the meliponario. The colonies of *Scp. pectoralis* are aggressive, and the smell of citral triggers a strong defensive response (pers. Obs.). When *L. niitkib* workers release citral during their attacks, *Scp. pectoralis* guards can detect and repel the intruders stopping the attack to colonies in the meliponario (González-Acereto and Medina-Medina 1998). A similar strategy has been proposed by Rech et al. (2013) in Brazil, using colonies of *Duckeola* and *M. fulva* to protect colonies in meliponarios of central Amazonia.

For the beekeeper, *Lestrimelitta* represents a pest and an enemy to eradicate Nogueira-Neto (1997). The number of colonies of *L. niitkib* destroyed annually in the Yucatan is unknown, but there is increased difficulty to find nests of this species (pers. Obs.), suggesting possible reduction in the population. Predators, as these cleptobiotic bees, possibly contribute to maintain the stability of stingless bee populations. In spite of the apparent destructive effect, *Lestrimelitta* could contribute to keep healthy populations and regulate their numbers (Roubik 1989).

Regarding other insect pests, interestingly, no wasp parasitoids or beetles are reported affecting stingless bees or their nests (Roubik 1989).

The intense transit of foraging bees in meliponarios attracts larger predators, like toads (*Bufo*) and lizards (*Anolis, Sceloporus*). These animals do not normally represent a great danger to colonies, as they only capture a few bees. However, we witnessed a temporarily high density of *Anolis* that were particularly keen on a nest of *L. niitkib* and destroyed it. The lizards waited at the nest's entrance tube and ate the bees coming in and out of the colony. To deal with this problem, we used vaseline around the base of the entrance tube. The lizards were repelled by the vaseline sticking to their feet. An alternative is also to use a large plastic cup, remove the bottom part, and place the device around the entrance tube so that lizards cannot climb directly on it. It is also important that the hives are not directly placed on the floor. Bases should be used to keep colonies at least 1 m above ground; this will also reduce the attacks by toads.

Mammals can have some impact on meliponarios. In some areas of Mexico, a small carnivore, *Tayra (Eira) barbara* (Mustelidae), can become a problem (De Araujo Freitas et al. 2010). Beekeepers frequently poison these creatures locally called "cabeza de viejo (*oldman head*)," but this practice has a negative impact on the ecosystems. One way to deal with this problem is to use traps to catch the animals and relocate them in areas far away from the meliponarios. The meliponarios can also be protected with some type of farm mesh around. This practice can reduce the incidence of attacks by *T. barbara* and other mammals like raccoons, possums, armadillos, and anteaters.

8.8 Hazards and Threats to Stingless Bees in Mexico

A priority today is the preservation of ecosystems and the services they provide to humankind. In this context, a central issue is the protection of keystone species that provide ecosystem stability. One of these important keystone groups is the pollinators (IPBES 2016). Sadly, it is well established that pollinators have experienced various degrees of decline worldwide (Biesmeijer et al. 2006; Freitas et al. 2009; IPBES 2016). Particularly important is a drastic reduction in the number of managed honey bee colonies in the Unites States and Europe, that triggered an international alert and discussion on the future of food production in relation to pollination (Potts et al. 2010). Concomitantly, the loss of honey bee colonies has helped to reveal the importance of preserving native bees as alternative providers of pollination services (Kremen et al. 2002; Garibaldi et al. 2013; Winfree et al. 2015). Nonetheless, declines in various species of native bees have also been documented in temperate areas (Biesmeijer et al. 2006). Regrettably, scant information is only available on the status of native bee populations in the tropics (Freitas et al. 2009). However, some tropical species are considered under threat, suggesting that bee declines are occurring in these areas too (Kerr 2002; Freitas et al. 2009; Vandame and Palacio 2010; Villanueva-Gutiérrez et al. 2013).

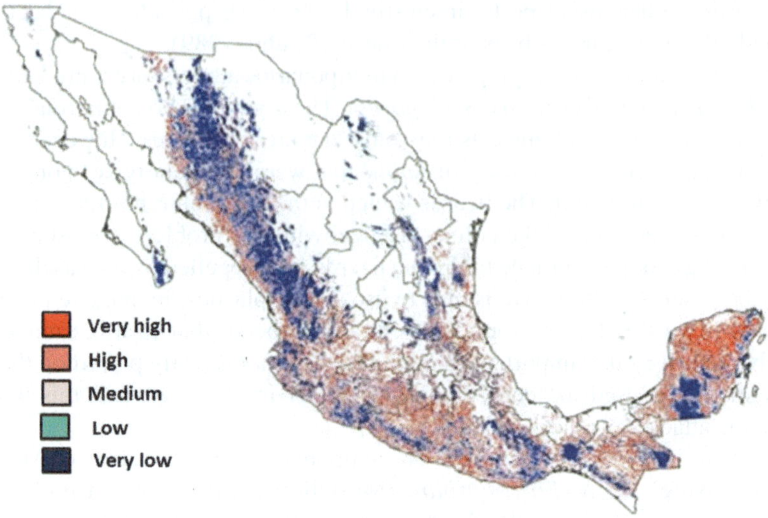

Fig. 8.17 Deforestation rates in Mexico (INEGI 2011)

Several drivers have been considered as major causatives of the decline of bees and other pollinators (Potts et al. 2010; DeGrandi-Hoffman and Chen 2015; García-Morales et al. 2016; IPBES 2016). Two of the most important drivers affecting pollinators in tropical areas are habitat loss and indiscriminate use of agrochemicals (Freitas et al. 2009).

Deforestation severely affects tropical biodiversity. In Mexico, agricultural expansion, road building, and urban development are the main causes of forest disappearance (Gibson et al. 2011; Laurence et al. 2014). Indeed, deforestation rates are high (Vidal et al. 2013; Rosete-Vergés et al. 2014). During the mid-1980s, the rate of deforestation in Mexico was 600,000 ha per year, by the mid-1990s it increased to 700,000 ha per year, and by the year 2000 that rate increased to 800,000 ha per year (Herrera and Herrera 2003). The Yucatan Peninsula is the region with the highest deforestation rate in the country (INEGI 2011) (Fig. 8.17).

The Yucatan has experienced a long history of deforestation since the Maya; large areas of rainforest were cleared down for agriculture and fuel (Abrams and Rue 1988). After the Spanish conquest, the development of large haciendas, with the cultivation of sugar cane (*Saccharum officinarum*) and henequen (*Agave fourcroydes*) further decimated the forests, in particular in the center and north of the Peninsula.

At the end of the last century (between 1980 and 1990), deforestation substantially increased as extensive agriculture expanded (Rueda 2010). More recently, urban developments have become an additional threat, with large sections of forest removed around Merida and Cancun. Uncontrolled man-made fires are also considered major hazards to the forest cover of this region (Ellis et al. 2017).

An overall impact of deforestation on Mexico's native apifauna has not been assessed. However, regional studies have shown that the diversity and abundance of bees decrease in altered landscapes compared with preserved fragments (Aguirre and Dirzo 2008; Meneses Calvillo et al. 2010). Similarly, the diversity of bees in crops varies with the degree of forest conservation around them. More diverse and abundant bee faunas are present in forests around *J. curcas* (Romero and Quezada-Euán 2013) and hot habanero pepper *C. chinense* (Landaverde-González et al. 2017). These studies suggest that deforestation in Mexico, as in other countries, is negatively affecting bee populations.

The consequences of small population size can be more severe in bees compared with other organisms. Small population sizes in Hymenoptera, carry the risk of an extinction vortex (Gilpin and Soulé 1986; Pellet et al. 2007). When colony numbers decrease, the numbers of alleles at the sex locus *csd* also reduce, increasing the potential for inbreeding and the production of diploid males, that represent reproductive dead ends (Crozier 1971; Crozier and Pamilo 1996; Zayed and Packer 2005).

Habitat degradation can also alter genetic flow by reducing the connectivity between regions (Brown and Albrecht 2001; Hanski 2011). In the case of stingless bees, characteristics like colony philopatry, and small body size of many species could further reduce the exchange of reproductives, increasing genetic isolation and inbreeding (Brown and Albrecht 2001; Quezada-Euán et al. 2012).

In general, moving towards more sustainable and pollinator-friendly agriculture and diversifying the agricultural landscapes could reduce the risks of pollinator decline. For this, three recommended key aspects to consider are (1) improving agricultural production and livelihoods while minimizing environmental damage; (2) strengthening diversified farming systems; and (3) investing in the protection and restoration of natural and seminatural habitats (IPBES 2016).

In Mexico, an important step towards the conservation of habitats is educating on the importance of pollinators. Presently, farmers are not aware of the economic value of pollinators and the importance of preserving them for the productivity of their farms. Farming practices could change through the economic incentive of increased crops (Morandin and Winston 2006; IPBES 2016).

Urban habitats could also be used for the conservation of bees and pollinators (Hernández et al. 2009). In general, people in cities are not informed of the importance of bees. Educating people is a key aspect to change perception on beneficial insects and reduce threats. Habitats for pollinators can enhance life quality, promoting and diversifying home and public gardens (Gaston et al. 2005). Large cities in Mexico are frequently limited in public gardens, but council programs could encourage "planting for bees", recommending and delivering flowering plants that can be used by bees and other insects. Artificial nest sites in parks (MacIvor and Packer 2015) and road edges could improve the number of bees in these areas too (Dicks et al. 2010). School gardens with bees nesting in them could be used to teach the importance of habitat conservation and pollinators to young generations (Decourtye et al. 2010; Dicks et al. 2010; Quezada-Euán and Ayala 2010).

Another major threat related with the expansion of agriculture is a heavy use of agrochemicals for pest and disease control in crops (IPBES 2016). All over the world, the extensive and uncontrolled use of pesticides (fungicides, herbicides, insecticides, rodenticides) is identified as a major factor threatening pollinators and wildlife (Phalan et al. 2013; Costantini 2015). Regarding bees, only a limited number of studies have assessed the effect of agrochemicals on non-*A. mellifera* species (Devillers and Pham-Delègue 2002; Fischer and Moriarty 2014; Johnson 2015). Regrettably, studies in tropical areas with native species are also scarce (Johansen and Mayer 1990; Belzunces et al. 1999). In addition, there is limited evidence of the interaction between agrochemicals and other drivers (multifactorial studies) (Motzke et al. 2015). Nonetheless, public awareness is rising and various initiatives to reduce the use of pesticides and promote alternative safer methods are increasing (IPBES 2016).

Toxicological studies on stingless bees are comparatively recent and the toxicity of different compounds has only been assessed on a few species (Malaspina and Stort 1983; Macieira and Hebling-Beraldo 1989; Moraes et al. 2000). Nonetheless, similar to the honey bee, many common pesticides are highly toxic to stingless bees, in accordance to the criteria of LD_{50} < 2 μg/specimen (Table 8.3; Devillers and Pham-Delègue 2002). However, differences exist in the toxicity of individual compounds between species (Table 8.3).

The first insecticides widely used belong to the organophosphate and carbamate family. These compounds are highly hazardous to bees, but their use is declining because of the risk to humans and increased resistance of target insects (Johansen and Mayer 1990). In Mexico, the use of DDT was prohibited in 1999 and other organochlorine compounds have been gradually banned, but organophosphate and carbamate insecticides are still in use (Waliszewski et al. 2013).

Neonicotinoids are a new generation of insecticides that have become popular because they seem potentially less harmful to humans and other vertebrates. However, neonicotinoids are highly toxic to insects, whose neuronal receptors present high affinity to such compounds (Casida and Durkin 2013). In insects, neonicotinoids block the neuronal receptors for acetylcholine, causing hyperexcitement and rapid death (Casida and Durkin 2013; Van der Slujs et al. 2013). In Mexico, the use of neonicotinoids started when imidacloprid was approved in 1993. Later on, in 2004 thiamethoxam was also endorsed. Both neonicotinoids are mainly used in Solanaceae (potato, tomato, and peppers) for the control of whitefly (*Bemisia tabaci*), a vector of viral diseases of economic interest (Gutiérrez-Olivares et al. 2007).

A potential problem of neonicotinoids on pollinators is their systemic mode of action. They are absorbed by the plant's roots and can reach the nectar and pollen (Stoner and Eitzer 2012). Neonicotinoids are also used as coat protection of seeds against the attacks of insect pests. However, they are active throughout the germination period, and can be potentially toxic to insects other than the target (Girolami et al. 2009). Laboratory tests have shown that neonicotinoids are among the most toxic compounds to stingless bees (Table 8.3; Valdovinos-Nuñez et al. 2009; Soares et al. 2015; Costa et al. Rosa et al. 2016).

Another class of insecticides gaining popularity is the so-called biopesticides. These compounds are derivatives from microorganisms or plants (Lasota and Dybas

Table 8.3 Relative toxicity of pesticides to stingless bees and honeybees, presented as lethal doses LD_{50}. Studies where LC_{50} is reported are labeled (*). Nontoxic doses are indicated in bold italics. All data are reported for worker bees, except where indicated

Pesticide group	Compound	Bee species	LD_{50} (μg/bee)	Reference
Organochlorines	Heptachlor	*Trigona spinipes*	0.016	Macieira and Hebling-Beraldo (1989)
	Dieldrin	*Trigona spinipes*	0.028	Macieira and Hebling-Beraldo (1989)
		A. mellifera	0.0006	Tasei (2002)
	Lindane	*Trigona spinipes*	0.133	Macieira and Hebling-Beraldo (1989)
	Endosulfan	*Trigona spinipes*	0.209	Macieira and Hebling-Beraldo (1989)
Organophosphates	Diazinon	*M. beecheii*	0.240	Valdovinos-Nuñes et al. (2009)
		M. beecheii (gyne)	0.190	Valdovinos-Nuñes et al. (2009)
		M. beecheii (male)	0.185	Valdovinos-Nuñes et al. (2009)
		Fr. nigra	0.240	Valdovinos-Nuñes et al. (2009)
		N. perilampoides	0.190	Valdovinos-Nuñes et al. (2009)
		A. mellifera	0.38	Sanchez-Bayo and Goka (2016)
	Parathion	*T. spinipes*	0.095	Macieira and Hebling-Beraldo (1989)
		A. mellifera	0.003	Tasei (2002)
	Malathion	*T. spinipes*	0.264	Macieira and Hebling-Beraldo (1989)
		Scp. tubiba	0.040	Moraes et al. (2000)
		A. mellifera	0.002	Tasei (2002)
	Trichlorfon	*Scp. tubiba*	0.080	Moraes et al. (2000)
		A. mellifera	0.024	Tasei (2002)
	Acephate	*T. spinipes*	0.423	Macieira and Hebling-Beraldo (1989)
	Methamidophos	*M. quadrifasciata*	0.066*	Del Sarto et al. (2014)

(continued)

Table 8.3 (continued)

Pesticide group	Compound	Bee species	LD_{50} (µg/bee)	Reference
Carbamates	Methomyl	*M. beecheii*	0.410	Valdovinos-Nuñes et al. (2009)
		M. beecheii (gyne)	0.290	Valdovinos-Nuñes et al. (2009)
		M. beecheii (male)	0.250	Valdovinos-Nuñes et al. (2009)
		Fr. nigra	0.150	Valdovinos-Nuñes et al. (2009)
		N. perilampoides	0.120	Valdovinos-Nuñes et al. (2009)
		T. spinipes	0.140	Macieira and Hebling-Beraldo (1989)
		A. mellifera	0.08 *	Drescher and Geusen-Pfister (1991)
	Carbaryl	*T. spinipes*	0.740	Macieira and Hebling-Beraldo (1989)
Pyrethroids	Permethrin	*M. beecheii*	0.098	Valdovinos-Nuñes et al. (2009)
		M. beecheii (gyne)	0.066	Valdovinos-Nuñes et al. (2009)
		M. beecheii (male)	0.072	Valdovinos-Nuñes et al. (2009)
		Fr. nigra	0.031	Valdovinos-Nuñes et al. (2009)
		N. perilampoides	0.014	Valdovinos-Nuñes et al. (2009)
		T. spinipes	0.072	Macieira and Hebling-Beraldo (1989)
		A. mellifera	0.19–0.28	Leahey (1985)
	Cypermethrin	*T. spinipes*	0.070	Macieira and Hebling-Beraldo (1989)
		A. mellifera	0.03–0.05	Inglesfield (1989); Greig-Smith et al. (1994)
	Fenvalerate	*T. spinipes*	***1.088***	Macieira and Hebling-Beraldo (1989)
	Deltamethrin	*Scp. tubiba*	0.730	Moraes et al. (2000)
		M. quadrifasciata	0.082*	Del Sarto et al. (2014)
		A. mellifera	0.010	Tasei (2002)

Neonicotinoids	Imidacloprid	*N. perilampoides*	0.001	Valdovinos-Nuñes et al. (2009)
		M. scutellaris	0.0024	Costa et al. (2015)
		M. quadrifasciata	0.024 *	Tomé et al. (2015a)
		S. postica	0.025	Soares et al. (2015)
		A. mellifera	0.018	Iwasa et al. (2004)
	Thiamethoxam	*N. perilampoides*	0.004	Valdovinos-Nuñes et al. (2009)
		A. mellifera	0.03	Iwasa et al. (2004)
	Thiacloprid	*N. perilampoides*	0.007	Valdovinos-Nuñes et al. (2009)
		A. mellifera	***14.6***	Iwasa et al. (2004)
Phenylpyrazole family	Fipronil	*M. scutellaris*	0.0004	Lourenço et al. (2012)
		Scp. postica	0.0005	Jacob et al. (2013)
		A. mellifera	0.013	Tasei (2002)
Biopesticides	Abamectin	*M. quadrifasciata*	0.015*	Del Sarto et al. (2014)
		A. mellifera	0.03	Sánchez-Bayo and Goka (2016)
	Spinosad	*M. quadrifasciata*	0.012*	Tomé et al. (2015a)
		M. quadrifasciata	0.114* (larvae)	Barbosa et al. (2015a,b)
		A. mellifera	0.078 0.063*	Miles (2003)
	Azadirachtin	*M. quadrifasciata*	0.210 * (larvae)	Barbosa et al. (2015a,b)
GM crops	Cry1Ac	*T. spinipes*	***1.8*** (larvae)	Lima et al. (2013)

1991; Sparks et al. 2001). In appearance, biopesticides seem more environmentally friendly, degrading rapidly, and because of this, are preferred in organic systems. However, laboratory studies have detected biopesticide levels of toxicity similar to conventional compounds (Xavier et al. 2010, 2015; Del Sarto et al. 2014; Barbosa et al. 2015a).

One of the first widely approved biopesticides is spinosad; it is obtained from the bacterium *Saccharopolyspora spinosa* (Sparks et al. 2001). This compound is a repellent to *Tr. fulviventris* and *Scp. mexicana* workers (Gómez-Escobar et al. 2014), and when fed to larvae of *M. quadrifasciata* caused deformities (Barbosa et al. 2015b; Tomé et al. 2015a). In studies using contact and oral application, spinosad resulted highly toxic to *Pt. helleri* and *Scp. xanthotricha* (Tomé et al. 2015b). Another popular biopesticide is abamectin, an avermectin product of the fermentation by *Streptomyces avermitilis* (Lasota and Dybas 1991). Abamectin resulted highly toxic to *M. quadrifasciata* (Table 8.3; Del Sarto et al. 2014).

Today azadirachtin is the most frequently used biopesticide. Azadirachtin is an extract from the neem (or nim) tree (*Azadirachta indica*). This plant was introduced in Mexico at the end of last century, specifically for the production of biopesticides (Koul 2004). In spite of initial optimism, recent assessments have shown that azadirachtin is highly toxic to most stingless bees (Barbosa et al. 2015a; Barbosa et al. 2015b; Lima et al. 2016). A significant antifeeding effect was observed when azadirachtin was fed to *M. quadrifasciata* and *Pt. helleri* workers (Bernardes et al. 2017).

Entomogenous fungi have been used to control insect pests by biological means (Ferron 1978). Various entomopathogenic fungi exhibit medium to high toxicity to Mexican *M. beecheii*, *Scp. mexicana*, and *Ttr. angustula* (Toledo-Hernández et al. 2016).

More recently, oil extracts from different plants have been used as biopesticides; among these, garlic, andiroba, citronella, eucalyptus, neem, and rotenone are more popular (Moreira et al. 2007; Mureithi 2008). All these compounds are toxic to honey bees (Xavier et al. 2015). However, in stingless bees contrasting results have been found. *N. testaceicornis* workers were not affected after contact with garlic extract, rotenone, andiroba, citronella, neem, or eucalyptus oils (Xavier et al. 2010). However, in *Ttr. angustula* andiroba, citronella, neem, and eucalyptus oils were toxic, but not the garlic extract nor rotenone. The most toxic compound to *T. angustula* was citronella oil, and the least toxic, andiroba oil. Eucalyptus oil and neem oil presented intermediate toxicity to *T. angustula* (Xavier et al. 2010).

These evidences show that biopesticides can be as lethal to bees as conventional pesticides, and should not be exempted from risk assessment (Xavier et al. 2015; Tomé et al. 2015a; Bernardes et al. 2017; Lima et al. 2016).

A growing tendency in agriculture is the use of genetically modified crops (GMC) (Morandin 2008). This type of crops has caused major concerns regarding their possible effect on bees and other pollinators. Today, two main groups of GMC crops exist, herbicide-tolerant (HT) and insect tolerant (IT) (Morandin 2008). HT crops are presently the most abundant class (70% of all GMC) and are potentially less harmful for bees (Huang et al. 2004). An argument raised in favor of HT crops is that the overall amount of chemicals used during the cycle of crop production could be reduced, representing less contamination of the environment (Morandin 2008). Nonetheless, HT crops can have an important indirect effect on bee populations by changing the botanical composition of plantations, and reducing the diversity of ruderals on which bees can feed (Haughton et al. 2003; Hawes et al. 2003). Thus, it seems, that paradoxically, the use of herbicides by eliminating ground vegetation can reduce pollination services to crops (Morandin and Winston 2005; Kremen and M'Gonigle 2015; Norfolk et al. 2016). Associated with the use of HT crops is the herbicide glyphosate. Honey bees exposed to concentrations of glyphosate normally found under field conditions were not affected, but there is a possibility that sensitivity to nectar and associative learning could be impaired (Herbert et al. 2014). The effect of glyphosate has not been tested in other bees.

In contrast to HT, IT crops can be potentially more harmful to bees. One group of IT crops is known as Bt because the inserts that confer resistance come from the bacterium *Bacillus thuringiensis* (Bt toxin). One important Bt toxin group is Cry, which in lab tests has proven lethal to Lepidoptera and Coleoptera (Benz and Joeressen 1994; Simpson et al. 1997). However, Bt toxins have not been lethal to Hymenoptera and their colonies (Malone and Pham-Delegue 2001; Hanley et al. 2003; Malone et al. 2004; Liu et al. 2005; Lima et al. 2011). A study conducted on honey bees showed that Cry proteins naturally occur in these insects, and that nurse bees foraging on Bt maize had no reduced life span, No change in the composition of gut bacteria was found either, indicating no indirect harmful effects on honey bees (Hendriksma et al. 2013). In Brazil, tests of Cry1Ac at an overall amount of 1.8 μg during the larval development of *Tr. spinipes,* revealed no harmful effects (Lima et al. 2013).

Another class of IT crops use protease inhibitors (PIs) as protectors (Morandin 2008). In contrast to Cry proteins, PIs can affect the development and longevity of worker honey bees (Broodsgard et al. 2003; Malone et al. 2004; Sagili et al. 2005), but no evidence of their effect on stingless bees is yet available. To date the main IT crops with PI protection are cotton, tobacco, potato, and wheat, whose flowers are not intensively visited by bees (Habib and Fazili 2007; Gatehouse 2011). However, more information is needed on IT crops, and their possible effect on bees and other pollinators.

Some fertilizers can also affect insects because they contain heavy metals (Johnson 2015; Rodrigues et al. 2016). Foliar fertilizers containing copper can be moderately toxic to the stingless bee *Frs. schrottkyi* (Rodrigues et al. 2016).

Apart from the lethal effect, most agrochemicals also have sublethal consequences on individual bees and colonies. Sublethal effects generally include impairment of the individual's physiology and behavior, resulting in a reduced life span and performance (Vandame et al. 1995; Cruz et al. 2010). Workers exposed to sublethal doses can carry contaminated pollen and nectar to their hives affecting whole colonies (Tomé et al. 2015a; Guedes et al. 2016). Thus, sublethal effects could be more detrimental on the long term for bee populations (Atkins 1993; Decourtye et al. 2005; Desneux et al. 2007; De la Rúa et al. 2009). Interestingly, apparently less harmful pesticides can have marked sublethal effects (Tomé et al. 2012; Jacob et al. 2013). For instance, azadirachtin causes little mortality in *Pt. helleri* and *Scp. xanthotricha*, but can affect their flight capacities (Tomé et al. 2015b).

Aerial control of insect pests (locusts and mosquitoes) can be particularly harmful because larger areas are sprayed at once. In Mexico, the main insecticide currently used for locust control is fipronil, a member of the phenylpyrazole family (Sagarpa 2008). Fipronil is highly toxic for Brazilian *M. scutellaris* (Lourenço et al. 2012), and can affect orientation and memory capacity in *Scp. postica* (Jacob et al. 2015). The effect of the aerial control of locust on beneficial insects has not been evaluated in Mexico.

Similarly, campaigns against mosquitoes to prevent vector-transmitted diseases, like dengue and Zika, are becoming more frequent during the Mexican rainy season and after hurricanes (Valdovinos-Nuñes et al. 2003). Permethrin is a commonly used insecticide in mosquito control, but it is highly toxic to *M. beecheii, Fr. nigra*, and *N. perilampoides* (Valdovinos-Nuñez et al. 2009). Normally, applications take place at dusk when they may be less harmful to bees. However, in the United States, it was found that pyrethroids for mosquito control can contaminate pollen (Long and Krupke 2016).

Some characteristics of the stingless bees could make them potentially more susceptible to the effect of pesticides. Several species have relatively small populations, and the development of individuals is comparatively slow (Moo-Valle et al. 2004). A sudden loss of workers could be, thus, more costly and detrimental to stingless bees compared to honey bees, for instance (Valdovinos-Nuñez et al. 2009). The stingless bee mass feeding system could increase the exposure of larvae to contaminated food (Lima et al. 2016). Indeed, an antifeeding effect in stingless bees seems common for various insecticides (Bernardes et al. 2017). In *Pb. droriana* the addition of chlorpyrifos to the larval food reduced consumption and produced deformities (dos Santos et al. 2016). In species in which female castes are determined by overfeeding, the production of queens can be compromised if larvae are not properly nourished (dos Santos et al. 2016).

Many species of stingless bee are small sized. Bees of small size can comparatively absorb more pesticide per body unit because of relatively large surface in relation to mass (Johansen et al. 1983; Lundin et al. 2015). In the Yucatan, Valdovinos-Nuñes et al. (2009) found that at similar dose, *M. beecheii* workers were less susceptible to permethrin and methomyl, compared with smaller *Fr. nigra* and *N. perilampoides*. Variables like pH of the hemolymph, level of fat reserves, and detoxification enzymes, have been found to affect the susceptibility of honey bees and leaf cutter bees to pesticides (Ahmad and Johansen 1973, Yu et al. 1984), but no data are available in Meliponini. Interestingly, detoxification enzymes in honey bees are less diverse than in other insects making them probably more susceptible to a range of chemicals (Johnson 2015). This aspect deserves evaluation in stingless bees.

Age and sex can also affect susceptibility to pesticides. Young workers are in general more susceptible than foragers, presumably because detoxification enzymes are still incipient (Meled et al. 1998; Guez et al. 2001; Xavier et al. 2015; Lima et al. 2016). In *A. mellifera* queens are more resistant to DDT than workers

(Graves and Mackensen 1965; Guirguis and Brindley 1975; Tasei 2002). In *M. beecheii,* sexually immature males were more susceptible than gynes to various insecticides (Valdovinos-Nuñes et al. 2009). It is important to document the effect of insecticides on mature male stingless bees because they live a solitary life, increasing the risk of contact with insecticides (Valdovinos-Nuñes et al. 2009).

Although pesticides have negative effects on bees, adequate management can reduce to a large extent the impact of application. Respecting the amounts and periodicity of application, together with early warnings and cooperation between beekeepers and farmer's can substantially reduce bee poisoning (Johansen and Mayer 1990). The effects of pesticides can also be reduced when combined with alternative forms of crop management to control pests and diseases. Integrated management of crops should be supported with education programs and policies developed to accomplish the farmers' needs (IPBES 2016).

Another potential threat to bees is the spread of pathogens and parasites (Paxton et al. 2007; Higes et al. 2009; van Engelsdorp et al. 2009; Neumann and Carreck 2010; Martin et al. 2013). So far, no evidence of major diseases or epidemics has been recorded in stingless bees (Nunes-Silva et al. 2016). An explanation could rely on the stingless bees' more efficient behaviors to fight diseases (Medina et al. 2009). However, extensive search of pathogens in stingless bees is still limited.

A latent problem regarding disease spread is the introduction of exotic bees, that may carry unknown pathogens (Colla et al. 2006; Plischuk et al. 2009). Interestingly, surveys of viruses in stingless bees have revealed little levels of cross infections with other Apids. In Brazilian *M. scutellaris* only evidence of acute paralysis virus was found (Ueira-Vieira et al. 2015). Similarly, in *Scp. mexicana*, only deformed wing and black queen cell viruses were found, out of seven candidates tested (Guzmán-Novoa et al. 2015). Apparently, virus transmission can occur from honey bees to stingless bees (probably by sharing food sources or robbing), but, curiously, no evidence of disease has been detected in colonies where viruses are present (Ueira-Vieira et al. 2015; Guzmán-Novoa et al. 2015). One possible explanation is innate or behavioral resistance of stingless bees to honey bee viruses and other pathogens. For instance, no molecular evidence of *Nosema* has been detected in workers of *M. beecheii* (Medina-Medina et al. 2012) and, artificial inoculation with spores failed to infect the midgut of workers (Soto-Vargas 2014). Similarly, no evidence of active replication of microsporidians and protozoans (Nosematidae and Leishmaniinae) was found in six species of Brazilian stingless bee (Nunes-Silva et al. 2016). Nonetheless, *Apicystis bombi*, a pathogen from bumble bees, has been detected in *Pb. emerina* and *Ttrg. fiebrigi*, albeit at low prevalence (3% out of 131 colonies) (Nunes-Silva et al. 2016).

Even though no major honey bee disease has been found affecting stingless bees, it is well documented that exotic bees can introduce novel pathogens and should be an important criterion against introduction (Otterstatter and Thomson 2008; Dafni et al. 2010; Goulson and Hughes 2015; Nunes-Silva et al. 2016). Moving colonies across regions may also represent a similar threat. The commercial exploitation of stingless bees is increasing and moving colonies across regions could scale. The evidence from other species and latitudes underlines the danger of such practice. A

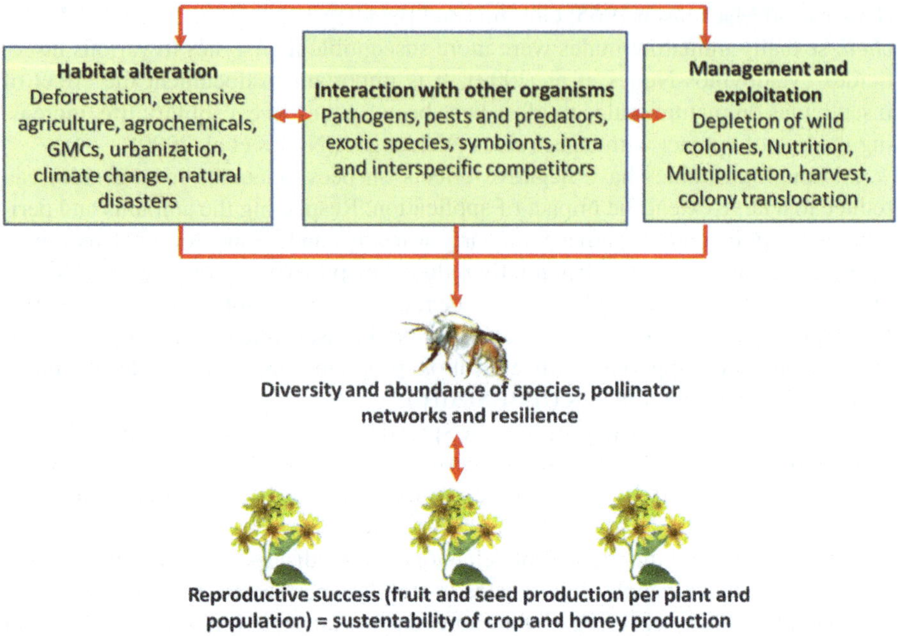

Fig. 8.18 Potential drivers of bee populations (first level), their impact on bee diversity and pollinator networks (second level), and the sustainability of crop and honey production (third level). The red arrows indicate possible interactions among drivers and effects

caution principle to stop the movement of colonies to areas where they are exotic would be the best option to avoid potentially disastrous consequences (Vandame and Palacio 2010; Quezada-Euán et al. 2012). An apparently harmless symbiont to one species can become a severe parasite when crossing over to a different species. An interesting example is the mite *V. destructor* that spread from a resistant host, *A. cerana*, to the susceptible *A. mellifera* causing major colony losses worldwide (Martin and Medina 2004).

In this section I pointed out the impact that several drivers can have on the diversity and abundance of bees in forest and agricultural ecosystems (Fig. 8.18). In the absence of these important pollinators, pollen flow, flower fertilization, and plant reproduction (fruit and seed) would be severely compromised (Potts et al. 2010; Hadley and Betts 2012; Fig. 8.18). For commercial crops, reduced bee richness and abundance means less productivity (Gemmill-Herren 2016). For the prosperous honey industry in Mexico, it is of central importance maintaining the vital link between nectar producing plants and pollinating bees. A reduction in the number and diversity of native plants because of low pollination rates and deforestation, would have a negative impact on the production of honey. Today, it seems that preserving natural habitats and heterogeneous agricultural landscapes is the best option to support large and diverse pollinator assemblages that sustain ecosystems, as well as crop and honey production (Novais et al. 2016).

References

Abrams EM, Rue DJ (1988) The causes and consequences of deforestation among the prehistoric Maya. Hum Ecol 16:377–395

Aguirre A, Dirzo R (2008) Effects of fragmentation on pollinator abundance and fruit set of an abundant understory palm in a Mexican tropical forest. Biol Conserv 141:375–384

Ahmad Z, Johansen C (1973) Selective toxicity of carbophenotion and thricloroform to the honey bee and the alfalfa leaf cutting bee. Environ Entomol 2:27–30

Alfaro-Bates RG, González-Acereto JA, Ortíz-Diáz JJ, Viera-Castro FA, Burgos-Pérez AI, Martínez-Hernández E, Ramírez-Arriaga E (2010) Caracterización palinológica de las mieles de la peninsula de Yucatán. UADY-CONABIO

Atkins EL (1993) Management strategies for protecting honey bees, *Apis mellifera*, from pesticides. In: Connor LJ, Rinderer TE, Sylvester HA, Wongsiri S (eds) Proceedings of the 1st international conference on the Asian honey bees and bee mites. Wicwas Press, Cheshire, pp 607–612

Barbosa WF, Smagghe G, Guedes RNC (2015a) Perspective pesticides and reduced-risk insecticides, native bees and pantropical stingless bees: pitfalls and perspectives. Pest Manag Sci 71:1049–1053

Barbosa WF, Tomé HVV, Bernardes RC, Siqueira MAL, Smagghe G, Guedes RNC (2015b) Biopesticide-induced behavioral and morphological alterations in the stingless bee *Melipona quadrifasciata*. Environ Toxicol Chem 34:2149–2158

Biesmeijer JC, Roberts SPM, Reemer M, Ohlemüller R, Edwards M, Peeters T, Schaffers AP, Potts SG, Kleukers R, Thomas CD, Settele J, Kunin WE (2006) Parallel declines in pollinators and insect-pollinated plants in Britain and the Netherlands. Science 313:351–354

Belzunces LP, Pélissier C, Lewis GB (1999) Hazards of pesticides to bees. Abstracts of the 7th International Symposium of the ICPBR Protection Group co-organized by INRA and ACTA, Université d'Avignon, France

Benz G, Joeressen HJ (1994) A new pathotype of *Bacillus thuringiensis* with pathogenic action against sawflies (Hymenoptera, Symphita). Bull OILB-SROP 17:35–38

Bernardes RC, Tomé HVV, Barbosa WF, Guedes RNC, Lima MAP (2017) Azadirachtin-induced antifeeding in Neotropical stingless bees. Apidologie 48(3):27–285. https://doi.org/10.1007/s13592-016-0473-3 (En prensa)

Broodsgard HF, Broodsgard CJ, Hansen H, Lovei GL (2003) Environmental risk assessment of transgene products using honey bee (*Apis mellifera*) larvae. Apidologie 34:139–145

Brown BV (2004) Revision of the *Melaloncha cingulata*-group of bee-killing flies (Diptera: Phoridae). Ann Entomol Soc Am 97:386–392

Brown BV (2005) Revision of the *Melaloncha* (*M.*) *furcata*-group of bee-killing flies (Diptera: Phoridae). Insect Syst Evol 36:241–258

Brown BV (2016) Two new bee-killing flies from Brazil (Insecta: Diptera: Phoridae: Melaloncha). Biodivers Data J 4:e7715

Brown JC, Albrecht C (2001) The effect of tropical deforestation on stingless bees of the genus *Melipona* (Insecta: Hymenoptera: Apidae: Meliponini) in Central Rondonia, Brazil. J Biogeogr 28:623–634

Casida JE, Durkin KA (2013) Neuroactive insecticides: targets, selectivity, resistance, and secondary effects. Annu Rev Entomol 58:99–117

Catzín-Ventura G, Delgado A, Medina-Medina L (2008) Evaluación de la actividad antibacteriana de la miel de *Melipona beecheii* y *Scaptotrigona pectoralis* de Yucatán, México. In: Memorias V Congreso Mesoamericano sobre abejas sin aguijón. Universidad Autónoma de Yucatán, Mérida, pp 55–62

Colla SR, Otterstatter MC, Gegear RJ, Thomson JD (2006) Plight of the bumble bee: pathogen spillover from commercial to wild populations. Biol Conserv 129:461–467

Colleto-Silva A (2005) Captura de enxames de abelhas sem ferrão (Hymenoptera, Apidae, Meliponinae) sem destruição de árvores. Acta Amazon 35:383–388

Costa L, Venturieri GC (2009) Diet impacts on *Melipona flavolineata* workers. J Apic Res Bee World 48:38–45
Costa LM, Grella TC, Barbosa RA, Malaspina O, Nocelli RCF (2015) Determination of acute lethal doses (LD50 and LC50) of imidacloprid for the native bee *Melipona scutellaris* Latreille, 1811 (Hymenoptera: Apidae). Sociobiology 62:578–582
Costantini D (2015) Land-use changes and agriculture in the tropics: pesticides as an overlooked threat to wildlife. Biodivers Conserv 24:1837–1839
Crozier RH (1971) Heterozygosity and sex determination in haplo-diploidy. Am Nat 105:399–412
Crozier RH, Pamilo P (1996) Evolution of social insect colonies. Sex allocation and kin selection. Oxford University Press, Oxford, UK
Cruz AS, da Silva-Zacarin EC, Bueno OC, Malaspina O (2010) Morphological alterations induced by boric acid and fipronil in the midgut of worker honeybee (*Apis mellifera* L.) larvae. Cell Biol Toxicol 26:165–176
Dafni A, Kevan P, Gross CL, Goka K (2010) *Bombus terrrestris*, pollinator, invasive and pest: an assessment of problems associated with its widespread introductions for commercial purposes. Appl Entomol Zool 45:101–113
De Araujo Freitas JC, González-Acereto JA, Marrufo-Olivares JC (2010) Apicultura práctica en la Península de Yucatán. Ediciones de la Universidad Autónoma de Yucatán, Mérida
DeGrandi-Hoffman G, Chen Y (2015) Nutrition, immunity and viral infections in honey bees. Curr Opin Insect Sci 10:170–176
De la Rúa P, Jaffé R, dall' Olio R, Muñoz I, Serrano J (2009) Biodiversity, conservation and current threats to European honeybees. Apidologie 40:263–284
Decourtye A, Devillers J, Genecque E, Le Menach K, Budzinski H, Cluzeau S, Pham-Delègue MH (2005) Comparative sublethal toxicity of nine pesticides on olfactory learning performances of the honeybee *Apis mellifera*. Arch Environ Contam Toxicol 48:242–250
Decourtye A, Mader E, Desneux N (2010) Landscape enhancement of floral resources for honey bees in agro-ecosystems. Apidologie 41:264–277
Del Sarto MCL, Oliveira EE, Guedes RNC, Campos LAO (2014) Differential insecticide susceptibility of the Neotropical stingless bee *Melipona quadrifasciata* and the honey bee *Apis mellifera*. Apidologie 45:626–636
Desneux N, Decourtye A, Delpuech JM (2007) The sublethal effects of pesticides on beneficial arthropods. Annu Rev Entomol 52:57–80
Devillers J, Pham-Delègue M (2002) Honey bees: estimating the environmental impact of chemicals. Taylor and Francis, London
Dicks LV, Showler DA, Sutherland WJ (2010) Bee conservation: evidence for the effects of interventions. Pelagic Publishing, Exeter
dos Santos CF, Acosta AL, Dorneles AL, dos Santos PDS, Blochtein B (2016) Queens become workers: pesticides alter caste differentiation in bees. Nat Scientific Rep 6:31605
Drescher W, Geusen-Pfister H (1991) Comparative testing of the oral toxicity of acephate, dimethoate and methomyl to honeybees, bumblebees and Syrphidae. Acta Hortic (288):133–136
Ellis EA, Romero Montero JA, Hernández Gómez IU (2017) Deforestation processes in the state of Quintana Roo, México: the role of land use and community forestry. Trop Conserv Sci 10:1–12
Ferron P (1978) Biological control of insect pests by entomogenous fungi. Annu Rev Entomol 23:409–442
Fischer D, Moriarty T (2014) Pesticide risk assessment for pollinators. SETAC-Wiley Blackwell, Oxford
Flores-Guido JS, Vermont-Ricalde R (2011) La vegetación de la peninsula de Yucatán y su miel. In: Echazarreta C (ed) La miel y las abejas, el dulce convenio del Mayab. Secretaría de Educación del Estado de Yucatán, Yucatánpp, pp 33–58
Freitas B, Imperatriz-Fonseca VL, Medina LM, Kleinter AMP, Galetto L, Nates-Parra G, Quezada-Euán JJG (2009) Diversity, threats and conservation of native bees in the Neotropics. Apidologie 40:332–346

García-Morales R, Moreno CE, Badano EI, Zuria I, Galindo-González J, Rojas-Martínez AE et al (2016) Deforestation impacts on bat functional diversity in tropical landscapes. PLoS One 11:e0166765

Garibaldi LA, Steffan-Dewenter I, Winfree R, Aizen MA, Bommarco R, Cunningham SA, Kremen K, Carvalheiro LG, Harder LD, Afik MO et al (2013) Wild pollinators enhance fruit set of crops regardless of honey bee abundance. Science 339:1608–1611

Gaston KJ, Smith RM, Thompson K, Warren PH (2005) Urban domestic gardens (II): experimental tests of methods for increasing biodiversity. Biodivers Conserv 14:395–413

Gatehouse JA (2011) Prospects for using proteinase inhibitors to protect transgenic plants against attack by herbivorous insects. Curr Protein Pept Sci 12:409–416

Gemmill-Herren B (ed) (2016) Pollination services to agriculture: sustaining and enhancing a key ecosystem service. FAO Routledge, Cornwall, UK

Gibson L, Lee TM, Koh LP, Brook BW, Gardner TA, Barlow J et al (2011) Primary forests are irreplaceable for sustaining tropical biodiversity. Nature 478:378–381

Gilpin M.E., Soulé M.E. (1986) Minimum viable populations: process of species extinction. En: Soulé M.E. (ed) Conservation biology: the science of scarcity and diversity. Sinauer, Sunderland

Girolami V, Mazzon L, Squartini A, Mori N, Marzaro M, Di Bernardo A, Greatti M, Giorio C, Tapparo A (2009) Translocation of neonicotinoid insecticides from coated seeds to seedling guttation drops: a novel way of intoxication for bees. J Econ Entomol 102:1808–1815

Gómez-Escobar E, Liedo P, Montoya P, Vandame R, Sánchez D (2014) Behavioral response of two species of stingless bees and the honey bee (Hymenoptera: Apidae) to GF-120. J Econ Entomol 107:1447–1449

González-Acereto JA (2008) Cría y manejo de abejas nativas sin aguijón en México. Ed. UADY-Secretaría Fomento Agropecuario y Pesquero Estado de Yucatán, México

González-Acereto JA, Medina-Medina LA (1998) La respuesta defensiva de *Scaptotrigona pectoralis*como un contundente escudo de protección contra las incursiones de *Lestrimelitta niitkib* dirigidas a otras especies de abejas sin aguijón. In: Memorias VI Congreso Ibero-americano de apicultura. Mérida, Yucatán, pp 171–173

González-Acereto JA, Quezada-Euán JJG, Medina-Medina LA (2006) New perspectives for stingless beekeeping in the Yucatan: results of an integral program to rescue and promote the activity. J Apic Res 45:234–239

Goulson D, Hughes WOH (2015) Mitigating the anthropogenic spread of bee parasites to protect wild pollinators. Biol Conserv 191:10–19

Graves JB, Mackensen O (1965) Topical application and insecticide resistance studies on the honey bee. J Econ Entomol 58:990–993

Greco M, Spooner-Hart R, Holford P (2010) A stingless bee hive design for a broader climate range. J Apic Res 49:290–292

Greig-Smith PW, Thompson HM, Hardy AR, Bew MH, Findlay E (1994) Incidents of poisoning of honeybees (*Apis mellifera*) by agricultural pesticides in great Britain 1981–1991. Crop Prot 13:567–581

Grüter C, von Zuben LG, Segers FHID, Cunningham JP (2016) Warfare in stingless bees. Insect Soc 63:223–236

Guedes RNC, Smagghe G, Stark JD, Desneux N (2016) Pesticide induced stress in arthropod pests for optimized integrated pest management programs. Annu Rev Entomol 61:1–20

Guez D, Suchail S, Gauthier M, Maleszka R, Belzunces LP (2001) Contrasting effects of imidacloprid in 7- and 8-day-old honeybees (*Apis mellifera*). Neurobiol Learn Mem 76:183–191

Guirguis GN, Brindley WA (1975) Carbaryl penetration into and metabolism by alfalfa leafcutting bees *Megachile pacifica*. J Agric Food Chem 23:274–279

Gutiérrez-Olivares M, Rodríguez-Maciel JC, Llanderal-Cázares C, Terán-Vargas AP, Lagunes-Tejeda Á, Díaz-Gómez O (2007) Estabilidad de la resistencia a neonicotinoides en *Bemisia tabaci* (Gennadius), biotipo B de San Luis Potosí, México. Agrociencia 41:913–920

Guzmán-Novoa E, Hamiduzzaman MM, Anguiano-Baez R, Correa-Benítez A, Castañeda-Cervantes E, Arnold NI (2015) First detection of honey bee viruses in stingless bees in North America. J Apic Res 54:93–95

Habib H, Fazili KM (2007) Plant protease inhibitors: a defense strategy in plants. Biotechnol Mol Biol Rev 2:68–85

Hadley AS, Betts MG (2012) The effects of landscape fragmentation on pollination dynamics: absence of evidence not evidence of absence. Biol Rev 87:526–544

Hanley AV, Huang ZY, Pett WL (2003) Effects of dietary transgenic Bt corn pollen on larvae of *Apis mellifera* and *Galleria mellonella*. J Apic Res 42:77–81

Hanski I (2011) Habitat loss, the dynamics of biodiversity and a perspective on conservation. Ambio 40:248–255

Haughton AJ, Champion GT, Hawes C, Heard MS, Brooks DR, Bohan DA et al (2003) Invertebrate responses to the management of genetically modified herbicide-tolerant and conventional spring crops: II. Within-field epigeal and aerial arthropods. Phylos Trans R Soc Lond B 358:1863–1877

Hawes C, Haughton AJ, Osborne JL, Roy DB, Clark SJ, Perry JN et al (2003) Responses of plants and invertebrate trophic groups to contrasting herbicide regimes in the farm scale evaluations of genetically modified herbicide-tolerant crops. Phylos Trans R Soc Lond B 358:1899–1913

Hendriksma HP, Küting M, Härtel S, Näther A, Dohrmann AB, Steffan-Dewenter I et al (2013) Effect of stacked insecticidal cry proteins from maize pollen on nurse bees (*Apis mellifera carnica*) and their gut bacteria. PLoS One 8:e59589

Herbert LT, Vázquez DE, Arenas A, Farina WM (2014) Effects of field-realistic doses of glyphosate on honeybee appetitive behaviour. J Exp Biol 217(Pt 19):3457–3464

Hérnandez JL, Frankie GW, Thorp RW (2009) Ecology of urban bees: a review of current knowledge and directions for future study. Cities Environ 2:1–15

Herrera y Herrera B. (2003) Análisis de la deforestación y dinámica territorial en México periodo 1980–2000. Universidad Autónoma de Chapingo, División de Ciencias Forestales, México

Higes M, Martín-Hernández R, Garrido-Bailón E, González-Porto AV, García-Palencia P, Meana A et al (2009) Honeybee colony collapse due to *Nosema ceranae* in professional apiaries. Environ Microbiol Rep 1:110–113

Hölldobler B, Wilson EO (1990) The ants. Harvard University Press, Cambridge

Huang ZY, Hanley AV, Pett WL, Langenberg M, Duan JJ (2004) Field and semifield evaluation of impacts of transgenic canola pollen on survival and development of worker honey bees. J Econ Entomol 97:1517–1523

INEGI (2011) Carta de uso de suelo y vegetación Serie IV. Instituto Nacional de Estadística, Geografía e Informática

Inglesfield C (1989) Pyrethroids and terrestrial non-target organisms. Pestic Sci 27:387–428

IPBES (2016) In: Potts SG et al (eds) Summary for policymakers of the assessment report of the intergovernmental science-policy platform on biodiversity and ecosystem services on pollinators, pollination and food production. Secretariat of the Intergovernmental Science-Policy Platform on Biodiversity and Ecosystem Services, Bonn, Germany, p 36

Iwasa T, Motoyama N, Ambrose JT, Roe MR (2004) Mechanism for the differential toxicity of neonicotinoid insecticides in the honey bee, *Apis mellifera*. Crop Prot 23:371–378

Jacob CRO, Soares HM, Carvalho SM, Nocelli RCF, Malaspina O (2013) Acute toxicity of fipronil to the stingless bee *Scaptotrigona postica* Latreille. Bull Environ Contam Toxicol 90:69–72

Jacob CRO, Soares HM, Nocelli RC, Malaspina O (2015) Impact of fipronil on the mushroom bodies of the stingless bee *Scaptotrigona postica*. Pest Manag Sci 71:114–122

Johansen CA, Mayer DF (1990) Pollinator protection: a bee pesticide handbook. Wicwas Press, Kalamazoo, Michigan

Johansen CA, Mayer DF, Eves J, Kious CW (1983) Pesticides and bees. Environ Entomol 12:1513–1518

Johnson RM (2015) Honey bee toxicology. Annu Rev Entomol 60:415–434

Kaspari M, O'Donell S (2003) High rates of army ant raids in the Neotropics and implications for ant colony and community structure. Evol Ecol Res 5:933–939

Kerr W. E. (2002) Extincão de especies: a grande crise biológica do momento e como afeta os meliponinios. En: Anais do V encontro sobre abelhas, pp 4-9. Riberao Preto, São Paulo

Kiatoko N, Raina SK, Langevelde F (2016) A vertical compartmented hive design for reducing post-harvest colony losses in three Afrotropical stingless bee species (Apidae: Meliponinae). Int J Develop Res 06:9026–9034

Koul O. (2004) Neem: a global prespective. En: Koul O, Wahab S (eds). Neem: today and in the new millennium. Kluwer Academic. Dordrecht

Kremen C, M'Gonigle LK (2015) Small-scale restoration in intensive agricultural landscapes supports more specialized and less mobile pollinator species. J Appl Ecol 52:602–610

Kremen C, Williams NM, Thorp RW (2002) Crop pollination from native bees at risk from agricultural intensification. Proc Natl Acad Sci USA 99:16812–16816

Leahey JP (1985) The pyrethroid insecticides. Taylor and Francis, London

Landaverde-González P, Quezada-Euán JJG, Theodorou P, Murray TE, Husemann M, Ayala R, Moo-Valle JH, Vandame R, Paxton RJ (2017) Sweat bees on hot chillies: provision of pollination services by native bees in traditional slash-and-burn agriculture in the Yucatan peninsula of tropical Mexico. J Appl Ecol 54(6):1814–1824

Lasota JA, Dybas RA (1991) Avermectina, a novel class of compounds: implications for use in arthropod pest control. Annu Rev Entomol 36:91–117

Laurence WF, Sayer J, Cassman KG (2014) Agricultural expansion and its impacts on tropical nature. Trends Ecol Evol 29:107–116

Lima MAP, Pires CSS, Guedes RNC, Nakasu EYT, Lara MS, Fontes EMG, Sujii ER, Dias SC, Campos LAO (2011) Does Cry1Ac BT-toxin impair development of worker larvae of Africanized honey bee? J Appl Entomol 135:415–422

Lima MAP, Pires C, Guedes R, Campos L (2013) Lack of lethal and sublethal effects of Cry1Ac Bt-toxin on larvae of the stingless bee *Trigona spinipes*. Apidologie 44:21–28

Lima MA, Martins GF, Oliveira EE, Guedes RN (2016) Agrochemical-induced stress in stingless bees: peculiarities, underlying basis, and challenges. J Comp Physiol A 202:733–747

Liu B, Xu CG, Yan FM, Gong RZ (2005) The impacts of the pollen of insect-resistant transgenic cotton on honeybees. Biodivers Conserv 14:3487–3496

Long EY, Krupke CH (2016) Non-cultivated plants present a season-long route of pesticide exposure for honey bees. Nat Commun 7:11629

Lourenço CT, Carvalho SM, Malaspina O, Nocelli RCF (2012) Oral toxicity of Fipronil insecticide against the stingless bee *Melipona scutellaris* (Latreille, 1811). Bull Environ Contam Toxicol 89:921–924

Lóriga-Peña W, Fonte-Carballo L, Demdio-Lorenzo J (2014) Reporte de *Aethina tumida* Murray (Coleoptera, Nitidulidae) en colonias de la abeja sin aguijón *Melipona beecheii* Bennett de Matanzas y Mayabeque. Revista de Salud Animal 36, no.3 La Habana sep.-dic

Lundin O, Rundlöf M, Smith HG, Fries I, Bommarco R (2015) Neonicotinoid insecticides and their impacts on bees: a systematic review of research approaches and identification of knowledge. PLoS One 10(8):e0136928

Macías-Macías JO, Otero-Colina G (2004) Infestation of *Pyemotes tritici* (Acari: Pyemotidae) on *Melipona colimana* (Hymenoptera: Apidae: Meliponinae): a case study. Agrociencia 38:525–528

Macieira OJD, Hebling-Beraldo MJA (1989) Laboratory toxicity of insecticides to workers of *Trigona spinipes* (F. 1793) (Hymenoptera: Apidae). J Apic Res 28:3–6

MacIvor JS, Packer L (2015) 'Bee hotels' as tools for native pollinator conservation: a premature verdict? PLoS One 10(3):e0122126

Malaspina O, Stort AC (1983) Estudo da tolerância ao DDT e relação com outros caracteres em abelhas sociais. Revista Brasileira de Biología 43:327–330

Malone LA, Pham-Delegue MH (2001) Effects of transgene products on honeybees (*Apis mellifera*) and bumble bees (*Bombus* sp.). Apidologie 32:287–304

Malone LA, Todd JH, Burgess EPJ, Christeller JT (2004) Development of hypopharyngeal glands in adult honey bees fed with a Bt toxin, a biotin-binding protein and a protease inhibitor. Apidologie 35:655–664

Martin S, Medina ML (2004) Africanized honeybees have unique tolerance to Varroa mites. Trends Parasitol 20:112–114

Martin SJ, Hardy J, Villalobos E, Martín-Hernández R, Nikaido S, Higes M (2013) Do the honeybee pathogens *Nosema ceranae* and deformed wing virus act synergistically? Environ Microbiol Rep 5:506–510

Medina LM, Hart AG, Ratnieks FLW (2009) Hygienic behavior in the stingless bees *Melipona beecheii* and *Scaptotrigona pectoralis* (Hymenoptera: Meliponini). Genet Mol Res 8:571–576

Medina-Medina L, Quezada-Euán JJG, Natsopoulou ME, Paxton RJ, Suraporn S (2012) Nosema in Mexican bees. Proceedings 5th European conference of Apidology, Halle der Salle 2012, p. 231

Meled M, Thrasyvoulou A, Belzunces LP (1998) Seasonal variations in susceptibility of *Apis mellifera* to the synergistic action of prochloraz and deltamethrin. Environ Toxicol Chem 17:2517–2520

Meneses Calvillo L, Meléndez Ramírez V, Parra Tabla V, Navarro J (2010) Bee diversity in a fragmented landscape of the Mexican neotropic. J Insect Conserv 14:323–334

Menezes C, Coletto-Silva A, Gazeta GS, Kerr WE (2009) Infestation by *Pyemotes tritici* (Acari, Pyemotidae) causes death of stingless bee colonies (Hymenoptera: Meliponina). Genet Mol Res 8:630–634

Menezes C, Vollet-Neto A, Imperatriz-Fonseca VL (2012) A method for harvesting unfermented pollen from stingless bees (Hymenoptera, Apidae, Meliponini). J Apic Res 51:240–244

Miles M (2003) The effects of spinosad, a naturally derived insect control agent to the honeybee. B Insectol 56:119–124

Moo-Valle H, Quezada-Euán JJG, Navarro J, Rodriguez-Carvajal LA (2000) Patterns of intranidal temperature fluctuation for *Melipona beecheii* colonies in natural nesting cavities. J Apic Res 39:3–7

Moo-Valle H, Quezada-Euan JJG, Wenseleers T (2001) The effect of food reserves on the production of sexual offspring in the stingless bee *Melipona beecheii* (Apidae, Meliponini). Insect Soc 48:398–403

Moo-Valle H, Quezada-Euán JJG, Canto J, González-Acereto JA (2004) Caste ontogeny and the distribution of reproductive cells on the combs of *Melipona beecheii* B. (Hymenoptera: Meliponini). Apidologie 35:587–594

Moraes SS, Bautista AR, Viana BF (2000) Avaliacão da toxicidade aguda (DL50 e CL50) de insecticidas para *Scaptotrigona tubida* (Smith) (Hymenoptera: Apidae): via de contacto. Anais da Sociedade Entomologica do Brasil 29:31–37

Morandin L. A. (2008) Genetically modified crops: effects on bees and pollination. James R.R. y Pitts-Singer T.L. (eds) Bee pollination in agricultural ecosystems, 203-218. Oxford University Press, NY

Morandin LA, Winston ML (2005) Wild bee abundance and seed production in conventional, organic, and genetically modified canola. Ecol Appl 15:871–881

Morandin LA, Winston ML (2006) Pollinators provide economic incentive to preserve natural land in agroecosystems. Agricult Ecosyst Environ 116:289–292

Moreira MD, Picanço MC, Martins JC, Campos MR, Chediak M (2007) Uso de inseticidas botânicos no controle de pragas. In: Zambolim L, Lopes CA, Picanço MC, Costa H (eds) Manejo integrado de doenças e pragas em hortaliças. Embrapa, Brasília, pp 577–606

Motzke I, Tscharntke T, Wanger TC, Klein AM (2015) Pollination mitigates cucumber yield gaps more than pesticide and fertilizer use in tropical smallholder gardens. J Appl Ecol 52:261–269

Mureithi JG (2008) Use plant pesticides to control crop pests. Kenya Agricultural Research Institute, Kenya

Neumann P, Carreck NL (2010) Honey bee colony losses. J Apic Res 49:1–6

Norfolk O, Eichhor MP, Gilbert F (2016) Flowering ground vegetation benefits wild pollinators and fruit set of almond within arid smallholder orchards. Insect Conserv Diver 9:236–243

Nogueira-Neto P (1997) Vida e criação de abelhas indígenas sem ferrão. Editora Nogueirapis, São Paulo

Novais SMA, Nunes CA, Santos NB, D'Amico AR, Fernandes GW, Quesada M et al (2016) Effects of a possible pollinator crisis on food crop production in Brazil. PLoS One 11:e0167292

Nunes-Silva P, Piot N, Meeus I, Blochtein B, Smagghe G (2016) Absence of Leishmaniinae and Nosematidae in stingless bees. Nat Sci Rep 6:32547

D'Oliveira APM, Venturieri GC, Contrera FAL (2013) Body size variation, abundance and control techniques of *Pseudohypocera kerteszi*, a plague of stingless bee keeping. B Insectol 66:203–208

Otterstatter MC, Thomson JD (2008) Does pathogen spillover from commercially reared bumble bees threaten wild pollinators? PLoS One 3(7):e2771

Paxton RJ, Klee J, Korpela S, Fries I (2007) *Nosema ceranae* has infected *Apis mellifera* in Europe since at least 1998 and may be more virulent than *Nosema apis*. Apidologie 38:558–565

Pellet J, Fleishman E, Dobkin DS, Gander A, Murphy DD (2007) An empirical evaluation of the area and isolation paradigm of metapopulation dynamics. Biol Conserv 136:483–495

Pernal SF, Baird DS, Birmingham AL, Higo HA, Slessor KN, Winston ML (2005) Semiochemicals influencing the host-finding behaviour of *Varroa destructor*. Exp Appl Acarol 37:1–26

Phalan B, Bertzky M, Butchart SHM, Donald PF, Scharlemann JPW, Stattersfield AJ, Balmford A (2013) Crop expansion and conservation priorities in tropical countries. PLoS One 8:e51759

Plischuk S, Martín-Hernández R, Prieto L, Lucía M, Botías C, Meana A, Abrahamovich AH, Lange C, Higes M (2009) South American native bumblebees (Hymenoptera: Apidae) infected by *Nosema ceranae* (Microsporidia), an emerging pathogen of honeybees (*Apis mellifera*). Environ Microbiol Rep 1:131–135

Potts SG, Biesmeijer JC, Kremen C, Neumann P, Schweiger O, Kunin WE (2010) Global pollinator declines: trends, impacts, and drivers. Trends Ecol Evol 25:345–353

Quezada-Euán JJG (1988) Estudio comparativo sobre la adaptación y desarrollo preliminar de colonias de *Melipona beecheii* B. en cuatro tipos de alojamiento en la zona centro del estado de Yucatán, México (Hymenoptera: Apídae). Tesis de Licenciatura, Universidad Autónoma de Yucatán, México

Quezada-Euán JJG, González-Acereto JA (1994) A preliminary study on the development of colonies of *Melipona beecheii* in traditional and rational hives. J Apic Res 33:167–170

Quezada-Euán JJG, González-Acereto JA (2002) Notes on the nest habits and host range of cleptobiotic *Lestrimelitta niitkib* (Ayala 1999) (Hymenoptera: Meliponini) from the Yucatán eninsula. México Acta Zool Mex 86:245–249

Quezada-Euán JJG, Ayala BR (2010) Abejas nativas de México, la importancia de su conservación. Ciencia y Desarrollo, Octubre 2010

Quezada-Euán JJG, May-Itzá W d J, González-Acereto JA (2001) Meliponiculture in México: problems and perspective for development. Bee World 82:160–167

Quezada-Euán JJG, May-Itzá WdJ, Rincón M, De la Rúa P, Paxton RJ (2012) Genetic and phenotypic differentiation in endemic *Scaptotrigona hellwegeri* (Apidae: Meliponini): implications for the conservation of stingless bee populations in contrasting environments. Insect Conserv Diver 5:433–443

Quezada-Euán JJG, Ramírez J, Eltz T, Pokorny T, Medina R, Monsreal R (2013) Does sensory deception matter in eusocial obligate food robber systems? A study of *Lestrimelitta* and stingless bee hosts. Anim Behav 85:817–823

Ramos EM, Medina ML, May-Itzá W (2003) Atracción del vinagre y el ácido acético sobre *Pseudohypocera kerteszi* (Diptera: Phoridae) en Yucatán, México. In: Memorias III seminario Mesoamericano sobre abejas sin aguijón. Tapachula, Chiapas, Mexico, pp 27–30

Rebelo KS, Ferreira AG, Carvalho-Zilse GA (2016) Physicochemical characteristics of pollen collected by Amazonian stingless bees. Ciência Rural 46:927–932

Rech RA, Schwade MA, Schwade MRM (2013) Abelhas-sem-ferrão amazônicas defendem meliponários contra saques de outras abelhas. Acta Amazon 43:389–394
Robinson GE (1981) *Pseudohypocera kerteszi* (Enderlein) (Diptera: Phoridae), a pest of the honeybee. Fla Entomol 64:456–457
Rickli M, Guerin PM, Diehl PA (1992) Palmitic acid released from honeybee worker larvae attracts the parasitic mite *Varroa jacobsoni* on a servosphere. Naturwissenschaften 79:320–322
Robroek BJM, Jong HD, Arce H, Sommeijer MJ (2003) The development of *Pseudohypocera kerteszi* (Diptera, Phoridae), a kleptoparasite in nests of stingless bees (Hymenoptera, Apidae) in central America. Proc Sec Experimental Appl Entomol Netherlands Entomol Soc 14:71–74
Rodrigues CG, Krüger AP, Barbosa WF, Guedes RNC (2016) Leaf fertilizers affect survival and behavior of the neotropical stingless bee *Friesella schrottkyi*. J Econ Entomol 109:1001–1008
Romero MJ, Quezada-Euán JJG (2013) Pollinators in biofuel agricultural systems: the diversity and performance of bees (Hymenoptera: Apoidea) on *Jatropha curcas* in Mexico. Apidologie 44:419–429
Rosa AS, Teixeira JSG, Vollet-Neto A, Queiroz EP, Blochtein B, Pires CSS, Imperatriz-Fonseca VL (2016) Consumption of the neonicotinoid thiamethoxam during the larval stage affects the survival and development of the stingless bee, *Scaptotrigona* aff. *depilis*. Apidologie 47:729–738
Rosete-Vergés FA, Pérez-Damián JL, Villalobos-Delgado M, Navarro-Salas EN, Salinas-Chávez E, Remond-Noa R (2014) El avance de la deforestación en México 1976-2007. Madera y Bosques 20:21–35
Roubik DW (1989) Ecology and natural history of tropical bees. Cambridge University Press, Cambridge
Rueda X (2010) Understanding deforestation in the southern Yucatán: insights from a sub-regional, multi-temporal analysis. Reg Environ Chang 10:175–189
SAGARPA (2008) Manual Operativo de la campaña contra la langosta, exploración, muestreo y control. https://www.gob.mx/cms/uploads/attachment/file/157270/Langosta.pdf
Sagili RR, Pankiw T, Zhu-Salzman K (2005) Effect of soybean trypsin inhibitor on hypopharyngeal gland protein content, total midgut protease activity and survival of the honey bee (*Apis mellifera* L.). J Insect Physiol 51:953–957
Sakagami SF, Roubik DW, Zucchi R (1993) Ethology of the robber stingless bee, *Lestrimelitta limao* (Hymenoptera: Apidae). Sociobiology 21:237–277
Sáncez-Bayo F, Goka K (2016) Impacts of pesticides on honey bees. In: Chambo ED (ed) Beekeeping and bee conservation—advances in research. InTech, Croatia. https://doi.org/10.5772/62487
de Silva-Barros JR (2006) Genetic breeding on the bee *Melipona scutellaris* (Apidae, Meliponinae). Acta Amazon 36:115–120
Simpson RM, Burgess EPJ, Markwick NP (1997) *Bacillus thuringiensis* Delta-endotoxin binding sites in two lepidóptera. *Wiseana* spp. and *Epiphyas postvittana*. J Invertebr Pathol 70:136–142
Soares HM, Jacob CRO, Carvalho SM, Nocelli RCF, Malaspina O (2015) Toxicity of Imidacloprid to the stingless bee *Scaptotrigona postica* Latreille, 1807 (Hymenoptera: Apidae). Bull Environ Contam Toxicol 94:675–680
Sommeijer MJ (1999) Beekeeping with stingless bees: a new type of hive. Bee World 80:70–79
Soto-Vargas C (2014) Efectos de la inoculación de esporas de *Nosema ceranae* (Microsporidia) y detección del virus de alas deformes (DWV) en la abeja nativa sin aguijón *Melipona beecheii* (Hymenoptera: Meliponini) en Yucatán, México. Tesis de Maestría Universidad Autónoma de Yucatán, México
Sparks TC, Crouse GD, Durst G (2001) Natural products as insecticides: the biology, biochemistry and quantitative structure activity relationships of spinosyns and spinosoids. Pest Manag Sci 57:896–905
Stoner KA, Eitzer BD (2012) Movement of soil-applied imidacloprid and thiamethoxam into nectar and pollen of squash (*Cucurbita pepo*). PLoS One 7:e39114
Tasei JN (2002) Impact of agrochemicals on non-*Apis* bees. In: Devillers J, Pham-Delègue MH (eds) Honey bees: estimating the environmental impact of chemicals. Taylor & Francis, London, pp 101–131

Toledo-Hernández RA, Ruiz-Toledo J, Toledo J, Sánchez D (2016) Effect of three entomopathogenic fungi on three species of stingless bees (Hymenoptera: Apidae) under laboratory conditions. J Econ Entomol 109:1015–1019

Tomé HVV, Martins GF, Lima MAP, Campos LAO, Guedes RNC (2012) Imidacloprid-induced impairment of mushroom bodies and behavior of the native stingless bee *Melipona quadrifasciata anthidioides*. PLoS One 7:e38406

Tomé HVV, Barbosa WF, Martins GF, Guedes RNC (2015a) Spinosad in the native stingless bee *Melipona quadrifasciata*: regrettable non-target toxicity of a bioinsecticide. Chemosphere 124:103–109

Tomé HVV, Barbosa WF, Corrêa AS, Gontijo LM, Martins GF, Guedes RNC (2015b) Reduced-risk insecticides in Neotropical stingless bee species: impact on survival and activity. Ann Appl Biol 167:186–196

Ueira-Vieira C, Almeida LO, Almeida FC d, Amaral IMR, Brandemburgo MAM, Bonetti AM (2015) Scientific note on the first molecular detection of the acute bee paralysis virus in Brazilian stingless bees. Apidologie 46:628–630

Valdovinos-Núñez GR, Quezada-Euán JJG, Marrufo-Olivares J (2003) Efecto de la aplicación aérea de permetrina en *Apis mellifera* y abejas nativas sin aguijón (Hymenoptera: Apidae) en Yucatán, México. In: Memorias XVII Seminario Americano de Apicultura. Aguascalientes, México, pp 147–149

Valdovinos-Nuñez GR, Quezada-Euán JJG, Ancona-Xiu P, Moo-Valle H, Carmona A, Ruiz-Sanchez E (2009) Comparative toxicity of pesticides to stingless bees (Hymenoptera: Apidae: Meliponini). J Econ Entomol 102:1737–1742

Vandame R, Meled M, Colin ME, Belzunces LP (1995) Alteration of the homing-flight in the honey bee *Apis mellifera* exposed to sublethal dose of deltamethrin. Environ Toxicol Chem 14:855–860

Vandame R, Palacio MA (2010) Preserved honey bee health in Latin America: a fragile equilibrium due to low-intensity agriculture and beekeeping? Apidologie 41:243–255

van der Sluijs JP, Simon-Delso N, Goulson D, Maxim L, Bonmatin JM, Belzunces LP (2013) Neonicotinoids, bee disorders and the sustainability of pollinator services. Curr Opin Environ Sustain 5:293–305

van Engelsdorp D, Evans JD, Saegerman C, Mullin C, Haubruge E, Nguyen BK, Frazier M, Frazier J, Cox-Foster D, Chen Y, Underwood R, Tarpy DR, Pettis JS (2009) Colony collapse disorder: a descriptive study. PLoS One 4:e6481

Venturieri GC, Oliveira PS, Vasconcelos MAM, Mattietto RA (2007) Caracterização, colheita, conservação e embalagem de méis de abelhas indígenas sem ferrão. Embrapa Amazonia Oriental

Venturieri GC (2008) A criação de abelhas indígenas sem ferrão. Embrapa Amazonia Oriental

Vidal O, López-García J, Rendón-Salinas E (2013) Trends in deforestation and forest degradation after a decade of monitoring in the monarch butterfly biosphere Reserve in Mexico. Conserv Biol 28:177–186

Villanueva-Gutiérrez R, Roubik DW, Colli-Ucán W, Güemez-Ricalde FJ, Buchmann SL (2013) A critical view of colony losses in managed Mayan honey-making bees (Apidae: Meliponini) in the heart of Zona Maya. J Kansas Entomol Soc 86:352–362

Villas-Bôas J (2012) Manual tecnológico Mel de abelhas sem ferrão. Brasilia, DF. Instituto Sociedade, População e Natureza (ISPN). In: Vit P, Pedro SRM, Roubik DW (eds) Pot-honey: a legacy of stingless bees. Springer, New York

Vit P, Pedro SRM, Roubik DW (eds) (2013) Pot-honey: a legacy of stingless bees. Springer, New York

Waliszewski SM, Caba M, Gómez-Arroyo S, Villalobos-Pietrini R, Martinez A, Valencia-Quintana R, Lozano-Flores ME, Regalado-Torres MA (2013) Niveles de plaguicidas organoclorados en habitantes de México. Rev Int Contam Ambiental 29:121–131

Watkins JF II (1982) The army ants of Mexico (Hymenoptera: Formicidae: Ecitoninae). J Kansas Entomol Soc 55:197–247

Winfree R, Fox JW, Williams NM, Reilly JR, Cariveau DP (2015) Abundance of common species, not species richness, drives delivery of a real-world ecosystem service. Ecol Lett 18:626–635

Wolff LF, Nava DE (2007) Ocorrência da mosca dos favos *Pseudohypocera kerteszi* (Diptera: Phoridae) em colméias de abelhas melíferas africanizadas no Rio Grande do Sul. Comunicado Técnico Embrapa Clima Temperado 178:1–4

Xavier VM, Message D, Picanço MC, Bacci L, Silva GA, Benevenute JS (2010) Impact of botanical insecticides on indigenous stingless bees (Hymenoptera: Apidae). Sociobiology 56:713–726

Xavier VM, Message D, Picanço MC, Chediak M, Santana Júnior PA, Ramos RS, Martins JC (2015) Acute toxicity and sublethal effects of botanical insecticides to honey bees. J Insect Sci 15:137

Yu SJ, Robinson FA, Nation JL (1984) Detoxication capacity in the honey bee, *Apis mellifera* L. Pest Biochem Physiol 22:360–368

Zayed A, Packer L (2005) Complementary sex determination substantially increases extinction proneness of haplodiploid populations. Proc Nat Acad Sci USA 102:10742–10746

Chapter 9
The Past, Present, and Future of Meliponiculture in Mexico

> *"There are two castes of bees and both are smaller than ours. The larger are kept in hives which are rather small; they do not make combs like our bees but small vesicles like wax hazelnuts, clustered tightly ... These bees do not sting and do no harm."*
>
> Fray Diego de Landa, 1553

A meliponario in the Yucatan at the start of the twentieth century

Accounts on ancient meliponiculture are not abundant for the Americas. Most pre-Columbian evidence comes from Mexico, in particular from the Yucatan Península. Early accounts from Spaniards indicate that outstanding levels of husbandry and large concentrations of colonies of *Melipona beecheii* ('Xunan kab' in Mayan) were present in the region. An extensive and probably monopolized trade of honey and

J. J. G. Quezada-Euán, *Stingless Bees of Mexico*,
https://doi.org/10.1007/978-3-319-77785-6_9

cerumen with distant territories may have supported the economy and the development of Mayan civilization. Rich manifestations of the intricate bond between the Maya and *Xunan kab* are still found in the complex cosmogony and mythology associated with the cultivation of these bees (Jax et al. 2013). The cultivation of *M. beecheii* still played a crucial role in the Yucatan Peninsula and southern Mexico by the end of the nineteenth century, being important in folk medicine, economy, and cosmology (Velednitsky 2011). However, recent changes in the economic system, accelerated a loss of indigenous knowledge and practices, including traditional stingless beekeeping. Meliponiculture, considered obsolete and non-profitable, gradually started to disappear (Stearman et al. 2008). Nonetheless, in recent decades there has been renewed interest in stingless bees and their conservation, coupled with important efforts by beekeepers, academics, and general public aiming at restoring meliponiculture (Quezada-Euán et al. 2001; Venturieri 2008; Reyes-González et al. 2014).

This chapter presents an account of the importance of meliponiculture for Mesoamerican civilizations, in particular the Maya (the past), discussing the different agents that caused a decline and also a recent surge of interest (the present), ending with an agenda of actions that can help to reduce current threats and sustain the activity (the future).

9.1 Meliponiculture in Pre-Columbian Mesoamerica

In Mayan cosmogony, honey is associated with fertility, rendering the bees that produce it, guarantors of life (de Jong 1999; López-Maldonado 2005). For the Maya, stingless beekeeping had similar importance to the cultivation of maize, a staple food for Mesoamerican civilizations (de Jong 1999).

As for other human groups, honey represented a significant source of carbohydrate and to obtain it, early Mesoamericans probably raided wild nests, much like it is still practiced today in other parts of the world (Crane 1990; Gonzalez-Acereto 2008). However, in Mesoamerica, a systematic cultivation of some species of stingless bee evolved, building a rich network between traditional knowledge and mythology (de Jong 1999; Quezada-Euán et al. 2001, 2011.

It is difficult to trace the exact origin of meliponiculture in Mesoamerica. However, in accordance to archaeological remains and its current geographic distribution, it is likely that it may have started in the region of the Yucatan Peninsula and north of Guatemala, probably around 1400–1900 years ago (Kent 1984; Crane 1992). It is believed that from this area, stingless beekeeping may have spread to other regions of Mesoamerica (Fig. 9.1).

Evidence of a Mayan origin of meliponiculture is, that apart from the highlands of Puebla and Veracruz, only rudimentary forms of stingless beekeeping exist outside the Yucatan and Guatemala (Posey 1983; Quintal and Roubik 2013).

Fig. 9.1 Some of the main Mesoamerican civilizations. The Mesoamerican region is shaded; the red line indicates the boundaries between the Maya and other cultures. Meliponiculture probably arose in the Maya region

It is likely that meliponiculture in the Mayan area evolved from a primitive type of honey hunting, with colonies maintained in the forest and harvested periodically. That type of practice still occurs today in the Yucatan, with species like *C. zexmeniae* (Gonzalez-Acereto 2008), that nests underground and does not adapt well to wooden hives. In the case of species that nest in hollow trees, it is possible that some colonies were moved from the forests near human settlements, to prevent others from stealing the honey. The Maya Chontal from Tabasco obtain colonies of stingless bees from the forest in this way. When a forest patch is cleared for agriculture, the nests of stingless bees found by the owner are left for some time in the field. One afternoon, when most worker bees have entered the nest, the log is transported to the land owner home (Vásquez-Dávila and Hernández 2011). Usually, the logs are hung horizontally with some rope, below the house eaves to protect them from direct sunlight and rain.

It is likely that species that produced little or less appreciated honey were probably less favored compared with more productive species that better adapted to man-made lodgings. The use of man-made lodgings probably ensured honey of good quality as well as colony survival (Quezada-Euán 2005). From wild nests established in backyards, it is possible that husbandry developed. For this, a crucial step occurred when people mastered the reproduction and division of colonies (Fig. 9.2). In the case of *M. beecheii,* the constant production of queens may have facilitated artificial division, as beekeepers did not need to learn how to rear them. Divisions may have been easier from colonies nesting in large cavities

Fig. 9.2 Traditional Mayan stingless bee keeping with *M. beecheii*. Major steps should have occurred when people learned how to reproduce the colonies (above) and systematically harvest the honey (below) from specially made hives in hollow logs or *jobones*

(as *M. beecheii*), allowing better manipulation of combs with less damage to the nest. Interestingly, no other stingless bee species was systematically cultivated in the Mayan region, despite their potential for beekeeping. This may also have to do with beliefs behind the cultivation of the different species, which will be explained in further paragraphs.

Methods to keep the new colonies developed. Interestingly, the use of hollow logs (*jobon,* pronounced "ho bon") to keep *M. beecheii* is widely spread and very similar across the Mayan region (Fig. 9.2). For example, the length of the modern Maya *jobon* is between 50 and 60 cm long, with a diameter of about 30 cm and walls of 4 cm in width (Bianco et al. 2017).

The pattern of the traditional meliponary (*Nahil-kab* in Mayan language) is another development widely used in the Mayan region. The orientation and storage of the *jobones* with their main axis from east to west are also remarkably similar across the Yucatan Peninsula (Fig. 9.3). The jobones are stacked on an A-frame under a thatched roof structure. The A-frame is made of two inclined wooden racks and is kept in the house lot (*solar*) (Bianco et al. 2017).

Fig. 9.3 Traditional meliponario in the Yucatan. The *jobones* are stacked on two opposite inclining groups forming an A shape

The adoption of meliponiculture techniques, especially colony division, must have triggered extensive trade of *M. beecheii* products in the Maya area and abroad (Quezada-Euán 2011). When the Spaniards arrived in the sixteenth century, meliponarios had colonies by the hundreds (de Landa 1553; Quezada-Euán et al. 2001). Large numbers of managed colonies probably meant a large and constant supply of honey and cerumen that could be traded. Indeed, it seems that the production of *M. beecheii* in the Yucatan Peninsula occurred as large-scale cottage industry, and was organized at household and neighborhood level (Batún Alpuche 2009). Thus, a potential accumulation of wealth could have derived from honey and cerumen trading (Quezada-Euán et al. 2001).

Although rational stingless beekeeping seems an ancient activity in Mesoamerica, only few archaeological evidences are found today. New techniques like the analysis of soil chemistry are promising methods to trace the origins of meliponiculture in this region (Bianco et al. 2017). Among the available evidence, round flat limestones (*U maak, tapas* or *panuchos*; López-Maldonado 2005), which were used to plug the extremes of the *jobon,* have been frequently found in Mayan preclassic settlements. Dating of these pieces suggests that meliponiculture was already well established in the Yucatan around 100–300 B.C. (Crane 1992). Interestingly, today *jobones* more frequently have *tapas* made of wood rather than stone.

The recent finding of a ceramic object representing a *jobon* in the building 99 of Nakum (Petén, Guatemala; Fig. 9.4) confirms the dates suggested by Crane (1992) for the development of meliponiculture. The ceramic piece has been assigned to the Proto-Classic period, around 100 B.C. (Zralka et al. 2014). The Nakum ceramic *jobon* is smaller than modern ones, measuring only about 30 cm long, with a maximum diameter of 18 cm. This is the oldest representation of a stingless beekeeping artifact.

Fig. 9.4 Ceramic representation of a Mayan stingless bee hive or *jobon* found in the Petén, Guatemala, ca. 100–300 A.D. (photo courtesy by Zralka et al. 2014)

Fig. 9.5 A ceramic representation of *Ah-Mucen-kab*, one of the Mayan god protectors of *M. beecheii* and meliponiculture. Originally exhibited at the Museo de Antropología e Historia in Palacio Cantón Mérida, Yucatan

Diverse objects related to stingless beekeeping have also been found at various sites of the Mayan region (Zralka et al. 2014). One representative incense burner comes from the island of Cozumel on the Eastern Caribbean Coast of the Yucatan Peninsula. This was an important center for stingless beekeeping in the late post-classic period (Díaz del Castillo 1568; Calkins 1974). The incenser, part of the Museo Cantón collection in Mérida, is a cylindrical vase of 34 cm of height representing *Ah-Mucen-kab*, the descending Mayan bee-god (Fig. 9.5). The image is holding what has been identified as a set of combs and, on each side, it carries a pair of small representations of *jobones*. Images of the descending god are also found in temples of Tulum from the postclassic around 1200 A.D. (Fig. 9.6). Another example of a guardian god of the bees is found in the National Museum of Anthropology in Mexico city, a vase in the shape of an old man, *Hobnil*, with a characteristic necklace carrying a *jobon* (Fig. 9.6).

Probably the best example of the importance of stingless beekeeping for the Maya is the Tro-Cortesianus or Madrid Codex (Figs. 9.7 and 9.8). This document, dated ca. 900–1500 AD, is one of the three only surviving pre-Columbian Mayan

Fig. 9.6 On the left, a ceramic representation of *Hobnil*, another guardian god of stingless bees. On the right, the descending god associated with stingless bees, at one of the temples of the Caribbean archaeological site of Tulum

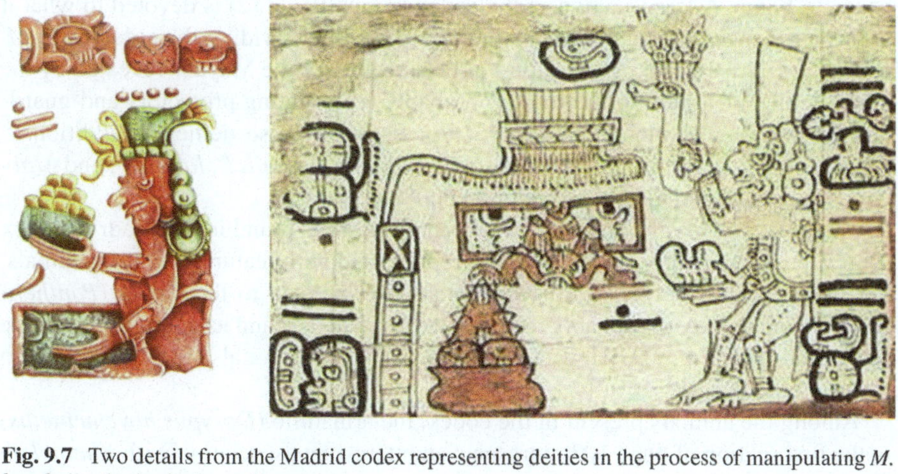

Fig. 9.7 Two details from the Madrid codex representing deities in the process of manipulating *M. beecheii* colonies in the meliponarios

Fig. 9.8 Pages 103–112 of the Madrid Codex representing stingless beekeeping with *M. beecheii* (*Xunan kab*). The images in Fig. 9.7 are taken from pages 104 and 106, respectively. Courtesy by Zralka et al. (2014)

codices and the longest one. The complete document, kept in the Museo de las Americas in Madrid, consists of 56 sheets painted on both sides, that is a total of 112 pages. The codex was named after Hernán Cortés, assuming that he brought the codex to Spain. A large section of the codex (pages 103–112) is devoted to what it seems a ritualized process of harvesting the honey and dividing the colonies of *M. beecheii* (Fig. 9.8). The sacred nature of beekeeping for the Maya is revealed by the many deities identified in the Codex, probably representing protectors and guardians of the bees or their huts (*Nahil-kab*). Some of these deities in addition to *Ah-Mucen-kab* have been identified as *Noh Yum-kab*, *Hobnil*, *Balam-kab*, and *Moc-chí* (Echazarreta and García-Quintanilla 2008).

Altogether 47 glyphs and representation of bees are found in the Madrid Codex (Fig. 9.6). Interestingly, all the bees are represented with features of other animals. For instance, the bee face with "fangs" has been related to the jaguar (*Panthera onca*, *Balam* in Maya), probably referring to the guard bee and its aggressive attitude (the *Balam-kab*; Fig. 9.9). However, the images of bees have also been suggested to represent the queen.

Among the animals present in the codex, the armadillo (*Dasypus novemcinctus*) in the first section of page 103 represents an interesting association with *Xunan kab*

Fig. 9.9 Detail of the Madrid Codex probably representing the guard (*Balam-kab*) or the queen bee of *M. beecheii*

in relation with fertility. It is believed that the armadillo, in its constant movement between earth and the underworld, carries the forces of fertility. Armadillos frequently have twin offspring, which may also link them with the metaphor of the Mayan Hero Twins, the brothers *Hunah Pu* and *Ix Balam Keh* of the *Popol Vuh* (López-Maldonado 2005).

In Mayan mythology, it is believed that *M. beecheii* was given to man directly by the main deity, *Kun' ku* o *Yumbil dios*. Thus, in essence, the stingless beekeepers do not own these creatures; they are only carers (de Jong 1999, 2001). The Mayans see *Xunan kab* colonies as extension of their family and considered them equal, calling them "gente," people. One aspect of this strong relationship is shown when the beekeeper dies. It is believed that the bees would perish unless another man in the hierarchical order of the family quickly informs the bees that he would look after them (Weaver and Weaver 1981). Today, it is still common that *jobones* are passed on to a male family member, but female beneficiaries are now accepted (Bianco et al. 2017).

The belief of a divine origin of *Xunan kab* determined that many of the activities related to the management of colonies were considered sacred. Today, for instance, the traditional harvest of the honey (*U hanli-kab* "the dinner of the bees") still involves a series of rituals that can only be performed by a priest shaman (*h-men*). The shaman acts as an intermediary to invoke the blessing and permission of deities, the real owners of the bees, before performing any type of harvest (Weaver and Weaver 1981). This Mayan belief system is based on the intrinsic connection among man, the bees, and the environment. Man interacts with nature in which all living and unanimated things had a soul and possess equal rights. In this way, humans can take goods from animals and the environment, but are also compelled to return part of the benefits to their divine owners or sacred winds (Echazarreta and García-Quintanilla 2008). Therefore, many Mayan rituals are aimed at obtaining permission and show gratitude for what has been obtained.

In Mayan ceremonies, the honey of *M. beecheii* is a basic ingredient because of its status as a vital fluid. It is used to prepare a drink called *balché,* also consumed in such ceremonies. This drink is a brew made specifically with *Xunan kab* honey mixed with other ingredients, the most important the bark of a local tree, the *balché* (*Lonchocarpus longistylus*). The Mayans may have recognized the enzymatic properties of the honey in accelerating the fermentation process. The alkaloids of the bark, released in the brew, may have induced hallucinations and trascendental states of conscience in the priests that consumed it (López-Maldonado 2005). *Balché* was only consumed during rituals and women were not allowed to drink it. Women consumed honey with *Sac-há,* a nonalcoholic drink made with ground maize. Presently, these drinks are still prepared and consumed during ceremonies related to care and harvest of the crops, the livestock, and the bees. Some *balché* is sprinkled on the hives during the *U hanli-kab* ceremony (Bianco et al. 2017).

Xunan kab is associated with the female gender; they live and work in the house, not in the forests, like men do (de Jong 2001). The temperature of the honey is an important feature for the Maya. The honey of *Xunan kab* is considered warm, as other vital fluids. In this regard, it is a carrier of life and fertility, like blood and semen. This vision relates to the Mayan concept of Mother Earth as a living entity made of spirit, blood, and flesh. The honey is obtained from the earth through the flowers by the sacred *Xunan kab* and is seen therefore as an earth fluid (de Jong 2001). That is why the honey from *Xunan kab* is also strongly associated with women fertility (de Jong 1999; González-Acereto et al. 2008). In association to the female cycle of fertility, the splitting of bee colonies is traditionally made during the full moon, when it is believed that the brood is mature (López-Maldonado 2005). Interestingly, other species of stingless bee are considered masculine (they inhabit and work in the forest), and their honey is regarded as cold (de Jong 2001).

In traditional Mayan medicine, the honey of *M. beecheii* is used to treat various diseases which are believed to result from imbalance of body temperature. One remarkable use of the honey is to stimulate conception (de Jong 1999). It is also used to speed up childbirth and applied to the baby's umbilical cord favors healing and detaching. The honey is used as sweetener for corn gruel drinks given to women after childbirth and tea drinks made from medicinal plants. In contrast to the wide use of honey, pollen has no use in traditional medicine because it is regarded as bee waste. During the harvest, pollen is traditionally buried in the ground (Gonzalez-Acereto 2008). Such a vision on the pollen may derive either from its strong and rancid smell or from the fact that when phorid flies invade colonies, the maggots are mostly seen in pollen pots.

Having the status of a sacred creature, the traditional cultivation of *M. beecheii* is strongly associated with various symbols, reminders of a divine origin and the creation of the world. For instance, the major axis of the jobones is normally oriented from East to West representing the daily journey of the sun (de Jong 1999, 2001). It is believed that a rectangular figure, frequently carved around the entrance (Fig. 9.8), represents the Earth and the four cardinal points, each protected by one of the gods called the *Balam ob.* A figure in the shape of a cross is also carved right above the entrance, which may represent the primordial tree, the *Yax-ché* (*Ceiba pentandra*). For the Maya, the highest point of the cross mirrors the level in the sky where the god creator resides (Fig. 9.10). An alternative purpose for the cross above the entrance may be to protect the colonies from evil eye. On a more practical view, it also indicates the position of the log to prevent placing it upside down after management (Weaver and Weaver 1981; Gonzalez-Acereto 2008).

Fig. 9.10 An engraved rectangle is frequently made around the entrance of the *jobon*, probably symbolizing the four cardinal points. Above the rectangle, a carved cross can be seen

Although trading of 'Xunan kab' honey and cerumen was important across Pre-Columbian Mesoamerica there are no data indicating the scale at which it occurred. It seems that Mesoamerica's trading functioned upon a system of regional specialization, meaning that the production of honey and cerumen, like other agrarian goods, was concentrated in certain areas. Such items were taken to distant locations where they could not be produced (Batún Alpuche 2009). It is known that stingless bee cerumen and honey from the Yucatan were traded for cacao seeds, precious stones and metals, highly valued by pre-Columbian people but, not present in the Maya area (Batún Alpuche 2009). In this way, the Yucatan Peninsula seems to have monopolized on the production of honey and cerumen in Mesoamerica. Honey and cerumen were also staple tribute or tax. In the Mendoza Codex images of towns delivering jars of honey to the Aztecs are depicted. From these evidences, it is clear that stingless beekeeping may have been one of the pillars of Mayan economy.

As described before, honey was mainly used in medicine and rituals rather than as everyday sweetener. In the case of cerumen, one use involded making of candles and figurines for use in rituals (Bianco et al. 2017). Archaeological evidence from ceramic and debris, suggests that lost-wax casting also formed an important component of postclassic-period Mayan metallurgical technologies (Tarkanian and Paris 2017). This activity must have used cerumen extensively.

After the Spanish conquest in 1521, the *encomienda* and *repartimiento* systems were a means of obtaining tax and goods from native communities, including honey and cerumen (Bianco et al. 2017). To control uprising and facilitate tax collection, Spanish authorities resettled Mayan people into few villages, called *congregación* (Bianco et al. 2017). Thus, traditional beekeeping practices were probably altered because of the increased concentration of beekeepers and restricted access to forests (Bianco et al. 2017). Colonization was accompanied by devastating exotic diseases, religious prosecution, enslavement, and destruction of Mayan cultural sites (Velednitsky 2011). However, despite substantial changes in the social system, stingless bee honey and cerumen continued playing a key role in colonial economy. From 1549 to 1551 alone, "the Spaniards took in 2182 *arrobas* (11-kg units) of cerumen and over 2035 *arrobas* of honey" from 173 towns in the province of Yucatan (Quezada-Euán 2011). Of these towns, 157 (or 90%) paid some part of their tribute in the form of honey, while 163 towns were required to pay some cerumen. Only ten communities in the Yucatan in 1549 did not pay any honey or cerumen taxes (Porter-Bolland 2001). From these figures, it is possible to estimate that the production of honey and cerumen was in the range of 7 tons per year, which assuming an average 2 l of honey produced per colony suggests that at least 3000–4000 colonies were maintained in Mayan villages of the Yucatan alone. Indeed, some references cite that meliponarios contained up to 500 colonies (de Landa 1553).

The large amount of honey and cerumen obtained from the Yucatan Peninsula may have prevented the Spaniards from introducing the honey bee into this region, despite having been successfully introduced since the sixteenth century into other areas of Mexico. Indeed, the introduction of honey bees was easily accepted in the central regions of Mexico, presumably because no significant stingless beekeeping existed there. On the other hand, the Maya accustomed to work with meliponines, were reluctant to work with a bee that stung (Labougle and Zozaya 1986). The introduction of the honey bee into the Yucatan occurred until the beginning of the twentieth century with stock from the United States. That is why the honey bee is locally known as the "American" bee (Quezada-Euán 2005).

During the Spanish colony, manuscripts recovering Mayan traditions and legends were still produced, among the most representative are the *Chilam Balam* of Chumayel and the ritual of the *Bacabes*. Both manuscripts were produced at the end of the sixteenth or start of the seventeenth centuries and include significant sections highlighting the importance of stingless bees. In the *Chilam Balam,* the cardinal points are described in association with a bee of different color. The great red bee belongs to the East, the great white bee to the North, the great black bee to the West, and the great yellow bee to the South. The different bees are probably associated with sacred winds and protectors of the earth (known as the 'Bacabes') that dwell at the four cardinal points. In the book of the *Bacabes*, descriptions of Mayan cosmogony and therapeutics include the use of the honey of *M. beecheii.*

It is interesting that the Mayan names presently given to *M. beecheii* (*Xunan kab* and *Colel-kab*) may have originated during the Spanish colony (González-Acereto, Pers. Inf.). In the descriptions made by de Landa (1553) no name is registered for the bees that he saw in Mayan meliponarios. This is surprising considering the detailed descriptions he made of the colonies, and ceremonies, using *M. beecheii* honey and cerumen. The words *Xunan* and *Colel* were used to refer to the Spanish ladies, and were also used to invoke the Virgin Mary. Given the divine nature granted to *M. beecheii,* it is possible that such title was used to separate this bee from other species. However, names given to other stingless bees by the Maya are well known and, probably date before the arrival of Spaniards (Fig. 9.11). The name *Pool-kab*, used in the southern part of the Peninsula, could be one of the original Mayan designations for *M. beecheii*.

Outside the Yucatan Peninsula, traditional stingless beekeeping only shows moderate levels of development. However, an exception is the northern parts of Puebla and Veracruz. In these regions, the Totonacas and Nahuas of the highlands cultivate large number of colonies of the species *Scp. mexicana* (Fig. 9.12), with levels of technification similar to those of *M. beecheii* (Quezada-Euán et al. 2001). It is puzzling that indigenous stingless beekeeping with *M. beecheii* is practically nonexistent outside the Yucatan Peninsula (Foster 1942; Calkins 1974), despite the extended natural distribution of the species in Mexico (Quezada-Euán et al. 2007). One explanation may be a monopolized cultivation by the Maya. However, one may ponder different reasons why this species was more prolific, or managed with greater success in some regions and not in others. For instance, I have seen that dividing and reproducing *M. beecheii* colonies are more difficult in the humid environments of tropical Chiapas compared to the dry lowland forests of the Yucatan.

In the northern highlands of Puebla and Veracruz native people cultivate *Scp. mexicana*, locally known as *Pisil-nek-mej* or *Taxkát*. These bees are reared in earthenware instead of hollow logs. Two pots of approximately 2 L united by their rims

Fig. 9.11 Mayan names (in red) given to the different species of stingless bees in the Yucatan Peninsula. *Denotes species whose products are used in traditional medicine

Fig. 9.12 Traditional meliponario on the northern highlands in Puebla. Colonies of *Scp. mexicana* are kept in earthenware. The bottom pot serves as brood chamber and the top as super

are used as brood chamber and super, respectively (Fig. 9.12). Not much is known about the cosmogony and ritual significance of *Pisil-nek-mej* and its honey, but it is possible that they had a cultural importance similar to maize. In the past, their honey may have been traded or paid as tithe to the Aztecs as explained in the codex Misantla (García-Flores et al. 2013). Today, the honey of *Scp. mexicana* is still used in local medicine.

Other indigenous groups like Zapotecos, Mixes, Zoques, and Popolucas of the Isthmus of Tehuantepec use the honey of *M. beecheii*, *Scp. pectoralis*, *Scp. mexicana*,

Ttr. angustula, and *Fr. nigra* in medicine. However, forest beekeeping is usually practiced and there is no rational management of stingless bees in this area (Vásquez-Dávila 2009). A similar situation occurs along the Mexican Pacific coast. In the highlands of Guerrero, the Tlapanecos keep colonies of *M. fasciata* (*Colmena real*) in hollow logs similar to the system used for *M. beecheii* (Gonzalez-Acereto 2008). However, it seems that the management of this species is only recent (Dixon 1987). Likewise, *Scp. hellwegeri* (*Abeja bermeja*) and *N. perilampoides* may have been cultivated to some extent on the Pacific coast but no evidence of ancient stingless beekeeping with these species is available (Bennett 1964).

As the number of settlements and the population grew during the Spanish colony, cattle ranching and agriculture expansion reduced natural forests. During this period, the introduction of sugarcane and the possibility to use cheap sugar for sweetening food, may have contributed to reduce the demand for stingless bee honey (Quezada-Euán et al. 2001).

9.2 Present Situation and Perspectives

After independence from Spain agriculture and cattle ranching continued expanding greatly across Mexico (Coerver et al. 2004). More recently, the cultivation of sisal or henequen (*Agave fourcroydes*) had a substantial impact on the local economy and landscape of the Yucatan. Henequen had been used since pre-Columbian times by the Maya. However, it gained economic relevance during the second half of the nineteenth century as a consequence of the large demand for fiber in the United States. The economic success of henequen was unprecedented, but being an extensive monocrop it took its toll on the forests and biodiversity of the Yucatan Peninsula. It has been estimated that about a third of the natural forests were logged down to cultivate henequen in the north of the Yucatan and adjacent regions that supplied wood needed for steam-powered extraction of the fiber (Russell 2010; Boyer 2012). The consequence of such extensive deforestation on stingless beekeeping, although not quantified, should have been significant.

More recently, the introduction of the honey bee and the Government's impulse on apiculture at the start of the twentieth century could have increased competition for floral resources, but, probably more negative, it decreased the interest and changed the local view on stingless beekeeping. Honey bee colonies produce at least ten times more honey and wax, and their products were in constant demand with a defined international market. Apiculture improved the income of peasant farmers and was eventually linked with modernity and prosperity (Quezada-Euán et al. 2001). Apiculture grew to such extent in the Yucatán Península that it became the most important honey producing area, with one of the highest densities of managed colonies (17/km^2) in México and the world (Quezada-Euán 2007). Such expansion was accompanied by a gradual abandonment of stingless beekeeping that was considered archaic and non-profitable.

In recent decades, a migration to urban areas reduced the interest in agronomic activities in general, and Mayan farmers have turned their back on the cultivation of indigenous plants and animals in favor of more cost-effective species or varieties. Recent surveys have shown a constant reduction in the number of colonies and people involved in meliponiculture (González-Acereto et al. 2006; Villanueva-Gutiérrez et al. 2013). For instance, in Quintana Roo, the number of stingless bee colonies in meliponarios declined by 93% between 1981 and 2004 (Villanueva-Gutiérrez et al. 2013).

Evidently, meliponiculture in Mexico had been continuously practiced from pre-Columbian times until the present, but its importance decreased significantly over time. By the last part of the last century it was considered at the verge of extinction (Quezada-Euán et al. 2001). Nonetheless, critical events together with the tenacity of determined promoters worked together in favor of the activity. A major change for beekeeping occurred by the end of the 1980s in Mexico, and particularly in the Yucatán Peninsula. Apiculture was mainly practiced in the backyards of rural homes. An approximate 15,000 beekeepers all over the Peninsula kept apiaries of 20–30 hives with European stock in their villages. However, in 1987 this situation changed. The Africanized honey bee, a hybrid of African and European honey bees, was first reported in Yucatan in August 1987 (Quezada-Euán 2007). Africanized honey bees are more defensive and stinging incidents prompted the relocation of apiaries outside towns, leaving Mayan villages with practically no honey producing bees. Africanization of the honey bees in the Yucatan abruptly ended the centennial

Fig. 9.13 Main indigenous groups practicing meliponiculture in twenty-first-century Mexico

tradition of beekeeping at home. Nonetheless, it also opened an opportunity for stingless beekeeping to take over. Stingless beekeeping was then almost forgotten, but with the help of academics and promoters it started to gain attention. Jorge Gonzalez Acereto had been in contact with Professors Virgilio de Portugal Araujo and Paulo Nogeuira Neto with the idea of using modern hives similar to those designed in Angola and Brazil to keep stingless bees in Yucatan. From 1986 to 1990 and with the support of UADY and the MacArthur foundation, he characterized the nests of *M. beecheii* and other species to design hives according to the volumes that they use in nature. With the use of modern hives, a program started to promote stingless beekeeping for the backyards of rural villages. Interestingly, Maya women from several villages got involved in this enterprise. Courses were given to groups of farmers who wanted to learn stingless beekeeping using modern methods or the traditional ones with *jobones*, respecting their culture. The popularity of stingless beekeeping increased slowly in the Yucatan and other areas of Mexico. One important catalyst was the contact among beekeepers and academics, impulsed by initiatives like the Seminario Nacional de Abejas sin aguijón in 1999, now Congreso Mesoamericano de Abejas Nativas, devised by Margarita Medina and other enthusiasts. Today, indigenous groups like *Ix-ek* from Campeche and *Tosepan Titataniske* from Puebla represent successful cases of new generation stingless beekeepers. Noteworthy, stingless beekeeping in twenty-first-century Mexico is still mainly practiced by indigenous peasant farmers. Thus, initiatives for rescuing and preserving meliponiculture must consider the rights of indigenous groups, local knowledge and governance systems, in relation to bees and land, to ensure effective protection of these resources, especially from plundering by outsiders. The main regions where stingless beekeeping is found today are the Yucatan Peninsula and parts of Tabasco and Chiapas (Maya), the highlands of Puebla and Veracruz (Nahua), the Huasteca in San Luis Potosi (Huasteco), and the mountains of Guerrero (Tlapaneco) (Fig. 9.13).

Today, meliponiculture has gained great popularity. It is seen as an activity important to maintain for cultural reasons, and stingless bee products are becoming trendy and high-priced. Paradoxically such great popularity is becoming one of the major threats to stingless beekeeping, particularly in the Yucatan Península. Only between 2016 and 2017 several projects to develop meliponiculture were Government approved and more than a dozen courses were advertised on the Internet at diverse social platforms. Unfortunately, many of those projects and courses are in the hands of inexperienced instructors resulting in countless numbers of colonies lost (Velednitsky 2011). Frequently, courses are only theoretical and the essential aspect of practice is absent, with no follow-up supervision to solve problems and flaws. Related with the former is a proliferation of hive models and sizes across the Yucatan Peninsula, many of excessive volume. It is not clear how exceeding the thermoregulation capacity of colonies has contributed to colony failure, but I suspect it is significant. Combined inexperience with lack of knowledge on the biology and management of stingless bees has proved disastrous. Certification of instructors and regulation of colony purchases and suppliers are urgent to avoid a man-caused extinction vortex.

One important aspect constraining economic revenues from meliponiculture is the lack of an established market for the products of stingless bees and, producers may find problems to sell them. It is also important to determine the composition of honey, pollen and resins and to assess their use as alternative health products. Schemes for accurate characterization of stingless bee honey are necessary to prevent adulteration (Quezada-Euán et al. 2001). Many efforts are being conducted internationally by groups of recognized researchers and standards are being established for different stingless bee honeys (Vit et al. 2013). The characterization of products and the development of beekeepers' networks may also serve to set product prices for a sustainable market (Jaffé et al. 2015).

Extension programs for indigenous communities should also consider traditions and beliefs as essential elements of sustainable production (Jaffé et al. 2015). This type of approach would preserve not only the activity but also the traditional knowledge associated with it (i.e., the use of honey and its medicinal properties) that is rapidly disappearing (Quezada-Euán et al. 2018).

Colonies are scarce and in high demand, boosting the price. The price of a colony of *M. beecheii* can reach $3000 MXN (the equivalent to ca. 150 USD). Unfortunately high colony demand is resulting in the decimation of populations too. We have seen that many of these colonies rapidly die because they have been badly divided, having meager population and sometimes queenless. Thus, the destruction of wild populations can become a server threat. In this regard, a way to prevent the destruction of wild colonies is by improving the methods of colony multiplication, including queen production (Quezada-Euán et al. 2001). It is essential to increase studies on different species and also to assess the possibility of controlling reproduction artificially.

The archaeological sites (Chichen-Itzá) and the Mayan Riviera have expanded their tourist offer. Hotels and resorts are now keeping *M. beecheii* colonies as a tourist attraction. This could be a potential way of preserving meliponiculture, provided that the colonies are well managed and propagated. Insects can be an important source for recreation and tourism (Lemelin 2012). Stingless bees can also be kept in urban areas (Hill et al. 2016). There is great potential for this activity and could be a way to preserve pollinators in the cities.

In the Yucatan, we have conducted a program to promote and rescue stingless beekeeping (González-Acereto et al. 2006). A keystone in this project is a committed participation of scientists, government, and beekeepers. The results are quite promising and we expect that this may have positive implications in other areas of the Peninsula and Mexico. The main objective was to maintain colony numbers and steadily increase them. For this, we followed these steps:

(a) Propose hive boxes as alternative to traditional hollow logs to facilitate management, feeding, and harvesting of colonies. This meant designing hive models for easy use and with the right size for the different species. However, the decision to continue working with hollow logs is respected, and does not interfere with trainining.
(b) Develop methods for controlling pests and parasites.

Fig. 9.14 Extension courses developed for groups of Mayan women. Hands-on work should be an important component of stingless bee courses

(c) Imparting hands-on-work courses with emphasis on methods for transference, reproduction, and management during dearth.
(d) Develop a stock of colonies that can be given on loan to beekeepers to start their own meliponarios.

A goal is to increase the number of skilled stingless beekeepers. To achieve this, a few beekeepers from strategic regions can be trained to later communicate their experience to other members in their communities. An important element of our teaching programs is a demonstrative meliponario. The university has a large stingless bee yard which serves this purpose, as well as a source of biological material for groups of beekeepers. This scheme has been useful in increasing the interest and promoting stingless bees even in areas where they had been unknown. Colonies of *M. beecheii* can be given on loan to potential beekeepers, but there is a compromise to also take one colony of the species *Scp. pectoralis* or *Fr. nigra*. The course is given to the recipients of colonies and constant supervision is provided for a reasonable amount of time, usually 1 year. After this period, the aim is to have skilled beekeepers and also daughter colonies that will remain in their property as well as their production. The mother colonies are returned to the university. A similar scheme of loan could be practiced from beekeepers that have large meliponaries to other members of the community or to different communities (González-Acereto et al. 2006) (Fig. 9.14).

It is crucial that farmers become aware of the potential problem of moving species and colonies across regions. As the interest in meliponiculture and the popularity of *M. beecheii* grows there have been attempts to bring colonies into areas where they are not native. There are many risks involved in the relocation of species ranging from genetic erosion to disease spillover (Quezada-Euán et al. 2012).

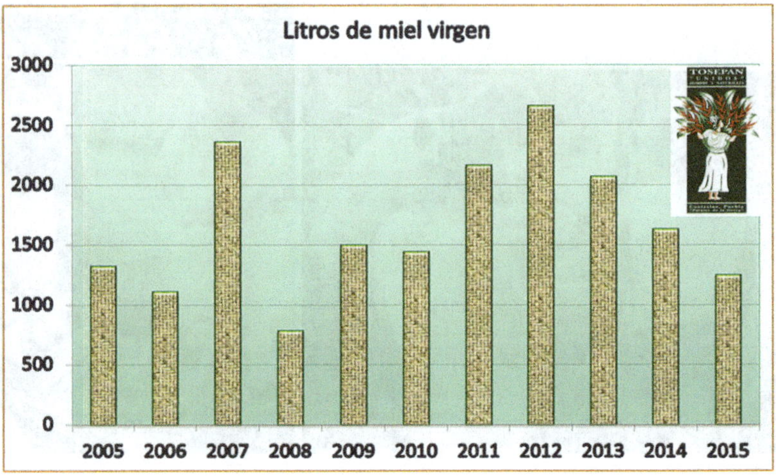

Fig. 9.15 Liters of honey (miel virgen) from *Scp. mexicana* harvested by the cooperative *Tosepan Titataniske* in Cuetzalan, Puebla, between 2005 and 2015

Another weakness of meliponiculture in Mexico is the limited number of stingless beekeepers' organizations. Working in groups could be key to impulse the activity, by improving the opportunities of financial support, marketing, and also to guarantee a fair trade (Jaffé et al. 2015). In Mexico, an example of a successful cooperative of stingless beekeepers is the Sociedad Cooperativa Agropecuaria Regional *Tosepan Titataniske* in the Nahua community of Cuetzalan, Puebla (Fig. 9.15). It is one of the largest cooperatives of indigenous people in Mexico and probably Latin America. Their main product is "*miel virgen*" obtained from thousands of colonies of *Scp. mexicana* (outside the Yucatan Peninsula the general term "*miel virgen*" is used for stingless bee honey). However, Tosepan has also diversified stingless bee products, producing sweets, eye drops, soap, shampoo, and other cosmetics using honey, pollen, and propolis as ingredients. The large metropolitan area around Mexico city has become their main market (Sociedad *Tosepan Titataniske* 2017).

To improve organization and conservation actions, it should be important to build up a data base of stingless beekeepers and their colonies, perhaps in different regions. This would be an essential tool to start monitoring the numbers of stingless beekeepers as well as their colony stocks. Regional and seasonal trends can be traced using this information and could eventually be used for the development of selection and reproduction programs.

Recently, stingless bee products have substantially increased in urban markets. However, one problem is that there is no certification of the origin and contents. Doses and effects on people are practically unknown, so tests should be implemented urgently. As a result of the increasing demand, it is also possible that adulteration will start occurring with more frequency. The value of *M. beecheii* honey can be up to $1200 MXN per liter (ca. 60 USD) and, less pricey *A. mellifera* honey is sometimes diluted to resemble stingless bee honey. The detection of unique components of stingless bee honey may prevent counterfeit (Vit et al. 2004, 2013; Catzín-Ventura et al. 2008). To stop adulteration, large-scale screenings could be made by the approved laboratories in Mexico as implemented for *A. mellifera* honey. In addition, it is crucial to obtain information on the medicinal properties of honey and resins from stingless bees. Recently, more evidence has been found for the antimicrobial, anti-inflammatory, anti-lipidemic, and antioxidant properties of various stingless bee honeys (Catzín-Ventura et al. 2008; Rao et al. 2016). Toxic properties of some honeys have been reported too (i.e., *Lestrimelitta;* Nogueira-Neto 1997), but it is possible that some of their components may have medicinal value. The confirmation of therapeutic effects could support the status of stingless bee products as alternative medicine and increase economic revenue (Quezada-Euán et al. 2001). Thus, the pharmacology of stingless bee products and its medicinal properties should be a promising field in the future.

Stingless bee colony production for commercial pollination is yet another way to promote the activity and provide additional income to rural families. Native stingless bees may be more efficient pollinators than introduced bee species especially for some neotropical crops like tomato, pepper, squash, and avocado because they evolved together with those plants (Cauich et al. 2004; Can-Alonzo et al. 2005; Palma et al. 2008). Promising species to use in commercial pollination are *Fr. nigra, Scp. pectoralis, Scp. mexicana, M. beecheii, M. colimana*, and *N. perilampoides* (Cauich et al. 2004; Palma et al. 2008; Romero and Quezada-Euán 2013). Efficient methods to reproduce colonies at large scales are urgently needed to fully develop this potential application (Quezada-Euán et al. 2001). Presently, there is a national program called "Sistemas Producto" (product system) devoted to agrarian products that have been identified as keystone for the national economy. The Sistemas Producto aim at increasing efficiency and competitiveness by the integration of the different actors (links) of the productive chain. Regrettably, out of 23 crop systems, none considers the contribution of pollination in their productive chain. This is an important aspect to consider in the short term, to promote pollination services in different regions of Mexico.

In general, in tropical America there is the lack of information on the biology and taxonomic status of pollinators and their services, which is particularly serious for bees (Freitas et al. 2009; Jaffé et al. 2015). Despite being one of the most biologically diverse countries, there are only a handful of taxonomic specialists and bee biologists in Mexico. Measures to preserve these insects should be supported with the development of scientific capacities (Freitas et al. 2009).

General aspects of programs for preserving pollinators can be applied for the conservation of stingless bees in Mexico (Dicks et al. 2010, 2016; Brown et al. 2016; IPBES 2016). Mexico is part of the North American Pollinator Protection Campaign and is developing joint schemes for the conservation of species shared with Canada and the United States. However, a national initiative including native species and considering the Mexican context is still needed. In Table 9.1, I summarize some important recommendations made by various groups of specialists for the conservation of pollinators worldwide and, include topics particular for Mexico.

Table 9.1 A list of problems and recommendations for the conservation of bees and stingless beekeeping in Mexico

Problem	Recommendation	Action
– Gaps in the knowledge of species: Taxonomy, biology, products, services, and management – Lack of standardization of management and technology	Promote the buildup of scientific and technical capacities and personnel Improve knowledge-sharing between scientists, farmers, public, stakeholders, and government	– Build a Mexican Initiative for the conservation and sustainable use of native bees. – Develop infrastructure and scientific capacity – Detect national needs of research prioritizing for different regions – Promote networks among scientific groups and farmers – Estimate the contribution of bees to pollination of different crops – Disregard hives of excessive size in favor of models of adequate volume – Incorporate the link of pollination services in "Sistemas Producto" – Assess the therapeutic value of stingless bee products – Monitoring bees and pollinators and their services
Extinction of indigenous and local knowledge	– Recognize traditions and beliefs as essential elements of sustainable meliponiculture	– Protect the rights of indigenous people over natural resources – Recover local knowledge on the use of bee products – Conduct surveys of stingless beekeepers and colonies – Prepare technicians from local groups as promoters of stingless beekeeping
Plunder of wild populations and loss of managed ones	Reinforce laws to prevent uncontrolled extraction of wild colonies Develop methods to sustainably reproduce colonies and preserve genetic material	– Apply regulations for extracting and selling colonies and their movement across regions – Establish a system for certification of technicians and extensionists – Monitor courses and new groups starting stingless beekeeping, ensuring long term supervision to avoid colony losses
Uncontrolled use of agrochemicals in rural and urban environments	– Revise the list of authorized chemicals, methods and doses – Educate on adequate use of chemicals – information-sharing between beekeepers and farmers on the application of chemicals – Propose alternative and integrated methods for pest control in rural and urban environments	– Regulate methods of application and doses in accordance to region and crop – Assess the effects of neonicotinoids and actual risks of GMO and chemicals used in conjunction (i.e., glyphosate) under field conditions – Promote organic and mixed-system agriculture – Avoid daytime applications in urban areas for control of insect vectors

Introduction of exotic species	– Estimate colony numbers needed for pollination in protected and open agriculture – Revise national regulations on importations of bees for pollination – Provide alternatives to exotic species based on the use of native bee species	– Strict import regulations of exotic bee species – Evaluate the efficiency and estimate colony/forager number for efficient pollination in different crops and systems – Develop methods to produce and manage native species in local and regional agriculture systems
Habitat fragmentation	– Plan for urban and agricultural expansion – Preserve natural reserves and apply legislation – Promote mixed agriculture schemes and methods to preserve bees and pollinators – Maintain forest corridors	– Control deforestation – Restore/increase diversity of bee plants in urban and rural areas (flowering ground vegetation) and agriculture systems – Maintain buffer areas of bare ground and strips of riparian vegetation around crops – Restore altered habitat/increase connectivity across fragments – Encourage farmers to preserve riparian land recognizing the benefits for crop productivity
Disinformation on the importance of bees and pollinators	– Use social networks and media to teach and increase public perception on bees and their importance – Develop school activities and for young people related to pollinator conservation – Promote citizen science	– Produce materials for publication in press and transmission on TV and social networks – Produce materials for schools and young children – Stimulate the diversity of bee plants in gardens and parks and the use of bee hotels – Consider bee tourism as a mean for conservation
Pests and diseases	Assess risks and conduct surveys on potential diseases	Obtain information on potential risks
Climatic change	Assess potential impacts of global warming on pollinators and pollination systems	Construct models and scenarios to develop plans to ameliorate potential effects of global warming on pollinator and pollination systems

The former list is by no means exhaustive, but the objective is to provide a base outline on which to develop a Mexican national initiative with actions that can be measured over time. The short-term application of policies and practices should minimize future threats and maximize future benefits (Brown et al. 2016).

It is possible that in the near future stingless beekeeping could express its full potential in benefit of farmers and society. This vision could only be possible with the participation, integration, and true commitment of stakeholders, beekeepers, academics, and government.

References

Batún Alpuche AI (2009) Agrarian production and intensification at a Postclassic Maya community, Buena Vista, Cozumel, Mexico. Ph.D. Dissertation, University of Florida, Gainesville

Bennett CF (1964) Stingless-bee keeping in western Mexico. Geogr Rev 54:85–92

Bianco B, Alexander RT, Rayson G (2017) Beekeeping practices in modern and ancient Yucatán, going from the known to the unknown. In: Mathews JP, Guderjan TH (eds) The value of things prehistoric to contemporary commodities in the Maya region. The University of Arizona Press, Tucson, pp 87–103

Boyer CR (2012) A land between waters: environmental histories of modern Mexico. The University of Arizona Press, Tucson

Brown MJF, Dicks LV, Paxton RJ, Baldock KCR, Barron AB, Chauzat MP et al (2016) A horizon scan of future threats and opportunities for pollinators and pollination. Peer J 4:e2249

Calkins C (1974) Beekeeping in Yucatan: a study in historical-cultural zoogeography. Ph.D. Thesis, University of Nebraska, Lincoln

Can-Alonzo C, Quezada-Euán JJG, Ancona-Xiu P, Moo-Valle H, Valdovinos-Nunez GR, Medina-Peralta S (2005) Pollination of 'criollo'avocados (Persea americana) and the behaviour of associated bees in subtropical Mexico. J Apic Res 44:3–8

Catzín-Ventura G, Delgado A, Medina-Medina L (2008) Evaluación de la actividad antibacteriana de la miel de *Melipona beecheii* y *Scaptotrigona pectoralis* de Yucatán, México. In: Memorias V (ed) Congreso Mesoamericano sobre abejas sin aguijón. Universidad Autónoma de Yucatán, Mérida, pp 55–62

Cauich O, Quezada-Euán JJG, Macias-Macias JO, Reyes-Oregel V, Medina-Peralta S, Parra-Tabla V (2004) Behavior and pollination efficiency of *Nannotrigona perilampoides* (Hymenoptera: Meliponini) on greenhouse tomatoes (*Lycopersicon esculentum*) in subtropical Mexico. J Econ Entomol 97:475–481

Coerver DM, Pasztor SB, Buffington RM (2004) Mexico: an encyclopedia of contemporary culture and history. ABC-CLIO, Mexico

Crane E (1990) Bees and beekeeping, science, practice and world resources. Heinemann Newnes, London

Crane E (1992) The past and present status of beekeeping with stingless bees. Bee World 73:29–42

de Jong HJ (1999) The Land of corn and honey: The keeping of stingless bees (Meliponiculture) in the Ethno-Ecological environment of Yucatan (Mexico) and El Salvador. Ph.D. Thesis, Utrecht University, Utrecht

de Jong HJ (2001) La meliponicultura en la cosmovisión Maya. Memorias II Congreso Mesoamericano de abejas sin aguijón. Universidad Autónoma de Yucatán, Mérida

de Landa D (1553) Relación de las cosas de Yucatán. Editorial Porrúa

Díaz del Castillo B (1568) Historia verdadera de la conquista de la Nueva España. Ed. Galaxia Gutenberg, España

Dicks LV, Showler DA, Sutherland WJ (2010) Bee conservation: evidence for the effects of interventions. Pelagic, Exeter

Dicks LV, Viana B, Bommarco R, Brosi B, Arizmendi MdC, Cunnigham SA, Galetto L et al (2016) Ten policies for pollinators. Science 354:975–976

Dixon CV (1987) Beekeeping in southern Mexico. Conference of Latinamericanist Geographers Yearbook 13:66–71

Echazarreta CM, García-Quintanilla A (2008) Las abejas y su guardián en la antigüedad maya. In: Memorias V congreso Mesoamericano sobre abejas sin aguijón. Universidad Autónoma de Yucatán, Mérida, pp 180–183

Freitas B, Imperatriz-Fonseca VL, Medina LM, Kleinter AMP, Galetto L, Nates-Parra G, Quezada-Euán JJG (2009) Diversity, threats and conservation of native bees in the Neotropics. Apidologie 40:332–346

Foster GM (1942) Indigenous apiculture among the popoluca of Veracruz. Am Anthropol 44:538–542

García-Flores A, de Rodriguez SM, Hernández-Colorado MR (2013) Taxkat, la abeja nativa de Mesoamérica. La ciencia y el hombre 25, enero-abril 2013

Gonzalez-Acereto JA (2008) Cría y manejo de abejas nativas sin aguijón en México. Universidad Autónoma de Yucatán. Secretaria de Fomento Agropecuario y Pesquero. Fundación Produce Yucatán, Mérida

González-Acereto JA, Quezada-Euán JJG, Medina-Medina LA (2006) New perspectives for stingless beekeeping in the Yucatan: results of an integral program to rescue and promote the activity. J Apic Res 45:234–239

González-Acereto JA, de Araujo Freitas C, Medina-Medina LA (2008) Aprovechamiento ancestral y sustentable de las abejas sin aguijón (Apidae: Meliponini) por campesinos mayas de Yucatán. In: Memorias V Congreso Mesoamericano de abejas sin aguijón. Ediciones de la Universidad Autónoma de Yucatán, Yucatán

Hill R, Kwapong P, Nates-Parra G, Breslow SJ, Buchori D, Howlett B, Le buhn G, Motta Maues M, Quezada-Euán JJG, Saeed S et al (2016) Biocultural diversity, pollinators and their sociocultural values. In: Potts SG, Imperatriz-Fonseca VL, Ngo HT (eds) IPBES, the assessment report of the intergovernmental science-policy platform on biodiversity and ecosystem services on pollinators, pollination and food production. Secretariat of the Intergovernmental Science-Policy Platform on Biodiversity and Ecosystem Services, Bonn, pp 275–360

IPBES (2016) Summary for policymakers of the assessment report of the Intergovernmental Science-Policy Platform on Biodiversity and Ecosystem Services on pollinators, pollination and food production. In: Potts SG, Imperatriz-Fonseca VL, Ngo HT (eds). Secretariat of the Intergovernmental Science-Policy Platform on Biodiversity and Ecosystem Services, Bonn

Jaffé R, Pope N, Torres Carvalho A, Maia UM, Blochtein B, Lopes de Carvalho CA, Carvalho-Zilse GA, Freitas BM, Menezes C, Ribeiro MdF, Venturieri GC, Imperatriz-Fonseca VL (2015) Bees for development: Brazilian survey reveals how to optimize stingless beekeeping. PLoS One 10:e0130111

Jax K, Barton DN, Chan KMA, de Groot R, Doyle U, Eser U, Görg C, Gómez-Baggethun E, Griewald Y, Haber W, Haines-Young R, Heink U et al (2013) Ecosystem services and ethics. Ecol Econ 93:260–268

Kent RB (1984) Mesoamerican stingless beekeeping. J Cult Geogr 4:14–28

Labougle JM, Zozaya A (1986) La apicultura en México. Ciencia y Desarrollo 69:17–36

Lemelin RH (2012) The management of insects in recreation and tourism. Cambridge University Press, Cambridge

López-Maldonado JE (2005) Ethnohistory of the stingless bees *Melipona beecheii* (Hymenoptera: Meliponinae) in the Mayan civilization, decipherment of the beekeeping almanacs Part I in the Madrid Codex and the study of their behavioral traits and division of labor. Ph.D. Dissertation, University of California, Davis

Nogueira-Neto P (1997) Vida e criação de abelhas indígenas sem ferrão. Editora Nogueirapis, São Paulo

Palma G, Quezada Euán JJG, Meléndez-Ramírez V, Irigoyen J, Valdovinos-Nuñez GR, Rejón M (2008) Comparative efficiency of *Nannotrigona perilampoides*, *Bombus impatiens* (Hymenoptera: Apoidea), and mechanical vibration on fruit production of enclosed habanero pepper. J Econ Entomol 101:132–138

Porter-Bolland L (2001) Landscape ecology of apiculture in the Maya area of La Montaña, Campeche, Mexico. Ph.D. Dissertation, University of Florida, Gainesville

Posey DA (1983) Keeping of stingless bees by the Kayapo' Indians of Brazil. J Ethnobiol 3:63–73

Quezada-Euán JJG (2005) Biología y uso de las abejas nativas sin aguijón de la Península de Yucatán, México (Hymenoptera: Meliponini). Universidad Autónoma de Yucatán, Mérida

Quezada-Euán JJG (2011) Xunan cab, la señora abeja de Yucatán. In: Echazarreta C (ed) La miel y las abejas, el dulce convenio del Mayab. Gobierno del Estado de Yucatán, Mérida, pp 15–29

Quezada-Euán JJG, de May-Itzá WJ, González-Acereto JA (2001) Meliponiculture in México: problems and perspective for development. Bee World 82:160–167

Quezada-Euán JJG, Paxton RJ, Palmer KA, May-Itzá WdJ, Tay WT, Oldroyd BP (2007) Morphological and molecular characters reveal differentiation in a Neotropical social bee, *Melipona beecheii* (Apidae: Meliponini). Apidologie 38:247–258

Quezada-Euán JJG, May-Itzá WdJ, Rincón M, De la Rúa P, Paxton RJ (2012) Genetic and phenotypic differentiation in endemic *Scaptotrigona hellwegeri* (Apidae: Meliponini): implications for the conservation of stingless bee populations in contrasting environments. Insect Conserv Divers 5:433–443

Quezada-Euán JJG, Nates-Parra G, Maués MM, Imperatriz-Fonseca VL, Roubik DW (2018) Economic and cultural values of stingless bees (Hymenoptera: Meliponini) for ethnic groups of tropical America. Sociobiology

Quintal B.R., Roubik, D.W. (2013) Melipona bees in the scientific world: western cultural views. En: Vit P., Pedro S.R.M., Roubik, D.W. (eds.) Pot-honey: a legacy of stingless bees, pp. 247–260. New York: Springer

Rao PV, Krishnan KT, Salleh N, Gan SH (2016) Biological and therapeutic effects of honey produced by honey bees and stingless bees: a comparative review. Rev Bras 26:657–664

Reyes-González A, Camou-Guerrero A, Reyes-Salas O, Argueta A, Casas A (2014) Diversity, local knowledge and use of stingless bees (Apidae: Meliponini) in the municipality of Nocupétaro, Michoacan, Mexico. J Ethnobiol Ethnomed 10:47

Romero MJ, Quezada-Euán JJG (2013) Pollinators in biofuel agricultural systems: the diversity and performance of bees (Hymenoptera: Apoidea) on *Jatropha curcas* in Mexico. Apidologie 44:419–429

Russell PL (2010) The history of Mexico: from pre-conquest to present. Routledge, New York

Sociedad Tosepan Titataniske (2017) http://www.slowfood.mx/2017/04/tosepan-pisilnekmej-el-modelo-unico-de-produccion/

Stearman AM, Stierlin E, Sigman ME, Roubik DW, Dorrien D (2008) Stradivarius in the jungle: traditional knowledge and the use of "black beeswax" among the Yuquí of the Bolivian Amazon. Hum Ecol 36:149–159

Tarkanian M, Paris E (2017) An evaluation of stingless bee wax as a pattern material in Mesoamerican investment casting. Proceedings 81st Annual Meeting of the Society for American Archaeology, Vancouver, British Columbia 2017

Vásquez-Dávila MA (2009) Las abejas nativas de los grupos étnicos del Istmo de Tehuantepec, Sur de México.En: *Memorias VI Congreso Mesoamericano Abejas Nativas*, Antigua Guatemala 2009

Vásquez-Dávila MA, Hernández EH (2011) Uso múltiple de los productos de la abeja nativa *Melipona beecheii* entre los yoko t'anob en Tabasco, México. En: Memorias VII Seminario Mesoamericano sobre abejas nativas. Ecosur, Unión de Cooperativas Tosepan. Cuetzalan, Puebla

Vednitsky S (2011) Transforming the colony: decolonial processes and outcomes in Melipona bee restoration. Bachelor dissertation, University of California, San Diego

Venturieri GC (2008) Criação de abelhas indígenas sem ferrão. Embrapa Amazônia Oriental, Belem

Villanueva-Gutiérrez R, Roubik DW, Collí-Ucán W, Güemes-Ricalde J, Buchmann SL (2013) A critical view of colony losses in managed Mayan honey-making bees (Apidae: Meliponini) in the heart of Zona Maya. J Kansas Entomol Soc 86:352–362

Vit P, Medina M, Enríquez ME (2004) Quality standards for medicinal uses of Meliponinae honey in Guatemala, Mexico and Venezuela. Bee World 85:2–5

Vit P, Pedro SRM, Roubik DW (eds) (2013) Pot-honey: a legacy of stingless bees. Springer, New York

Weaver N, Weaver EC (1981) Beekeeping with the stingless bee *Melipona beecheii*, by the Yucatecan Maya. Bee World 62:7–19

Zralka J, Koszkul W, Radnicka K, Sotelo-Santos LE, Hermes B (2014) Excavations in Nakum structure 99: new data on protoclassic rituals and precolumbian Maya beekeeping. Estudios de cultura Maya XLIV:86–117

Annex 1: List of Abbreviations of the Genera of Stingless Bees Used in the Book and Recognized Species for Mexico

Presently, there are two accepted systems of nomenclature used in stingless bees (Michener 2000; Camargo and Pedro 2007; Moure et al. 2007; Rasmussen and Cameron 2010). In the book, I preeminently use the genera proposed by Camargo and Pedro (2007) and Moure et al. (2007). Stingless bee genera appear abbreviated in the book; however, I use a non-conventional system of abbreviation when different genera start with the same letter. Here is the list of genera abbreviations used in the book:

Apis= A.
Austroplebeia= Au.
Cephalotrigona= Cph.
Friesella= Frs.
Frieseomelitta= Fr.
Lestrimelitta= L.
Melipona= M.
Nannotrigona= N.
Partamona= Pt.
Plebeia= Pb.
Scaptotrigona= Scp.
Scaura= Scr.
Schwarziana=Sch.
Tetragonilla=Ttga.
Tetragonisca= Ttr.
Tetragonula= Ttgn.
Trigona= Tr.
Trigonisca= Tgca.

In Mexico in accordance to Camargo and Pedro (2007) and Moure et al. (2007) 16 genera of stingless bees are reported and 46 species are formally recognized (Ayala 1999):

J. J. G. Quezada-Euán, *Stingless Bees of Mexico*,
https://doi.org/10.1007/978-3-319-77785-6

1. *Cephalotrigona eburneiventer*
2. *Cph. oaxacana*
3. *Cph. zexmeniae* **
4. *Dolichotrigona schulthesi*
5. *Lestrimelitta chamelensis*
6. *L. niitkib* **
7. *Melipona beecheii***
8. *M. belizae.*
9. *M. colimana.*
10. *M. fasciata.*
11. *M. lupitae*
12. *M. solani*
13. *M. yucatanica***
14. *Nannotrigona perilampoides***
15. *Oxytrigona mediorufa*
16. *Paratrigona guatemalensis*
17. *Partamona bilineata***
18. *Pt. orizabaensis*
19. *Plebeia cora*
20. *Pb. frontalis* **
21. *Pb. fulvopilosa*
22. *Pb. jatiformis*
23. *Pb. llorentei*
24. *Pb. manantlensis*
25. *Pb. melanica*
26. *Pb. mexica*
27. *Pb. moureana***
28. *Pb. parkeri***
29. *Pb. pulchra***
30. *Scaura latitarsis*
31. *Scaptotrigona hellwegeri*
32. *Scp. mexicana*
33. *Scp. pectoralis***
34. *Frieseomelitta nigra***
35. *Geotrigona acapulconis*
36. *Tetragona dorsalis*
37. *Tetragonisca angustula*
38. *Trigona corvina*
39. *Tr. fulviventris***
40. *Tr. fuscipennis***
41. *Tr. nigerrima*
42. *Tr. silvestriana*
43. *Trigonisca azteca*
44. Tgca. maya**
45. *Tgca. mixteca*
46. *Tgca. pipioli***

*Asterisks indicate species present in the Yucatan Peninsula

Annex 2: Nectar- and Pollen-Producing Plants in the Yucatan Peninsula

PLANTS OF THE PRE-HARVEST

Merremia aegyptia (Convolvulaceae)

Mayan name: *tso' ots' aak'*

Nectar production

Wide Naotropical distribution

Type: Climbing plant

Turbina corymbosa (Convolvulaceae)

Mayan name: *xtabentún*

Nectar production

Found from Mexico to Colombia and Ecuador

Type: Climbing plant

Jacquemontia azurea (Convolvulaceae)

Mayan name: *yaax ak*

Nectar and pollen production

Found from souther US to South America

Type: Climbing plant

J. J. G. Quezada-Euán, *Stingless Bees of Mexico*,
https://doi.org/10.1007/978-3-319-77785-6

PLANTS OF THE HARVEST

Viguiera dentata (Asteraceae)

Mayan name: *tah, tajonal, tah ché*

Nectar and Pollen production

Widely distributed in the Neotropics

Type: Herb

Gymnopodium floribundum (Polygonaceae)

Mayan name: *ts'iits'ilché, dzi dzil ché*

Nectar production

Distributed from Mexico to Central America

Type: Small tree

Piscidia piscipula (Leguminosae)

Mayan name: *ja'abin*

Nectar production

Tropical areas from the south of the US to Honduras

Type: Tree

PLANTS OF THE POST-HARVEST

Bursera simaruba (Burseraceae)

Mayan name: *chakaj*

Pollen production

Wide Neotropical distribution

Type: Tree

Senegalia gaumeri (Leguminosae)

Mayan name: *box katzin*

Nectar and pollen production

Distributed in southern Mexico and Guatemala

Type: Tree

Lysiloma latisiliquum (Leguminosae)

Mayan name: *tsalam*

Pollen and nectar production

Fournd from Southern US to Central America and the Antilles

Type: tree

PLANTS THAT CAN BE USED TO FEED THE COLONIES DURING DRAUGHT

Aloysia virgata (Verbenaceae)

Common name: lipia, residán

Nectar production

Native to the Pacific islands but widely spread across Mexico

Type: Bush

Antigonon leptopus (Polygonaceae)

Common name: *makal,* Flor de San Diego

Nectar production

Distribution: Native to Mexico but present in tropical America

Type: Climbing plant

Bidens pilosa (Asteraceae)

Mayan name: *k'an tumbuub*

Nectar production

Native from tropical America, now found in many parts of the world

Type: Herb

PLANTS THAT CAN BE USED TO FEED THE COLONIES DURING DRAUGHT

Bauhinia divaricata (Leguminosae)

Mayan name: *ts' ulub took'*

Nectar production

Distribution: South of Mexico, Central America and Greater Antilles

Type: Bush

Apoplanesia paniculata (Leguminosae)

Mayan name: *chulúul, cholul*

Pollen production

Distribution: South of Mexico down to Venezuela

Type: Tree

Cassia fistula (Leguminosae)

Common name: lluvia de oro

Nectar and pollen production

Native from Asia but widely spread in tropical México.

Type: Tree

PLANTS VISITED BY *M. beecheii*

Thouinia paucidentata (Sapindaceae)

Mayan name: *k'an chuunup*

Nectar and pollen production

Found in Southern Mexico, Belize and Guatemala

Type: Tree

Senna atomaria (Leguminosae)

Mayan name: *Tu ché*

Nectar and pollen production

Neotropical distribution

Type: Tree

Psidium guajava (Myrtaceae)

Common name: *pichi*, guayaba

Nectar and pollen production

Neotropical distribution

Type: Tree

Annex 3: Record Sheet Template, Nest Dimensions, and Hive Sizes for Stingless Bees from the Yucatan Peninsula

Annex 3.1: An Example Record Template for a Meliponario

Name of Meliponario: Kanlol Locality: Izamal Bee species: *Melipona beecheii*						Date of record: 13/May/2017		
Colony/ hive number	Queen presence	Strength	New combs	Pupae combs	Total	Number of pots	Phorids or pests	Observations
2	Yes	Strong	7	6	13	21	No	No
6	No	Weak	0	3	3	10	A few	Possibly queenless, feed and provide phorid traps
9	Yes	Medium	4	4	8	17	No	Feed

Annex 3.2: Nest Dimensions of Stingless Bee Species from Yucatan

J. J. G. Quezada-Euán, *Stingless Bees of Mexico*,
https://doi.org/10.1007/978-3-319-77785-6

Species	Entrance		Cavity			Brood chamber		Combs				
	Length	Diameter	Length	Diameter	Wall thickness	Length	Diameter	*n*	new	Sc	D Major	D minor
Melipona beecheii ($n = 15$)												
Mean ± standard deviation		1.15 ± 0.18	59.07 ± 5.89	13 ± 0.18	4.02 ± 1.12	16.92 ± 3.57	12.5 ± 1.38	8.7 ± 2.59	3.7 ± 1.48	0.62 ± 0.17	9.85 ± 2.8	5.2 ± 2.0
Range		0.9–1.5	50–69	10–15.5	2.5–8	10–21	8–14.5	5–13	2–7	0.6–0.9	8.0–17.0	4.0–6.2
Scaptotrigona pectoralis ($n = 39$)												
Mean ± standard deviation		1.24 ± 0.97	58.8 ± 24.7	9.31 ± 3.13		16.47 ± 5.23	8.18 ± 2.86	9.59 ± 3.25	3.75 ± 1.63	0.34 ± 0.6	6.77 ± 2.91	4.51 ± 2.33
Range		0.46–1.6	25–126	4.5–17.5		11–29	4.25–15	3–18	2–8	0.3–0.4	4.7–8.33	2.5–6.5
Frieseomelitta nigra ($n = 4$)												
Mean ± standard deviation		0.52 ± 1.08	56.76 ± 13.1	7.06 ± 2.41	7.87 ± 1.43							
Range		0.4–0.7			6–10							
Plebeia frontalis ($n = 1$)												
		0.45	50	14				11		0.45	5	3
Cephalotrigona zexmeniae ($n = 7$)												
Mean ± standard deviation		0.61 ± 0.25	75.28 ± 17.8	14.71 ± 1.7		17.4 ± 4.39		21 ± 16.08				
Range		0.6–0.65	60–101	13–18		13–23		10–52				
Lestrimelitta niitkib ($n = 3$)												
Mean ± standard deviation	7.6 ± 2.23	7 ± 1.41	103.3 ± 4.7	12.3 ± 0.47		33.3 ± 4.71	12	31 ± 11.3	12.5 ± 4.5	0.36 ± 1.6	9 ± 0.8	7
Range	5.5–8.4	5–8	100–110	12–30		30–40	0	23–47	8–17	0.3–0.8	8–10	0

	Cells		Involucrum		Pots		Bitumen	Gallery		Queen cell	
Species	Height	Width	Number of layers	Thickness	Height	Width	Thickness	Length	Diameter	Height	Width
Melipona beecheii (*n* = 15)											
Mean + standard deviation	1 ± 0.02	0.48 ± 0.35	1.2 (1 − 2) ± 0.41	0.12 ± 0.85	3.38 ± 0.14	2.37 ± 0.2	1.21 ± 0.9	14.05 ± 4.6	1.93 ± 0.57	–	–
Range	1.0–1.1	0.4–0.5	1.0–2.0	0.1–0.15	3.0–3.5	2.1–2.7	0.6–1.5	7.1–20	1.6–2.5		
Scaptotrigona pectoralis (*n* = 39)											
Mean + standard deviation	0.523 ± 0.52	0.306 ± 0.3	1.98 ± 0.79	1.54 ± 1.58	1.82 ± 0.4	1.86 ± .38	0.99 ± 1.57	9.14 ± 3.6	1.3 ± 0.1	1.92 ± 1.09	2.85 ±0
Range	.5–6	.275–.4	1–4	1			1–8	4–19	1	1–2	2.85
Frieseomelitta nigra (*n* = 4)											
Mean + standard deviation	0.5 ± 0	0.34 ± 0.29			1.75 ± 0.27	1.2 ± 0.1	3.67 ± 1.02				
Range	0.3–0.5	0.32–0.39			1.6–1.8	2.2–1.3	2.65–4.7				
Plebeia frontalis (*n* = 1)											
Mean	0.3	0.15			1.5	1	4.5	6.5	1		
Lestrimelitta niitkib (*n* = 3)											
Mean + standard deviation	0.5 ± 0	0.3 ± 0			2.36 ± 0.18	1.76 ± 0.12				0.65 ± 0.5	0.45 ± 0.5
Range	0	0			2.1–2.5	1.6–1.9				0.6–0.7	0.4–0.5

D Diameter, *Sc* Space between combs

For *Cephalotrigona zexmeniae* (*n* = 7) a mean of 1252 (±1828) brood cells was obtained and a scutellum length of 25.2 cm (±14.6) and width of 11.6 (±7) cm were recorded

Recommended internal dimensions of wooden hives (wood thickness at least 2.5 cm) for stingless bees from the Yucatan. Volumes (V) were calculated from data on cavity size (diameter and length) in Annex 3.2 and using an approach to the cylinder volume

($V = \pi r^2$ length):

Horizontal model TIBGA (measures are for the brood chamber, the super should be half the height):

M. beecheii

Natural volume = 7.8 L
9 cm height × 15 cm width × 55 cm length

Scp. pectoralis

Natural volume = 4 L
9 cm height × 12 cm width × 40 cm length

Fr. nigra

Natural volume = 2.2 L
9 cm height × 10 cm width × 30 cm length

N. perilampoides and *P. frontalis*

8 cm height × 9 cm width × 25 cm length

Vertical models as for *Portugal Araujo-1*:

M. beecheii

Brood chamber: 30 cm height × 16 cm width × 16 cm length
Super: 5 cm height × 16 cm width × 16 cm length

Scp. pectoralis

Brood chamber: 30 cm height × 12 cm width × 12 cm length
Super: 5 cm height × 12 cm width × 12 cm length

Fr. nigra

Brood chamber: 25 cm height × 10 cm width × 10 cm length
Super: 4 cm height × 10 cm width × 10 cm length

N. perilampoides and *P. frontalis*

Brood chamber: 20 cm height × 9 cm width × 10 cm length
Super: 3 cm height × 9 cm width × 10 cm length

It is important to note that the measures for horizontal hives are recommended based on measures taken from natural nests. Colonies may grow in the hives and would need additional space. In such case supers with the same length and width dimensions but with half the height can be provided.

In the case of *Cephalotrigona zexmeniae* colonies generally do not adapt well to wooden hives unless wood of at least 4 cm thick is used. Styrene could also be used as an additional outer cover. Hive sizes for *Scp. pectoralis* or *M. beecheii* would be suitable for this species.

Index

J. J. G. Quezada-Euán, *Stingless Bees of Mexico*,
https://doi.org/10.1007/978-3-319-77785-6

I

The manufacturer's authorised representative in the EU is Springer Nature Customer Service Centre GmbH, Europaplatz 3, 69115 Heidelberg, Germany. If you have any concerns regarding our products, please contact ProductSafety@springernature.com

Printed and bound by CPI Group (UK) Ltd, Croydon, CR0 4YY
30/09/2026
02225655-0008